AMERICAN BRIDGE

Lemelson Center Studies in Invention and Innovation

Joyce Bedi, Arthur Daemmrich, and Eric S. Hintz, general editors

Arthur P. Molella and Joyce Bedi, editors, *Inventing for the Environment*

Paul E. Ceruzzi, *Internet Alley: High Technology in Tysons Corner, 1945–2005*

Robert H. Kargon and Arthur P. Molella, *Invented Edens: Techno-Cities of the Twentieth Century.*

Kurt Beyer, *Grace Hopper and the Invention of the Information Age*

Michael Brian Schiffer, *Power Struggles: Scientific Authority and the Creation of Practical Electricity Before Edison.*

Regina Lee Blaszczyk, *The Color Revolution.*

Sarah Kate Gillespie, *The Early American Daguerreotype: Cross-Currents in Art and Technology*

Matthew Wisnioski, Eric S. Hintz, and Marie Stettler Kleine, editors, *Does America Need More Innovators?*

Kathryn D. Sullivan, *Handprints on Hubble: An Astronaut's Story of Invention*

Joris Mercelis, *Beyond Bakelite: Leo Baekeland and the Business of Science and Invention*

Eric S. Hintz, *American Independent Inventors in an Era of Corporate R&D*

David H. DeVorkin, *From the Laboratory to the Moon: The Quiet Genius of George R. Carruthers*

Matthew Wisnioski, *Every American an Innovator: How Innovation Became a Way of Life*

Gregory Dreicer, *American Bridge: Reinventing Building, Making History*

AMERICAN BRIDGE

REINVENTING BUILDING, MAKING HISTORY

GREGORY DREICER

The MIT Press
Cambridge, Massachusetts
London, England

in association with
The Lemelson Center
Smithsonian Institution
Washington, DC

The MIT Press
Massachusetts Institute of Technology
77 Massachusetts Avenue, Cambridge, MA 02139
mitpress.mit.edu

The MIT Press would like to thank the anonymous peer reviewers who provided comments on drafts of this book. The generous work of academic experts is essential for establishing the authority and quality of our publications. We acknowledge with gratitude the contributions of these otherwise uncredited readers.

Furthermore:
a program of the J.M.Kaplan Fund

Support for *American Bridge* was provided by a grant from Furthermore: a program of the J. M. Kaplan Fund

Front cover:
Grandfey Viaduct (W. Nordling, built 1857–62), Prosper Macherel, photographer, April 3, 1908, Source: Peter and Ruth Herzog, Basel, Swiss National Museum

This book was set in Bembo Book by Westchester Publishing Services. Printed and bound in the United States of America.

Library of Congress Cataloging-in-Publication Data is available.

ISBN: 978-0-262-55211-0

EU Authorised Representative: Easy Access System Europe, Mustamäe tee 50, 10621 Tallinn, Estonia | Email: gpsr.requests@easproject.com

CONTENTS

PREFACE

Welcome to *American Bridge*, an exploration of the stories that shape our understandings of invention, design, and each other. I hope that this book will inspire you to consider why and how we might change these stories. This preface begins with a concise chapter guide, followed by acknowledgments and a few clarifications.

GUIDE

American Bridge is arranged roughly chronologically, stretching from early nineteenth to early twentieth centuries, but you can read the chapters, each organized in a different way, in any order. Here's a quick description of what they're about. The introduction, "Making History," surfaces the evolutionism and nationalism that permeate our understandings of technology and history. Acknowledging and questioning these shared imaginings leads us to "Fabricating Materials" (chapter 1), where we see how tales about wood and iron signal the status of innovators and nations and determine how designers use these materials. In "Assembling the Parts" (chapter 2) we'll investigate how new ways of putting structures together enabled construction on a mass scale. "Building Future States" (chapter 3) tracks techno-utopian aspirations and industrial archetypes that together propel the construction of national infrastructures. "Transforming Structures" (chapter 4) portrays a web of technologists making a new model the new normal, as "American" inventions inspire the creation of novel bridges and new methods of analysis. "Reinventing Building" (chapter 5) revolves around a transformative moment in construction that exposes the soul of industrialization. In "Engineering Nations" (chapter 6), engineers developing bridge types and forming professional communities make public diverse understandings of, and hide ambiguities inherent in, materials, structures, and national status. Then, in "Global Positioning" (chapter 7), we observe elite America-makers producing architecture and infrastructure intended to define national identities. We conclude by "Getting High" (chapter 8) on an examination of why high-rise building systems were developed long before the "invention" of the skyscraper. We also touch on height-activated national elation. At the very end of our journey, "Restoring the Story" invites you to build new bridges to collective pasts and futures. Confronting and dismantling the divisive identity tales endlessly pushed upon us can help us instead envision narratives that bring people together. Has there ever been a better time to scrutinize the construction of bridges, stories, and communities?

THANKS

There are an enormous number of individuals and institutions for me to thank. They supported a research project whose development spread across many years, places, and disciplinary domains.

The advice, encouragement, wisdom, and patience of Amy Slaton were foundational for this project and enabled me to complete it. I will never be able to thank her enough! Almost from the beginning, I benefited from the guidance, encouragement, and friendship of Helmut Hilz, who later became director of the Deutsches Museum Library, a crucial source for this book. Jacques Leboeuf, librarian at the Conservatoire National des Arts et Métiers, enabled me to spend weeks immersed in the stacks, surveying hundreds of journal volumes. Catherine Bishir's generous and careful assistance was invaluable as I made a foray into the history of the antebellum South. The title of this book was bestowed upon me during a coffee with Christena Nippert-Eng, when she declared, "Call it *American Bridge*!"

These were only a few of the uncounted interactions that shaped this book. Some people I met once or twice; others I intermittently consulted; others provided book-saving interventions. I regret that, rather than adding another chapter with the details, I am able only to provide this incomplete, alphabetical acknowledgement. *Thank you! Merci!* and *Danke!* to Lynne Ambrosini, Jessica Blaustein, Hans Joachim Braun, Mike Chrimes, Susan Colangelo, Michel Cotte, Brian Crockett, Joe Dahmen, Françoise Domas, Dario Gasparini, André Guillerme, Katie Hafner, David Hounshell, Dana Hutt, Bruno Jacomy, Paul Jaskot, Andreas Kahlow, Anousheh Karvar, Kate Keleman, Gary Kulik, Karl-Eugen Kurrer, Bertrand Lemoine, Edward Lifson, Gunhild Lischke, Steven Lubar, Joanna Merwood-Salisbury, Brigitte Schroeder-Gudehus, Cecil O. Smith, John Ochsendorf, Jonathan Ochshorn, Ann Ostrander, Georges Ribeill, Brandy Savarese, Henry Schmidt, Ellan Spero, Michel St. Pierre, Eva Schöffel, Anna Slafer, John Staudenmaier, Jeffrey Stine, Martin Trautz, Will Truax, Jeff Ubois, Geert Verbong, Amy Vlastelica, Robert Vogel, Ulrich Wengenroth, and Denis Wornoff. Thanks to the five anonymous reviewers of the draft manuscript, who will be glad to know that I've revised it. Several times.

Special thanks to my *Doktorväter*, an old-school term for my PhD professors, who supported me in various ways over the years: Tom F. Peters, who brought me to Cornell and inspired and oriented my work; Ronald Kline, who immersed me in the complexities of the history of technology, and Mark Jarzombek, who guided me into theoretical and historiographic concerns. Before I embarked on the PhD, Norman Weiss provided life-changing encouragement and launched my research and career.

This project has taken so long to achieve that it includes a necrology. So I'm sending my gratitude into worlds where we all soon will reside. The enthusiasm for this project of Richard Sanders Allen, the seminal covered-bridge expert, meant so much. (Sanders' personal archive, which I was unable to consult, awaits researchers at the Hudson Mohawk Industrial Gateway and the National Society for the Preservation of Covered Bridges.) HAER founder Eric DeLony was a supporter; Ellen Wells offered me a research spot in the Dibner Library; Knut Gabriel welcomed me in Stuttgart; David Fischetti provided encouragement and expertise, as did Daniel Fink, who was intending to focus on the Portage Bridge.

As a public historian with a full-time job in curatorial strategy and management, I lacked the time, research assistance, and colleague and student interactions available to university professors, but my ability to develop public engagement projects including *Barn Again!*, *Between Fences*, *Architecture of Segregation*, *Me, Myself and Infrastructure*, and *Unbelievable*, enabled me to investigate and hone this book's themes. I'd like to thank my many collaborators.

Almost every preface includes an acknowledgment of a long-suffering family. Heartfelt thanks to Heather, Alula, Kai, and Ginger for bearing with me and offering unexpected perspectives and sly encouragement. I expect soon to be working with you on "American Bridge: The Musical." Special thanks to Mona for heroic support in image and rights wrangling. Eternal thanks to my parents.

Amazingly generous institutional backing, along with the trust behind it, enabled this research and book to happen. I am indebted to Cornell University, whose monumental support is the basis for all. This included the Andrew D. White Fellowship, several research grants, and backing of the former Program in the History and Philosophy of Science and Technology and its successor, the Technology Studies Department. Essential also was the support of the German Academic Exchange Service (DAAD), the Bourse Chateaubriand of France's Ministère de la Culture, the Deutsches Museum Library, and the Smithsonian Libraries' Dibner Library Resident Scholar Program. I am beyond grateful to the Lemelson Center for the Study of Invention and Innovation at the Smithsonian Institution for supporting my research with a fellowship and agreeing to copublish this book; I'm indebted to Art Molella and my editor Joyce Bedi, who stuck with me over many years. And thanks to the MIT Press team including Justin Kehoe, Suraiya Jetha, and Katie Helke. A beautiful grant from Furthermore: a program of the J. M. Kaplan Fund, enabled the inclusion of more of the images this publication required. Like our postprint, posttext world, the nineteenth century was awash in images.

Most of the research behind this book was done via "immersive research methodologies," otherwise known as spending a lot of time with paper documents.[1] I'd like to thank the libraries, archives, and museums who preserve and share the materials that otherwise would have disappeared or become inaccessible. Often, publications are the only surviving traces of people, structures, and places, prominent or not. Digital networks and archives provide awe-inspiring access and search capabilities but often cannot substitute for actual documents and artifacts. Although most images in this book are in the public domain, I've been careful to credit, that is, show deep appreciation, for the institutions that safeguard irreplaceable originals. Organizations that were particularly important include Deutsches Museum Library, Library of the Conservatoire National des Arts et Métiers, Smithsonian Libraries at the National Museum of American History (thanks so much to staff including Nancy Gwinn, Leslie Overstreet, and Jim Roan), Library of Congress, New York Public Library, École Nationale des Ponts et Chaussées, and Bibliothèque Nationale de France. University libraries essential for this project include Cornell, Stuttgart, Munich, Columbia, Harvard, Yale, British Columbia, and Berkeley. I have frequently taken advantage of the invaluable, free card that enabled me to access public libraries in New York, Chicago, Vancouver (BC), Berkeley, and Oakland.

CLARIFICATIONS

This book's title more accurately would be *Les ponts américains* or *Amerikanische Brücken* because the "American bridges" were not identified by that name in the United States any more than baguettes were called "French bread" in France. Moreover, the title should not be taken as an assertion that the bridges were American—that's for you to determine. Inventors did patent bridges in the United States that were widely perceived as American. Of course, "America" refers to a territory and political entity that changed dramatically during the nineteenth century; it's a nation that long had a narrow conception of who might enjoy all the benefits of citizenship and it's a place whose meanings and potentials we continue to change.

Although the fundamental, formative role of evolutionism and nationalism in Western thought and far beyond have been noted and analyzed in depth, their pervasiveness often enables scholars and popular storytellers to render them invisible. Like all infrastructure, people take narrative framings for granted—until things go wrong or catastrophe occurs. During my initial phase of research, people were amused when I asked what "iron bridge" meant or why there was so much interest in the first one; now there are entire books on "firsting." Historians once focused on "national technological cultures"; though such stereotyping appears still to be acceptable, increased attention to what's now called "othering" may make it less so.

American Bridge welcomes readers without experience in engineering or physics or history. I hope that my concise clarifications will be helpful for both humanities- and technology-based readers.[2] Please note that measurements are given in either meters or feet, according to the source; this will accommodate readers in both mensural camps. Although the length of a bridge span is an important factor for bridge designers, I have deliberately not played the numbers game common to historical accounts of engineering, in which longer is deemed superior. I've provided sources that enable anyone who wants the numbers to find them. It can be challenging to determine, even for well-known structures, the identities of the key designers. Many projects were team efforts, though often only the engineer responsible is named; regarding designers of the construction process, the task is even more difficult. When a range of dates is given for a bridge, this indicates the construction period; if just one date, this indicates the opening. I have tried to confirm all data with multiple sources. I am responsible for translations (pre-artificial-intelligence) from the French and German.

Because much here challenges two centuries of nationalist accounts, ample evidence is critical, though it won't be necessary for many readers. Notes were condensed for publication, so reach out if you'd like further information.

Parts of this book had previous lives. Chapter 6 is an overhauled version of "Building Bridges and Boundaries: The Lattice and the Tube, 1820–1860," *Technology and Culture* 51, no. 1 (2010): 126–163. Bits and pieces of the following may have found their way into several chapters: "History of Engineering, Reverse-Engineered," Fifth International Congress on Construction History, Chicago, 2015; "Influence and Intercultural Exchange: Engineers, Engineering Schools and Engineering Works in the 19th Century," *History and Technology* 12 (1995): 163–177; "Building Myths: The 'Evolution' from Wood to Iron in the Construction of Bridges and Nations," *Perspecta* 31 (2000): 130–141; "Nouvelles inventions: L'interchangeabilité et le génie national," *Culture Technique* 26 (1992): 213–220; "Technology, Biology, and History," Technological Development and Science in the 19th and 20th Centuries, a conference in Eindhoven, Holland, 1990.

INTRODUCTION: MAKING HISTORY

Stories are our bridges to the future. They connect us across place and time. They tell us who's in our community—who we'll remember and who we'll forget. For two centuries, we've heard tales of inventors as evolutionary agents fulfilling national destinies in a struggle-to-survive world. Today, it's challenging to think about history and innovation in any other way. Can we change the story?

American Bridge explores the shared narratives behind the reinvention of building, which enabled the construction of nations and their infrastructures. Our centerpiece is the lattice bridge, internationally known as the "American bridge" that crystallized the industrialization of building (figure I.1). Patented in 1820 by engineer-architect Ithiel Town (1784–1844), the lattice stood for a new way of putting things together and organizing the people doing the work. In the long view, it rivals the Model T as transformer. Almost a century before Ford's car, this iconic structure was a buildable articulation of an aspiration that innovators had long been working toward: the making of long, level structures out of small, uniform parts, on a mass scale, and fast. More than just a bridge type, it demonstrated the reimagining of a basic building element: the beam. The lattice, which energized inventors, engineers, and industrialists, was a bridge to the future. It can serve that purpose for us, too: as inspiration for rethinking the interwoven construction of technology, story, community, and place.

A rethink will help us understand transformations in building, one of the world's oldest professions and still essential to the wellbeing of individuals and nations. Our focus on the invention of process and processes of invention will reveal a historical double whammy: most storytellers have reinforced, instead of investigating, nineteenth-century technologists' narrative framings of materials, structures, and places. Rather than continuing to amplify oft-told tales and raising a nostalgic toast to the covered bridge, we'll exploit the explosive power of simple questions: Who did it? Why? Where? How do we know? We'll blaze alternate paths to our pasts in order to construct stories that have not yet been told.

Figure I.1
George Durrie, *Ithiel Town Truss Bridge*, 1853. Durrie's nostalgic American scenes soothed an urbanizing nation. The bucolic aura of this image belies its subject: the quintessential industrial product. This lattice (Ithiel Town, 1823) spanned the Mill River at Eli Whitney's gun factory near New Haven, Town's and Durrie's hometown. Durrie was a student of Nathaniel Jocelyn, who created the first lattice-in-landscape image (fig. 3.4). Courtesy and Collection of Mattatuck Historical Society.

TELLING TALES

We know a little about the origins of the structures that shape our environments and perhaps a little more about the narratives that explain those origins, but rarely is the impact of their mutual making examined. Widely shared tales about engineering reveal many things, but they infrequently highlight far-flung designers sharing knowledge, manipulating meanings, and making history along with their creations. The predominant story, which tells us about individuals within a hierarchy of invention-molding nations, obscures the shifting, transborder spaces of innovative activities that precede, construct, and outlive the nations and empires they flow through.[1]

Why focus on a bridge? Ideas about infrastructure and landscape, and nation and community, are inextricably bound to bridges (figure I.2). People, goods, and information all flow across them. But bridges are more than essential links within the communication and transportation networks on which our lives and communities depend; bridges determine the layout of roads that lead to them and the routes we choose to travel.[2] They have the ability to make a reputation by intensifying place and strengthening identity.[3] Bridges are transitional structures; the journey across can transport individuals and groups from one state to another. They bring people together into community, leading to relationships that are the basis of invention. Bridges are local artifacts that enable and support ever-changing local, national, and international interests. They give a physical form to the political and economic might required for their design and construction, which is to say, they represent the potential of human connection to transform environments and lives.[4] They advertise the power and wealth of communities with the resources to build and maintain them.[5] Bridges are crucial, exposed connectors, wartime targets that span identity-shaping political and geographical divides. The rise of industrial capitalism depended on the bridges that conveyed cotton from field to factory to consumer. Communities that lack bridges pay an enormous toll, measured in prolonged travel times, financial disadvantage, and lives lost.[6] Bridges are platforms for the public performance of trust: your life depends on the security of every bridge you pass over or under. We might honor bridges as a promise kept by the people who design, build, and care for them, sometimes at the cost of their lives. By taking as an example something as seemingly solid as a bridge type and examining the stories told about it, we'll explore the apparent permanence and real vulnerability of infrastructures, nation-states, and landscapes (figure I.3).

A bridge can take us above the landscape, allowing us to see our surroundings in unique and surprising ways. Yet, in a train or car, sometimes we're unaware we're crossing a bridge. Infrastructure—technological or narrative—that functions reliably functions

Figure I.2
Franklin, New Hampshire. Bridges and roads, like religious and governmental buildings, help define community. This photograph shows the Republican Bridge, built in 1802 and later replaced with a lattice. It no longer exists, but the church still stands. Collection of the author.

Figure I.3
Railroad bridge on North Anna River, Virginia, 1865, destroyed during the Civil War to prevent Union troops from advancing. Infrastructure and its condition are inseparable from the nation and its apparent permanence and vulnerability. War Photograph & Exhibition Company, "Destruction of a Railroad Bridge," photographed May 26, 1864, by Timothy H. O'Sullivan, photographer. Library of Congress, Prints and Photographs Division.

invisibly. Only when something goes wrong does it become visible. Lack of maintenance, short-term planning, inflexible public policy, and climate change mean that engineering infrastructure will become increasingly, perhaps catastrophically, visible. At the same time, political extremism, social and governmental dysfunction, and the normalization of misinformation seems to require a public questioning of the timeworn narrative infrastructure that determines our pasts and futures. *American Bridge* responds to these conditions by exploring why and how innovators, dispersed in different regions, developed processes and structures as they participated in the construction of nations.

At mid-nineteenth century, observers informally employed the phrase "American system" to refer to the mechanization of manufacturing in the United States; a few decades earlier, the phrase "American bridge" had referred, a bit more consistently, to novel structures that demonstrated concepts that would shape the manufacture of consumer goods. These included simplified, uniform parts, rapidity of production, and the lowering of labor skill. In the 1920s, the term "mass production" was applied to interchangeable parts, mechanization, and the assembly line of Ford factories. But more metal and more machinery were not essential to the reconception of a foundational industry, construction. The Ford idea was only one avenue of industrialization, rather than a model for all industrializations.[7] Mass construction was different than mass production, but no less momentous.

A significant inspiration was the lattice, an open-source solution to builders' longstanding challenges. Its perceived novelty elicited published reactions that will enable us to survey a variety of perspectives. But silence surrounds the lattice's chief value: how it was built. Ithiel Town never publicly revealed the process, which could not be patented (figure I.4). An even deeper silence surrounds how Town developed his version of structural, constructional, and managerial changes that constituted a reinvention of building. In light of the impact of the "American bridge" as a constructional prototype, we might consider this to be one of the most consequential mysteries in the history of industrial capitalism. We'll attempt to solve it by tracking the lattice, including its earliest iterations, and visiting a variety of little known and renowned structures with which it shared industrial imperatives, including the Portage Bridge, Britannia Bridge, Crystal Palace, and Quebec Bridge. It is not the intention of this book to elevate the status of Ithiel Town, though that may be an inevitable byproduct. He was a highly successful member of elite social and technological networks who after his death was mostly forgotten.

Town's bridge and subsequent variations were a Rorschach test. How did designers read the lattice? They might see a beam. Or a truss. They might see a nation, America—which could trigger a surge of expectations or anxieties. The assemblage of simple parts was hard to comprehend and seemed as if it should not have functioned. Observers interpreted its level, tubular form and its uniform, interlaced members as a

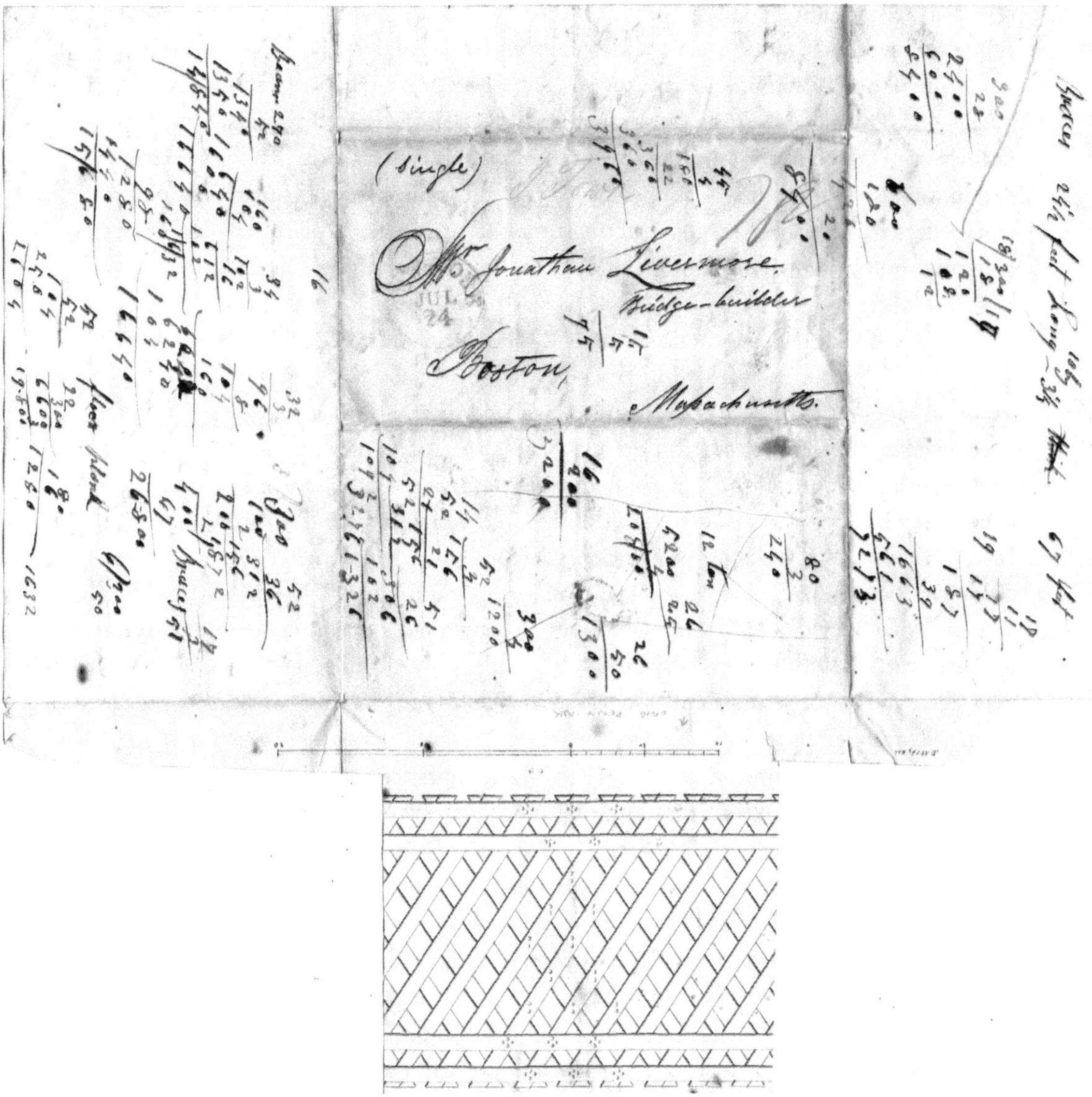

Figure I.4
Ithiel Town sent a drawing with instructions to Boston bridge builder Jonathan Livermore, July 23, 1836. Marjorie and John Geiser, II Collection, Southeastern Architectural Archive, Special Collections Division, Tulane University Libraries.

repudiation of tradition and an invitation to the construction of tomorrow. *American Bridge* embraces ambiguities and contradictions as it ponders: Who reinvented construction? How? What can a beam reveal? How does infrastructure define place—and vice versa? This means asking questions about the questions usually asked about technology.[8] So let's consider who's been answering the questions, how, and why. Embedded in long-shared narratives are perspectives you may not share. Understanding how storytellers make history by constructing order across time and place will enable us to recognize and bridge divides in our understandings of technology and culture.

A HISTORICAL ROMANCE: ENGINEERING AND EVOLUTION

To the extent that there is an enduring historical understanding of invention in civil engineering (an anachronistic category that originally referred to nonmilitary activities), it is evolutionist. "Evolutionism" in *American Bridge* signifies the integration of evolutionary ideas, often but not always allied with thinkers in the biological and anthropological sciences, to human social history. Evolutionism represents a romantic recounting of technological change that links artifacts in a chain, usually implying superiority with respect to past artifacts, and usually associated with specific places or cultures—which means it often overlaps with nationalist beliefs.[9] It is so deeply embedded and pervasive that it may seem odd to pause and think about it. When a storyteller evokes evolution, the idea often is so ambiguous that its implications are not clear, even to the person employing it. This brief introduction is unable to review all the mutations and permutations of two hundred years of evolutionist thought, but in order to explore a vast gap in the historical record—the reimagining of construction—we must understand how the designers of history did their work; in this case, that means investigating how they incorporated evolution and its closely allied partner, progress, into the conception of stories about materials, bridges, and inventors. Evolutionism contains a blend of beliefs about civilization, nationality, and race (which might refer to people sharing a national territory or a set of visible physical characteristics), as it highlights competition, reinforces difference, and authorizes exclusion. Its role as a rationale for, if not a driver of, dehumanization, aggression, and destruction is reason enough to identify the ramifications of this worldview and seek alternatives.[10]

Just discussing a myth can affirm it in the mind of believers, which might include you! With this hazard in mind, let's briefly explore how evolutionist and nationalist perspectives reinforce each other; indeed, they share a genesis and serve as the basis for stories that feel factual, even when supporting facts are absent. These perspectives will repeatedly surface in *American Bridge*, as they operated in the minds of historical figures *and* the storytellers who reported on those figures and their creations, generating the evolutionist double whammy that presents a formidable challenge to our task of historical reconstruction.

Darwin remains the public face of evolution, but his work was plugged into a network of thinkers and popularizers; he believed that his key contribution was showing how species changed.[11] Many ideas labeled "Darwinian" originated with others, such as naturalist Jean-Baptiste Lamarck, scientist Alfred Russel Wallace, and philosopher Herbert Spencer, who worked for several years (around 1840) as an engineer. Spencer's widely read evolutionary philosophy and often-evoked 1864 phrase "survival of the fittest" made him a major force in the application of evolution to fields outside biology; he believed that evolutionary laws should replace traditional historical accounts that emphasized human agency.[12]

Anthropologists who studied the US patent system—which defines novelty and classifies objects—fortified the melding of innovation and evolution as they narrated technological development through a genealogy of inventions. Some crafted linear stories that described a universal evolution of decontextualized objects; others focused on specific social or political conditions within an evolutionary scheme.[13] These models of culture map approximately onto evolutionism and nationalism, which have remained the key narrative frameworks aiming to explain invention. Within these perspectives, bridge building, iron making, and urbanization became criteria for identifying human groups furthest along in the civilization process. During the nineteenth century, the inhabitants of a place lacking a substantial collection of supposedly advanced technological artifacts might be labeled "savages" or "barbarians." Technologies denoted a group's status as potential collaborator, commercial threat, or target for conquest.[14]

The idea of evolution is as much biological as technological and cultural in its origins—in its DNA, per the popular expression. Biological analogies have pushed understanding of technology and vice versa. Technological metaphors based on computing and coding have shaped understanding of how the brain functions. People thinking about engineering and industry apply terms such as "ecosystem" and "meme" to nonbiological realms, no matter how misleading biopoetry may be.[15] Darwin, for example, was reading about technological practices and contemplating nature as a technological system while forming his thoughts. When he wrote, "Almost every part of every organic being is so beautifully related to its complex conditions of life that it seems as improbable that any part should have been suddenly produced perfect, as that a complex machine should have been invented by man in a perfect state," he was considering factors contributing to invention as he built his understanding of the natural world.[16]

"Evolution" appears once in the first edition of Darwin's *Origin of Species*; since the seventeenth century, the word had implied only a sequence of events.[17] It frequently surfaces in writings about bridges that aim to explain how the structures came to be, what they are, and what they will become.[18] The concept endows invention with momentum. Engineering professor Jules Gaudard began his 1892 text on bridge construction by proclaiming, "Under the hand of man, things seem to come to life, join with his aspirations, and, like him, struggle for existence." Engineer-historian Henry Petroski claimed a century later that "with the increasing production and application of iron in the nineteenth century, trusses naturally evolved into a plethora of types and styles employing the new material." He believed that the development of lighter and more flexible designs for tall buildings could be understood because "such evolutionary developments are natural in engineering."[19]

When Darwin drew a parallel between his study of animal breeding techniques and natural forces, he did not thoroughly examine the implications. Natural selection made biology machine-like and thereby scientific, but the use of the word "selection" gives the impression that nature, like a breeder, makes choices, rather than illuminating nature's autonomous processes of variation, which was Darwin's intention. As Karl Marx put it, "Darwin has interested us in the history of Nature's Technology."[20] The evolutionary idea of change in small steps may help some people understand innovation; in this respect, it might be better than no model at all.[21] But applied to a social process, evolution transforms technological development into a supernatural phenomenon: individual designers become enactors of a force that molds their behaviors. This force, called "progress," is often integral to accounts in which technology is always getting better or more sophisticated or that is beyond humanity's power to control. Darwin wrote of "progress towards perfection" in nature; this idea, dating at least from the eighteenth century, informs an understanding of technological development shared by many to the present day.[22] Evolution is often portrayed as an antireligion, but technological progress thinking is rooted in religious thought: it is a faith-based account of the world that taps into "nature" while distracting from an absence of evidence. Evidence may be beside the point, however, as the objective is to communicate a perspective on the status of nations or races, technological and industrial change, and the people who profit most from it.

What does evolution tell us about bridge building? When Petroski writes that "the evolution of bridges can be traced back to primitive man felling a log across a brook," he recalls hierarchies of human civilization that were once widely believed.[23] It was all inevitable, he explains: "While each invention and artifact has its unique aspects, there is also a certain sameness about the evolutionary way in which a stylus develops into a pencil, a sketch into a palace, or an arrow into a rocket."[24] In engineering's just-so stories, structures depicted by Renaissance architect Andrea Palladio (1508–1580) are frequently cited (figure I.5). Palladio did appear to be focusing on issues that two centuries later would

DEL PONTE DEL CISMONE. Cap. VII.

IL Ciſmone è vn fiume,ilquale ſcendendo da i monti,che diuidono l'Italia dalla Germania entra nella Brenta alquanto ſopra Baſſano; e percheegli è velociſſimo,e per lui i montanari mandano giù grandiſſima quantità di legnami,ſi preſe riſolutione di farui vn ponte,ſenza porre altrimenti pali nell'acqua. Percioche le traui,che vi ſi ficcauano, erano dalla velocità del corſo del fiume, e dalle percoſſe de i ſaſſi,e de gli arbori,che da quello cōtinuamente ſono portati all'in giù,moſſe,& cauate: Onde faceua biſogno al Conte Giacomo Angarano,il quale è patrone del ponte,rinouarlo ogni anno. La inuentione di queſto ponte,a mio giudicio,è molto degna di auertimento,perche potrà ſeruire in tutte le occaſioni, nellequali ſi haueſſero le dette difficoltà; perche i ponti coſi fatti vengono a eſſer forti,belli,e commodi: forti,perche tutte le loro parti ſcambieuolmente ſi ſoſtentano; belli,perche la teſſitura de' legnami è gratioſa;& commodi, perche ſono piani,e ſotto vna iſteſſa linea col rimanente della ſtrada. Il fiume nel luogo oue ſi ordinò queſto ponte,è largo cento piedi. Si diuiſe queſta larghezza in ſei parti eguali,& oue è'l termine di ciaſcuna parte(fuor che nelle ripe,le quali ſi fortificarono con due pilaſtri di pietra)ſi poſero le traui,che fanno il letto,e la larghezza del ponte;ſopra lequali,laſciatoui vn poco di ſpacio nell'eſtremità loro,ſi poſero altre traui per il lungo,lequali fanno le ſponde: ſopra queſte,al diritto delle prime ſi diſpoſero dall'vna, e l'altra parte i colonnelli (coſi chiamiamo volgarmente quelle traui, che in ſimili opere ſi pongono diritte in piedi.)Queſti colonnelli ſi incatenano con le traui, lequali ho detto,che fanno la larghezza del ponte,con ferri, che nominiamo Arpici,fatti paſſare per vn buco fatto a queſto effetto nelle teſte delle dette traui, in quella parte, che auanza oltra le traui che fanno le ſponde. Queſti arpici,perche ſono ne la parte di ſopra a lungo i detti colonnelli diritti, e piani, e forati in piu lochi, & nella parte di ſotto vicino alle dette traui groſſi,e con vn ſol foro aſſai grande;furono inchiodati nel colonnello, e ſerrati poi di ſotto con ſtanghette di ferro fatte a queſto effetto,onde rendono in modo vnita tutta l'opera,che le traui, che fanno la larghezza, e quelle ſponde ſono come di vn pezzo con i colonnelli, & in tal modo vengono i colonnelli a ſoſtentar le traui,che fanno la larghezza del ponte;e ſono poi eſſi ſoſtentati dalle braccia, che vanno da vn colonnello all'altro: onde tutte le parti l'vna per l'altra ſi ſoſtentano, e tale viene a eſſer la lor natura, che quanto maggior carico è ſopra il ponte,tanto più ſi ſtringono inſieme,e fanno maggior fermezza dell'opera. Tutte le dette braccia,e l'altre traui,che fanno la teſſitura del ponte non ſono larghe più di vn piede,nè groſſe più di tre quarti. Ma quelle traui che fanno il letto del ponte, cioè che ſono poſte per il longo,ſono molto più ſottili.

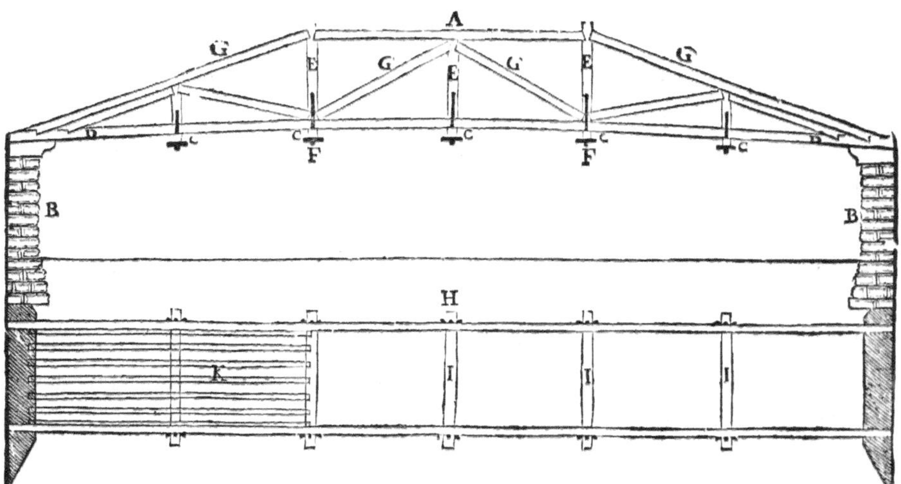

Figure I.5
Andrea Palladio, Bridge over Cismon River. Evolutionists typically present Palladio's designs as the source of the truss bridge. Andrea Palladio, *I quattro libri dell'Architecttura* 3 (Dominico de' Franceschi, 1570), 15 (image from 1601 ed.). The Getty Research Institute.

occupy bridge designers: standard connections and parts and how to assemble them into rigid structures. His work may have intrigued nineteenth-century bridge designers, but at best this has a tangential connection to long-span bridges, railway loads, and strategies for building on a mass scale; experts question how well the structures Palladio portrayed would have functioned even in his own time. Some included very long members—which would have been challenging to obtain—and a truss-like appearance did not mean truss-like function; historian David Yeomans points out that Palladio's designs were not the outcome of thinking of trusses as engineers later conceived of them.[25]

The Palladian example underscores the pitfalls of evolutionist thinking: a prior event may have no connection to a later one, even if both fall within a category ("bridge"). But historians, like nineteenth-century anthropologists, wanted to highlight progress through a demonstration of an "unbroken continuity that follows tools back to embryonic forms."[26] Physical resemblance, however, may or may not indicate a relationship. Structures that look alike may not function alike, and physical dissimilarity does not indicate the absence of a relationship. This means that a technology that looks nothing like a truss bridge and may not be a bridge at all might have been essential to the development of bridge structures. In biological terms, a bridge is as likely to be related to other "species" (such as ships or shoes, as discussed in chapter 2) as to other bridges. While it's unlikely that Palladio's designs contributed meaningfully to nineteenth-century engineering, they did bestow ancient Greek and Roman cachet—classical ideals and forms being the mark of Eurocentric "civilization." Invoking Palladio established a lineage of Western achievement that provided a cultural sheen for the industrialist engineers.[27]

When evolutionists construct tales, they align artifacts of the past with contemporary cognates, which they deem the most highly evolved. An arrow becomes a rocket. The railroad becomes a stepping-stone on the path to automobility. We still hear in popular publications stories that link Stone Age tools to information technology or explain technologist's behavior by referring to that of animals; digital tools represent the apogee of humankind's innovative trajectory.[28] In the evolutionary tale of industrialization, a peak is the mass-production factory. This encourages the devaluation of places of production such as the bridge-building construction site, which does not conform to the standard progress narrative. Although building processes were as important as the structures that were their outcomes, historians typically have kept their focus on tangible, drawable, patentable artifacts.[29] Process, an abstraction, resists the objectification and codification necessary for easy evolutionary narrative. To an evolutionist who considers steam and electricity as progress benchmarks, people- and horsepower, which were sometimes essential to the building of vast bridges, cannot represent industrial progress.[30] But, as we'll see, human power advanced bridge construction, literally and figuratively.

Can evolutionists explain the behaviors of individual innovators, whose stories they purport to tell? Many scientists believe that the human mind is adapted to conditions faced for two million years by our ancestors, the Pleistocene hunter-gatherers, rather than being the product of quick alteration during the few thousand years since the development of agriculture.[31] A link between, for example, changes in people's brains over the last 250 years and technological development has never been shown, although cooperation—which would include inventors' social networks—is so significant in human society that some consider it as important as natural selection.[32] Cooperation, whether collaborative or competitive or somewhere in between, is the heart of this book's story. While collaborative behaviors, whose outcomes include bridge building and religion and cities, may be the result of thousands of years of evolution, experts caution against interpreting specific inventive activities as instances of evolution. As zoologist Steven Vogel observed, "the familiar act of human creation is harder to explain in acceptably scientific terms than the way nature creates."[33] There is sharp disagreement about how one might distinguish human inventive behaviors that are due to evolution from those that are not.[34] Nest-building birds may be appealing models for understanding the development of building technology, but natural history accounts for neither human inventors' decisions nor the cultural and technological extinctions and resuscitations that these decisions set in motion and that are impossible in nature.[35]

This quick overview of evolutionism's profound impact highlights how we understand invention: we wrap it in stories whose implications we may not understand. If the stories create a believable reality, facts may be irrelevant. In any case, evolution enabled biologists and engineers to become "scientific" and boost their professional status. It helped historians, too, by delivering a grand narrative that appeared to provide natural, scientific evidence for human history. It provided order, essential for the work of engineers and historians.[36] Most significantly, evolutionist metaphors provide a shared narrative; individuals who use them forge identity and become members of a community.[37] Whether it's an evolutionary tree, timeline, or Bildungsroman, familiar, easily graspable narrative templates give the appearance of providing a coherent explanation of how the world works, along with a rationale for predicting the future. For example, an evolutionary panorama (figure I.6), created not long after the 1848 joining of cantons into a state, Switzerland, portrays a progression from untamed forest to humble wooden footbridge to stone arch to rectilinear iron lattice, upon which a puffing icon of technological supremacy passes. All this unfolds in front of the Alps, symbol of national identity. Railway and viaduct were badges of industrial competitiveness, enablers of unity, and markers of national strength and pride.[38]

Figure I.6
Sitter Viaduct (Carl Etzel, built 1853–56), Sitter valley, Bruggen, St. Gallen Railroad, Switzerland. Arch bridge in foreground: Kräzern Bridge (Ulrich Haltiner, 1812). Print, c. 1860. Collection of the author.

Evolutionist narratives appeal to popular audiences: happy endings help sell magazines and movies. But even within the scientific community the pleasure of a speculative creation story is hard to resist: biologists have been known to attribute changes to adaptation without evidence that adaptation is actually the mechanism behind the change.[39]

In *American Bridge*, we'll remember that a new form of storyteller, the historian, helped a new genre of designer, the civil engineer, build the conceptual and physical infrastructure of the nation-state. We'll keep in mind that (evolutionist) narrative determines if, how, and where we seek facts, which ones we'll embrace, and which ones we'll never uncover. Technologists and the people who told their tales knew that successfully inventing structures and strategies meant inventing stories—an activity shared with clients and colleagues, competitors and critics, consumers and citizens.

FROM NATION TO NATION

Chroniclers of bridges and engineering artifacts of all types often locate their evolutionist understandings of design within a divided landscape. The conceptual bifurcations include wood versus iron, craft versus industry, and practice versus theory, with the last of each of these severings the more "civilized" to storytellers who sort them always with an eye on the crucial divide: nation versus nation. They align their tales so that, for example, "practical," "wooden," and "American"—mythic categories that helped drive personal, institutional, industrial, and national competition in the nineteenth century—became historical fact in the twentieth. But the great divides do not explain; they must be explained.[40] Doing so allows us to see how national divides facilitate the fabrication of innovator relationships within a supposedly national community while obscuring relationships among communites.

The puzzle of why bridges seem to be different in various places of the world is as intriguing as the differences noted by Darwin between European and American weeds.[41] Conditions in specific places always shape technologists' work, but places don't invent, people do. And place itself is territory that people define and redefine. Nationalism, "a celebration of the nation" that takes many forms and intensities, can bind people together and be a force for public good, though it's often a stimulus for exclusion. Nationalism merges or submerges the diversities of language, race, place-based or political loyalties, ethnicities, and traditions of all kinds that were (and are) present in the nation-states formed during the last couple centuries. The "nationalization of humanity," in geographer Peter Taylor's apt phrase, has encouraged the prioritizing of "places over flows."[42] Flipping that perspective, that is, reorienting ourselves to flows among people, is what we'll be doing here. This will allow us to see individuals pursuing personal and collective interests and working with and against each other within and across boundaries.

One of the earliest lattice bridges, built over the South Yadkin River near Salisbury, incited a North Carolinian to exclaim in a local newspaper in 1825: "There is nothing splendid or magnificent about it: the plan is *indigenous* to our free country—being the invention of Mr. *Town*, of Connecticut; and for simplicity, strength and symmetry, corresponds with the character of our government."[43] Eye-catching infrastructure typically inspires citizen-observers to express national values.[44] In this case, the novelty of the lattice inspired a blending of ideas about Americanness, appearance, innovation, liberty, and the political system of a new nation. It made an implicit contrast to other nations. And it was an intriguing assertion, as enslaved people were making substantial contributions to the "free country," preparing lumber, building bridges along with the roads that led to them, and clearing rivers that flowed beneath them. Over the succeeding two centuries, however, storytellers fortified a narrative infrastructure capable of sustaining such profoundly contradictory facts. "Nature's progress" helped justify or paper over inconsistencies, disputes, persecutions, and gaps in knowledge about historical events.[45]

Historians for whom national boundaries are foundational have often explored the early United States and its interactions by wielding a sedentarist concept, "transfer."[46] It has inspired some to regard the United States as a "technological colony" and "debtor culture."[47] Transfer theorists detail the importation and adaptation of a technology, treating it as a taking, a one-way event between two points, rather than an ongoing interaction within and among networks of technologists acting within and across a diversity of nations, cultures, and communities. Inside the transfer scenario, the borrower is usually an adaptor of a completed artifact rather than manipulator of a changeable thing formed and reformed within dynamic relationships.[48]

Let's take a quick look at how narrators envision the sites of transfer launch and landing. Ithiel Town built his earliest lattice bridges in North Carolina (see chapter 5), where the Raleigh and Gaston, a risky, high-tech undertaking completed in 1840, became one of the earliest railroads to include a lattice bridge. How do we understand the state, in two senses, of North Carolina, in relation to other states and nations? In 1836, a local bemoaned:

> It is a lamentable fact, that there exists a mutual jealousy all over our State, not only between East and West, but between the Cape Fear, the Neuse, the Roanoke, &c. And the consequence is, whenever any improvement is proposed in one section, an opposition arises in another, and a resort must be had to that system, known by the name log rolling, to carry any point.[49]

North Carolina's government wanted a railroad in order to promote commerce within the state, but various parts of the state had business relationships with parts of neighboring states, which had comparable internal issues of their own.[50] This is similar to conditions today, when relationships among individuals and groups within a state may not align with or might contradict broader goals, which themselves are subject to change. Closely located urban and rural areas might have vastly different perspectives, while cities dispersed across the globe might share commonalities as meaningful as any that we might dream exist within a state.[51] While the transfer concept does draw our attention to flows essential to technological development, it isolates a place and moment in an ongoing process. In transfer, the nation is the story. That's why technology is not "transferred," for example, from North Carolina to Connecticut, or from one region of North Carolina to another, but it can be "transferred" from Britain to the United States.[52] (An agent of transfer might travel and learn and then decide to do something completely different, but this is unlikely to be recognized as transfer at all.)

Historians and politicians monumentalized the concept of nation-based innovation, in which the concept of transfer is grounded, with the Britishness of the Industrial Revolution, an event said to have occurred during the latter part of the eighteenth century and the beginning of the next. A peek into its national roots, however, complicates the story:

> The conventional narrative of the Industrial Revolution . . . has often mistaken the first introduction of a device imported from abroad for a product of native ingenuity. . . . Foreign contact, either through the migration of men or ideas, has always played an important part in English industrial advancement. That English industrialists or craftsmen borrowed extensively and persistently from their European neighbors, and that the genius of the country has lain not so much in originality as in transforming and blending native and foreign traits, is acknowledged by many. . . . When parallel and concurrently-running local histories [in England] are compared, there is at least one event, the intrusion of craft-bearing strangers, which is recurrently, if not always, antecedent to recurrent events of innovation.[53]

Craft-bearing strangers drive technological development. During the nineteenth century, technologists working across the industrializing nations deliberately sought and imported ideas, technologies, and workers. They shared common industries (including textiles, iron, railways, mining, and chemicals) and developed technologies through colonization, collaboration, and competition.[54] Recognition of transborder exchange as status quo and the establishment of new trades in Britain as "inexplicable except through the . . . settlement of Flemish, French, Dutch, and Italian workmen" means the United States might be regarded as less of a debtor and Britain less British than has been portrayed.[55] Conspicuous is the fact that empires that stretched around the globe diversified the creative mix with craft-bearers from places well beyond European borders. Enslaved Black people were enmeshed in the development of industrial materials, tools, and management strategies.[56] But the (mythic or real, per your belief) Industrial Revolution endowed Britain with special status because the concept itself was a product of British nationalism, even if the term was coined in France.[57] The enslaved have not been visible on the transfer map; storytellers saw them as inhabitants of an inferior civilization, noncitizens who contributed nothing across the evolutionary divide. The idea that knowledge exchange among peoples of various Western and non-Western lands, regions, and backgrounds is in some way unusual, or that the borders of nation-states align with cultural or technological communities, is irreconcilable with transnational flows of people and products. That ideas and information might be contained by a political border—that is a hypothesis that requires copious proof.

Tales of firsts and transfers deepen nation-versus-nation divides as they erase participants involved in ongoing exchange. The presumption that invention has a starting point (within a group deemed a race, culture, or nation that embeds its "character") and then spreads to other places is called "diffusionism" by political geographer J. M Blaut; in this scenario, "most humans are imitators, not inventors."[58] This is why it can be startling for people steeped in Eurocentric histories and nationalisms to discover that development of "British" iron, so long a symbol of industry and empire, likely capitalized on the minds of enslaved Africans. The big surprise would be if it had not.[59] Likewise, India's textile production made a marked impression on British inventors and industrialists, whose efforts subsequently created an industrial revolution that destroyed cotton manufacture in India, along with the lives of millions of Indians. This phenomenon is encapsulated in the word "factory," which comes from European merchants' trading centers in places including Africa and India.[60] Though people subjugated in colonized lands or not granted citizenship in Western nations participated in world-changing technological and cultural practices, narrative makers omitted them and they received no meaningful share of the profits.[61]

WORLDVIEW

The grand historical narrative of the nineteenth-century United States rested on two pillars: the idea of progress in the West, with Western institutions representing a cultural apex, and the belief that the United States would be a wellspring of that progress. On the ground, the advance of civilization equaled the decimation of Indigenous peoples, unbounded exploitation of natural resources, and the transformation of nature through technologies including mapping, deforestation, wetlands destruction, agriculture, and the construction of transportation infrastructure—all eventually organized within a railway-driven standardization of time.[62] In evolutionist eyes, the creation and use of tools that remade environments were evidence of "human industry [as] a natural expression of the development of the earth itself," in the words of historian John Tresch.[63]

Foundational to environmental transformations was the building of infrastructure. Within the process of national self-construction, bridges served as indispensable physical links—and offered awe-inspiring nation-making experiences. Novel large-scale structures invited citizens and visitors to come together and experience national dominion over natural, rural, and urban landscapes. In 1800, for example, the future 8th arrondissement of Paris was countryside; by the 1870s it was a neighborhood marked by the bustling St. Lazare train station near a six-limbed lattice bridge, the Pont de l'Europe, which enabled vehicles and people to continue through the streets that converged above the tracks. Artists including Gustave Caillebotte and Claude Monet produced depictions of the bridge in artworks that showed the lattice structuring and symbolizing changes in big city life (figure I.7). The starkly geometric imprint in the

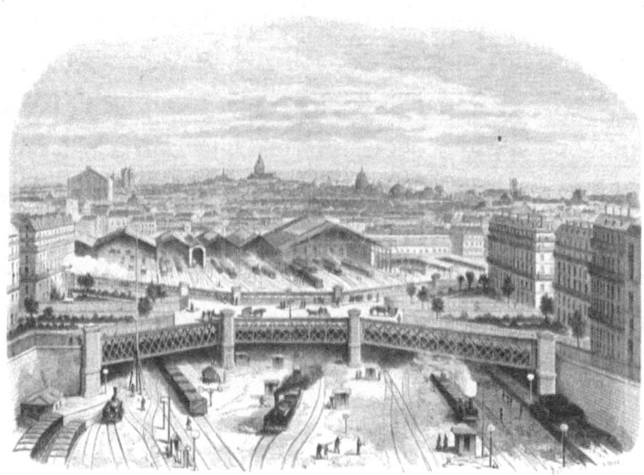

Figure I.7
Pont de l'Europe (Julien, built 1865–1868, demolished 1930). Caillebotte's precise rendering of the iron lattice Pont de l'Europe in Paris portrayed a social space that provided views of tracks and trains, St. Lazare station, and a new neighborhood in a city transformed under Haussmann. Top: Gustave Caillebotte, *Pont de l'Europe*, 1877. Petit Palais, Geneva, Switzerland/Bridgeman Images. Middle: Pont de l'Europe, ca. 1900. Collection of the author. Bottom: Pont de l'Europe, Pierre-Auguste Lamy, ca. 1868. *L'Illustration*, April 11, 1868, 236. Musée Carnavalet, Histoire de Paris.

street plan made the metallic hexapod emblematic of Haussmannization, a transfiguration more controversial than it might appear today.[64] Such imposing, unclad structures in busy urban settings were new, though in less populated, more rural locations, bridges such as the Portage, Kinzig, and Grandfey, all of which we explore in *American Bridge* and which no longer exist, had long been industrial culminations and popular attractions. National infrastructural experiences reached climaxes with the pedestrian walkway of the Brooklyn Bridge and the observation decks of the Eiffel Tower, a supersized bridge pier. Citizens and visitors contemplated, climbed on, and gazed out through structures that framed mind-blowing views of landscapes, intensified perceptions of national power, and carried people through shared experience into national community (figure 8.16).[65]

Understanding the worldview embodied in the national-infrastructural experience—whose sharing occurred in person and through the endless reproduction of images sent around the world—may enable us to change the story and build new bridges to the future. Recognizing, as we will, the interactions of bridge builders who worked in places including Schaffhausen, North Carolina, Connecticut, Prussia, Paris, and Wales—not to mention shifting national boundaries, imperialist expansion, and industrial firms' multinational work—leads us to wonder: Is the Eiffel Tower French? Though its première was deliberately drenched in French nationalist rhetoric, this question was salient at the time of its construction, when this supposed symbol of democratic society, lacking apparent connection to its surroundings or Paris' past, was labeled American, foreign, and, by anti-Semites, Jewish. Popular prejudice, especially during periods of heightened anxiety about cultural, social, and economic change, portrays Jews as cosmopolitans who threaten national identity.[66] The reinventors of building, connected far and wide, were the epitome of border-crossing cosmopolitans. With this in mind, *American Bridge* explores design, identity, and history through conceptual and material flows that enabled people to transform nations with bridges, towers, and stories.

1

FABRICATING MATERIALS

> When we no longer look at an organic being as a savage looks at a ship, as at something wholly beyond his comprehension; when we regard every production of nature as one which has had a history; when we contemplate every complex structure and instinct as the summing up of many contrivances, each useful to the possessor, nearly in the same way as when we look at any great mechanical invention as the summing up of the labour, the experience, the reason, and even the blunders of numerous workmen; when we thus view each organic being, how far more interesting, I speak from experience, will the study of natural history become!
>
> —Charles Darwin, *Origin of Species*, 1859

Darwin's recognition of the significance of the innovative interactions of "workmen" aided his investigations of the natural world.[1] With technology as a lens, he contrasted "savages" with civilized people, his readers; this racialized perspective on who is capable of knowledge and who is not, endures in ways obvious and hidden.[2] Anthropologist Alfred Russel Wallace, who developed a theory of natural selection at the same time as Darwin, had a different view of evolutionist hierarchies: he referred to "our modern civilization—more correctly barbarism" when recognizing the plight of industrial workers under capitalism: "A million people in London without sufficient food and clothing and fire for a healthy life."[3] Certainly, most of us today would qualify as "savages" with limited understanding of the technological mechanisms that shape our lives and seem to provide unlimited capacity to control them.

As storytellers sorted humanity into savages, barbarians, and civilized citizens, their evolutionist worldview transformed materials and structures into actors and status symbols. Historians diminished the significance of the "numerous workmen" as they cast iron in a starring role in a national-technological epic. In this chapter we'll dive into material facts and fictions in order to reset our minds. This may encourage us to break free of standard framings as we explore the reinvention of building. We'll begin de-evolutionization by disentangling materials, stories, structures, and nations, along with the people who were in the businesses of designing, manufacturing, building, and exporting bridges. We'll consider codependency rather than hierarchy in the cellulosic-ferric landscape. We'll ask basic questions: What is a "wooden bridge?" How do we know? Who decides? Was the "first iron bridge" a watershed? Whose interests do such categories serve? This quick recalibration will indicate how and why evolutionism, steeped in ideas about progress and national culture, delimits our views of past and future.

A branching diagram or "simile," as Darwin called the sole image in *Origin of Species*, became a model for establishing categories and tracing relationships. Subsequently, tree pictograms confirmed life forms' common ancestry while portraying subsequent ramifications. But a tree is unable to depict the unnatural human-made interconnections behind technological development. Even as a biological tool, the diagram can falsely communicate that the top is the most evolved, or that there is only one way to classify things, or that one branch is more important than another (figure 1.1).[4] A tree of bridges doesn't reveal the thoughts of the inventors and engineers who imagined and realized new possibilities.[5] While biologists can reconstruct millennia-long histories through artifacts, merely examining the artifacts will hardly suffice for understanding material choices, structural ideas, and constructional strategies in bridge building.

Wood, as we'll see, embodied what historians usually call "revolution," but they did not celebrate nonmetallic breakthroughs. They did not highlight wood and iron's intimate codependency. They may have been unable to perceive the wooden choice, so relentless were the progress-focused legends of industry and technology, in which wood plays at best a supporting role as the material of underresourced, unevolved, or uncivilized peoples. In this narrative, wood exists primarily to be replaced by metal. Storytellers made the first metal bridge into a symbol of civilization, a break from a supposedly unscientific, preindustrial past, an era still inhabited by (non-Western) people stalled in a previous phase of cultural evolution.[6] Such beliefs were the offspring of an informal partnership of scientific theorists and industrial capitalists: anthropologists theorized an evolutionarily advanced, science-based civilization that iron epitomized, while engineering entrepreneurs promoted new, advanced metals they were seeking to sell.[7]

Industrialization, however, meant an increase in the production and use of lumber, a new, standardized product, along with the ability to ship it via expanding transportation networks that enabled ever-greater consumption. In Britain and Ireland between 1840 and 1870, 268 railroad companies built about 2,500 wooden bridges, half between 1855 and 1870 (well into Britain's iron age). The factors behind this modest but not insignificant number include cost, ease of transportation of materials, speed of construction, and durability. Remembering a past in which wood is less than civilized means remembering (that is, envisioning) a future in which wood remains so. It means forgetting that designers have and can employ wood in structural roles on a par with metal.[8]

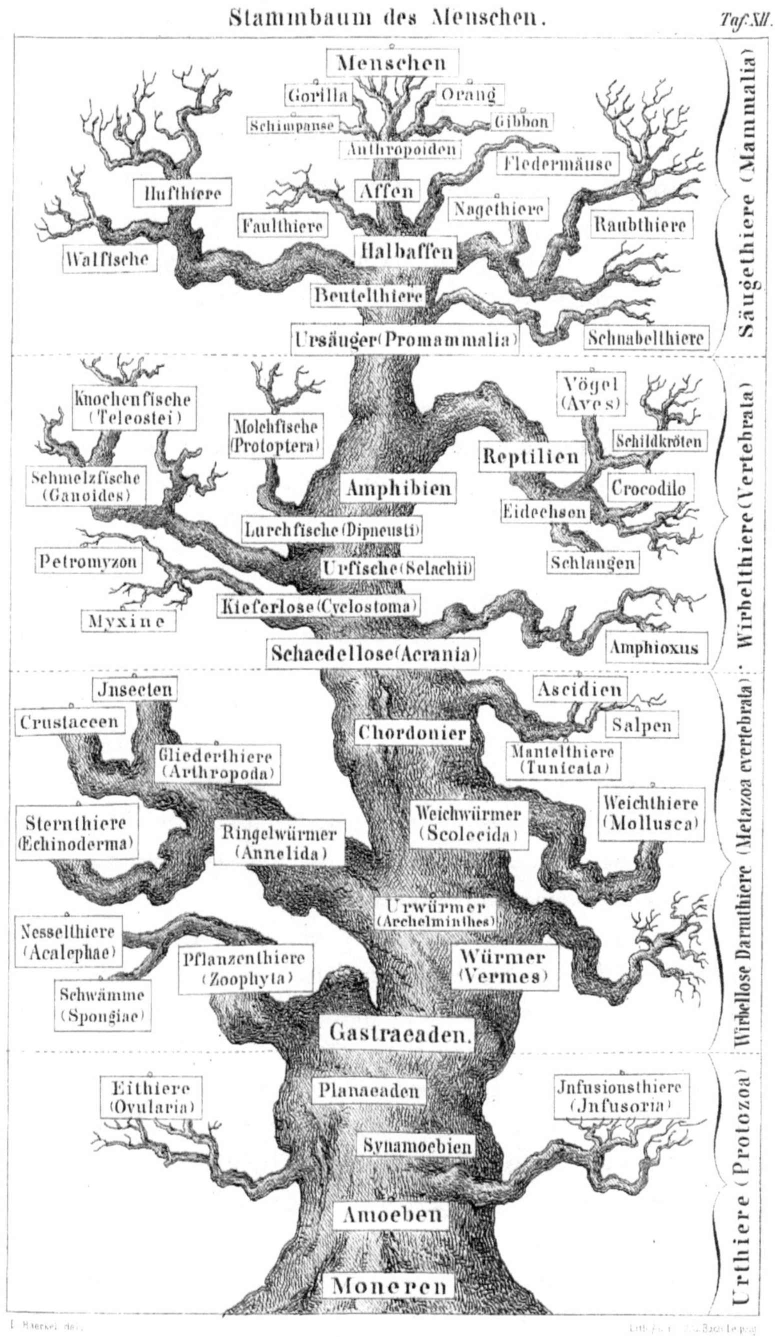

Figure 1.1

Ernst Haeckel, Family Tree of Man. Tree pictograms laid out and justified evolutionary, human-centered hierarchies. From 1866, zoologist and Darwin advocate Ernst Haeckel published hundreds of such images, which provided scientific support worldviews based on race and nation. In the English edition, this image was entitled, "Pedigree of Man." Ernst Haeckel, *Anthropogenie oder Entwickelungsgeschichte des Menschen* (Engelmann, 1874), table 12. Deutsches Museum Library.

IRON FIRST

Wood and iron products each challenge designers to proceed with specific factors in mind. Neither material has inherent moral, scientific, economic, or technological superiority. Many qualities of wood compare favorably to metal. In railway stations, wood displayed better resistance to corrosive fumes and flammability than steel. Wood is lightweight, strong, easily worked, and elastic. Wood's properties change in relation to its moisture content; in a structure, this can necessitate readjustment over time.[9] Both wood and iron are vulnerable to weather conditions, fire, water, vibrations, and a lack of care. Under some conditions, one will be preferred; sometimes it's a toss-up. But the interdependent, almost clandestine, relationship of wood and iron—with wood serving as silent partner, model, foundation, scaffolding, and fuel, while iron is glorified—forms the context for our investigation.

What is a wooden bridge? Many of the best-known wooden bridges of the eighteenth century (and later) contained tons of iron, in the form of critical parts such as bolts, straps, and rods (figure 1.2).[10] The renowned 234-foot-tall, 800-foot-long Portage Bridge (Buffalo & New York City Railroad, which became part of the Erie Railroad in 1861; Silas Seymour, engineer) in New York State, built between July 1851 and August 1852 with fifteen towering piers interconnected to create a trestle, was considered the largest "wooden" bridge in the world. It contained 135,500 cubic feet of timber, 92,000 cubic yards of masonry (for the abutments and bases of its fifteen towers), and 49 tons of wrought iron—bolts and rods.[11]

The use of iron connectors and members in "wooden" bridges and roof structures was not uncommon in Europe by the 1600s; hardware plays an important role, for example, in Palladio's Cismon bridge design (figure I.5). The earliest printed depictions of bridges with wooden members reveal the integration of iron components. The internationally known "wooden" Grubenmann bridges (figure 2.16) of the mid-eighteenth century were held with iron tension rods.[12] In "iron" bridges, wood was sometimes used as a deck material; wood and cast and wrought iron were sometimes used together in composite structures such as the popular Howe bridge (see chapter 4). Complex temporary wooden scaffolding, known as centering, most likely held together with metal connectors, was often used during the erection of bridges. Even when wood was not part of a finished (stone or metal) product, it was essential to construction.

So why have historians celebrated the "iron bridge," with rapturous attention to the "first"? Let's take a look at Abraham Darby III, manufacturer and builder of the so-called first iron bridge, over the River Severn in Coalbrookdale, England in 1779. He erected a cast-iron arch whose design was modeled after stone construction, with

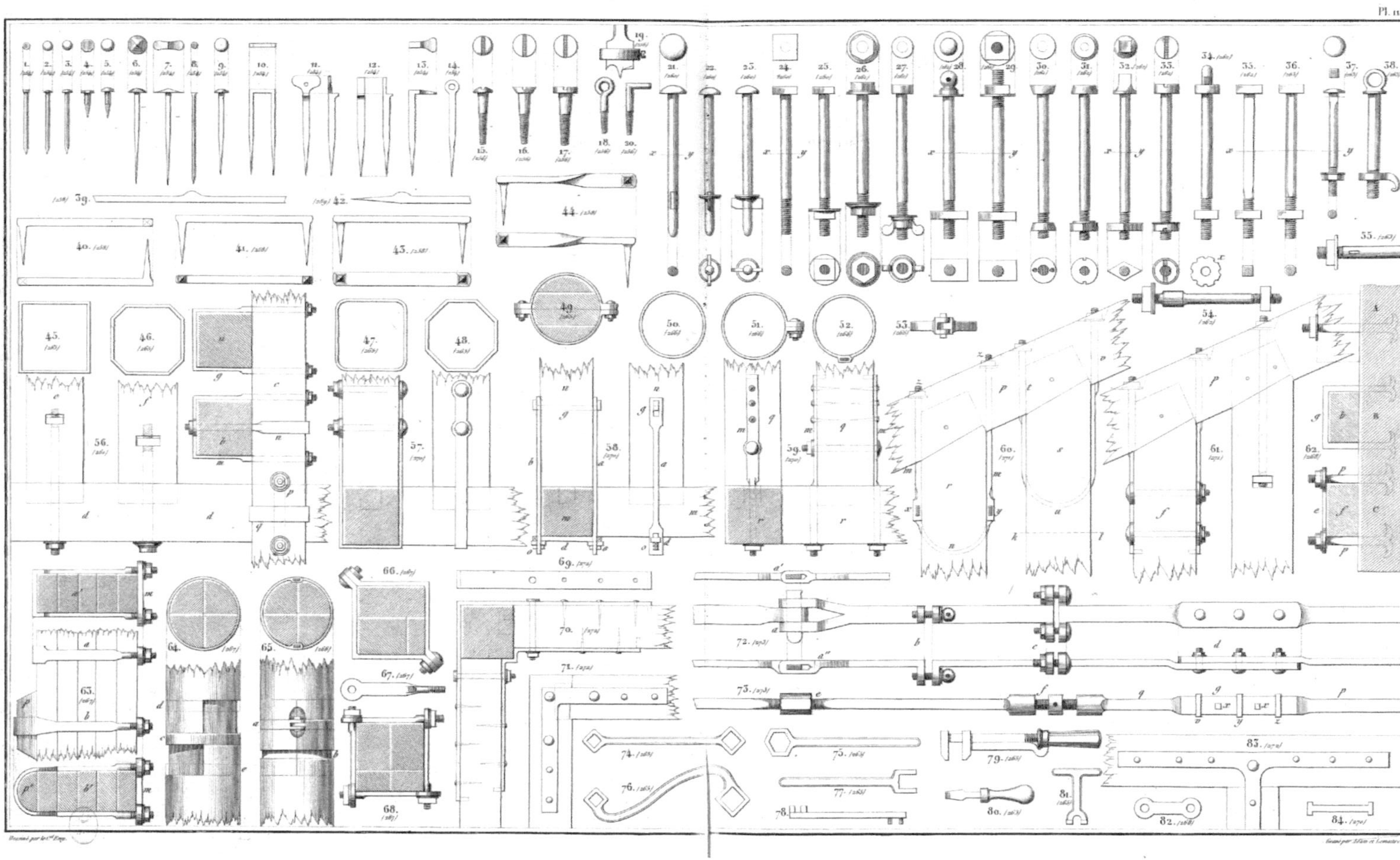

Figure 1.2
So-called wooden structures depended upon metal connectors, displayed here like natural history specimens. Amand-Rose Émy, *Traité de l'art de la charpenterie* (Carilian-Goeury et Anselin, 1841), pl. 116. Bibliothèque Nationale de France.

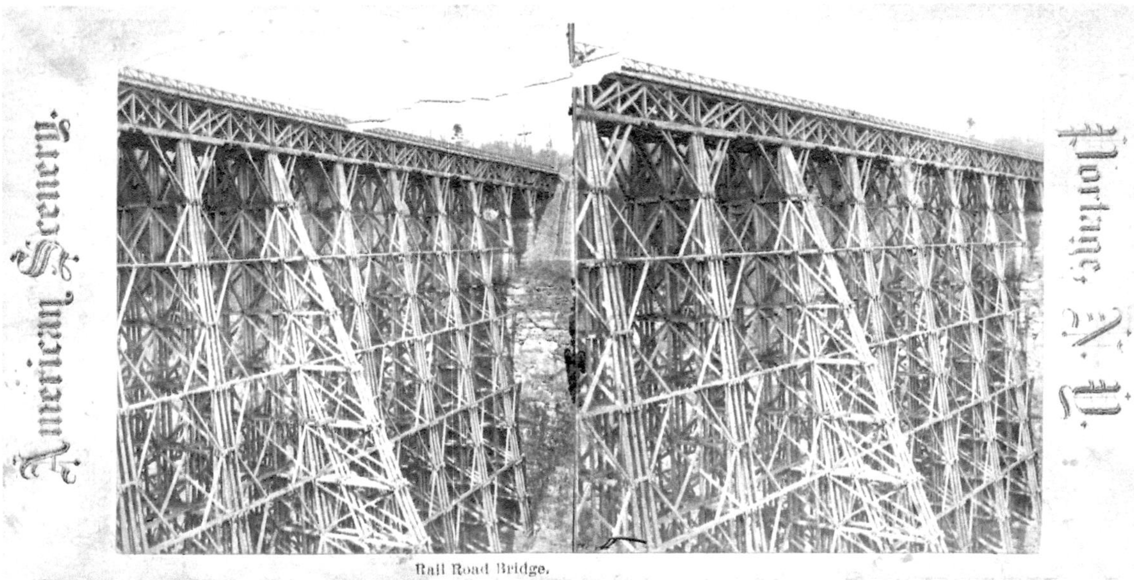

Figure 1.3
Portage Bridge (Engineer, Silas Seymour; Contractor, Lauman, Rockafellow & Moore, built 1851–52) over the Genesee River, Livingston/Wyoming Counties, New York. One of the world's largest "wooden" bridges contained tons of iron. Collection of the author.

connections inspired by wooden structures. This was an advance in some respects similar to the substitution of metal for wood in doorknobs—a translation rather than a transformation—made possible thanks to iron manufacturing methods. The building of the Coalbrookdale bridge itself, along with paintings of it that Darby commissioned, were promotional efforts for his foundry (figure 1.4). Although, the artists who for decades produced images of the bridge helped make it an indelible "first" in Britain's industrial revolution story, the structure did not become a model for subsequent design or construction.[13] Nor did the artists and historians reveal that the slave trade had been a source of wealth essential for the Darby business or hint at the role of enslaved Africans in the development of ironmaking.[14] And the structure at Coalbrookdale was a "first" preceded by an iron bridge over the Rhone in Lyon, France, a project aborted, because of cost, after the erection of its first arch.[15] (Firsts happen in a variety of forms and chance often determines their locations). Beyond its role as a Shropshire connector, however, the "first iron bridge" is most significant as an assertion of empire.

The material composition of the bridge at Coalbrookdale was novel. For evolutionist accounts of innovation, novelty (rather than functionality or maintenance) is usually the defining issue. But as historian Amy Slaton notes, apparent newness depends on who

Figure 1.4
William Williams, *The Iron Bridge*, 1780. Cast-iron bridge (Thomas Pritchard, design; Abraham Darby III, manufacturer and builder, built 1777–1781) over the River Severn, Coalbrookdale, Shropshire, England. This image was commissioned as an advertisement for the foundry that produced the iron. Ironbridge Gorge Museum Trust.

controls the knowledge, who is investing in production, and whose identity newness is linked to.[16] The person or community accumulating associated money or power typically identifies the new. So "firsting" identifies heroes and celebrates the place associated with them. It downgrades or erases what came before. (For example, the notion that science was suddenly "applied" to engineering erases a long history of technologists employing a variety of knowledge types in their work.[17]) This was a particularly meaningful narrative strategy in the United States, where inventors had to establish novelty in order to access the patent system's potential to generate wealth—in a place where Europeans' claims of being first was the rationale for taking possession of the continent.[18]

In the case of Coalbrookdale, the British government was determined to make iron British. The Iron Act of 1750 was a (not very effective) attempt to limit the North American colonies to producing only raw metal (pig iron) for export to Great Britain and to prevent the colonial expansion of the manufacture of iron products, in order to keep the colonies as major consumers of British-made products, such as nails and agricultural tools.

Submerged in the complexity of the wood-iron relationship is the fact that wood fed the fires of industrialization—the smelting of iron was dependent upon wood until the Darbys and others helped make the transition to coal. Through the nineteenth century till today, wood has been used as fuel; in 2022 in Europe, 50 percent of harvested wood is burned for heat or electricity.[19] Metal's supremacy over wood was ideological and tangible: its manufacture has devastated forests and environments.

A crux of the invention of the lattice bridge was that the structure required no iron.[20] This wood-metal divorce enabled a breakthrough: the construction of simply assembled long-span beams. As a prominent observer exclaimed in 1837, American bridges "are all remarkable in that they require almost no iron."[21] Iron was considered a disadvantage: it was expensive, difficult to manufacture, it could fail without warning, and it rusted. Eighteenth- and nineteenth-century structural innovators who developed structural systems only with wood were the industrial avant-garde.[22] Wood was the medium through which a rethinking of construction and structure took place.

BRIDGE OF WOOD, BRIDGE OF IRON

Let's continue our attempt to shatter over two centuries of metal mythmaking by recalling that structural iron at mid-nineteenth century was an experimental material. There was no inexorable rush to embrace it. And for bridges, iron, like wood, was long considered inferior to stone. Engineer Charles Goschler, for example, was doubtful about the use of iron; reporting in 1856 to the Société des Ingénieurs Civils on the long-span iron lattice viaducts of Carl Etzel (see chapter 5) he said:

> What strikes the observer is the almost exclusive use of iron as the principal structural material in the bridges. If one looks for the cause of the Swiss engineers' preference for iron to the exclusion of other materials, it can only lie in the desire of the builders to put their [railroad] lines into operation in the shortest possible time. Isn't it to be feared that here again, as in the case of the choice of railway routes, the companies have sacrificed their future interests for those of the present? Experience will tell.[23]

Material perceptions reflect the ever-shifting contexts and relationships that determine material choice.

Designers in France, as well as the United States and German states, considered stone a superior, even noble, material. Despite "America's wonderful plenty of wood," it appears that, in 1822, thirteen of the twenty longest bridges in Pennsylvania were built of stone, and two more were partly stone.[24] There were debates among Baltimore and Ohio Railroad engineers about stone and wood bridges, how much they should cost, and who had the authority to select the materials. When the B&O built the Washington-Baltimore line (1833–35), all fifteen bridges were stone (figure 1.5), as was pointed out in the *American Railroad Journal*'s long rebuttal of engineer David Stevenson's false assertion that "there is not a stone bridge in the whole of the United States or Canada."[25] In the United States, stone was sometimes publicly proclaimed as the superior material even as it began to be rejected in favor of wood. The durability and heaviness of stone was not desired for structures whose design lives fit the values of railroads that were under continual reinvention and rebuilding. Still, engineers replaced the renowned wooden lattice bridges of Moncure Robinson, built in the 1830s on the Philadelphia and Reading Railroad (see chapter 3), with stone arches in the 1840s and 1850s, even though wooden structures could be built more quickly.[26]

Even if stone railway bridges in the United States were "rare occurrences," North America was not poor in stone. Material abundance, however, was not necessarily the deciding factor in design. Economic interests, blended with inventors', builders', entrepreneurs', and managers' attitudes toward each material, were as important as what historian Edwin Layton Jr., has called the "intractable nature of materials."[27] How people test, use, and interpret materials—the stories we tell about them—establishes their "nature," as does the wealth that their manufacture and use generates. Materials are tractable stuff.

In 1830, an editor of *Dingler's Polytechnisches Journal* responded huffily to an engineer's proposal in the *Journal of the Franklin Institute* to use wooden ties under rails in railroads. (The German term "auf dem Holzweg" literally translates as "on the wooden path" and means "on the wrong track."[28]) He wrote:

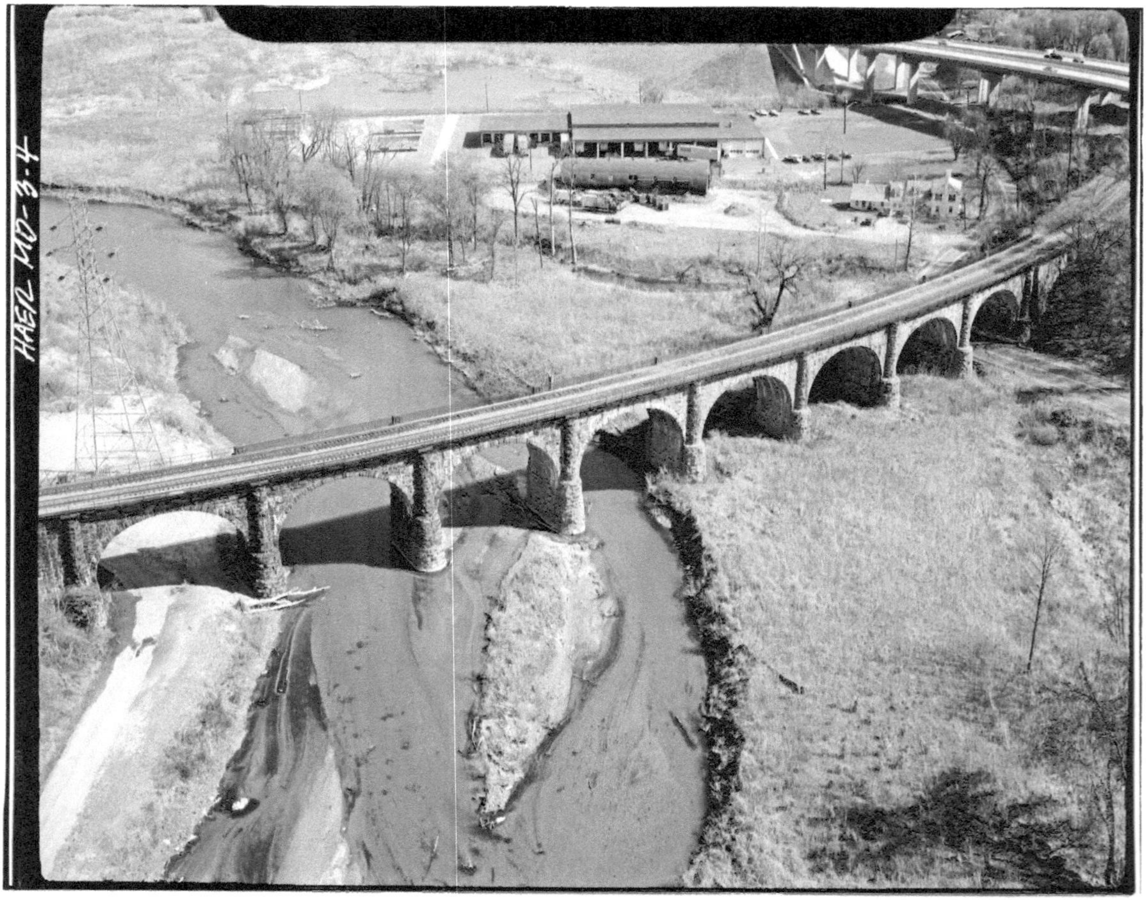

Figure 1.5
Thomas Viaduct (Benjamin H. Labrobe II, built 1833–1835), Patapsco River, Baltimore and Ohio Railroad, Elkridge, Howard County, Maryland; Jack Boucher, photographer, 1977. Maintenance and reconstruction enabled this granite bridge's survival. Library of Congress, Prints and Photographs Division, HAER MD-14-ELK, 1.

> Although North America still has more oak trees than people, it seems to us that what must be made of stone and iron to be good, could not be better made of wood. Even the students in Germany believe that their misfortunes can't be better described than with the expression "Holzweg." One may think of them as children or fools; at least children and fools tell the truth. Do something right, or don't do it all.[29]

Wood, "more perishable [than stone] in the abstract, but more enduring as a component part of a railway structure," became the standard tie material, the foundation of the rails.[30]

But as children and fools—and many engineers, entrepreneurs, and historians—believed, wood was not the material of progress. Wooden rails capped with iron, used in the earliest railway systems, proved inadequately resilient and durable.[31] The terms for "railroad" in French and German, "chemin de fer" and "Eisenbahn" (both, literally, "iron path") helped establish iron as the material of the future. Telegraph wires of metal appeared to confirm this—though they were carried, like the rails, on wood. In 1880 alone, 60 million ties were installed under US railroad tracks.[32] But storytellers celebrated neither ties nor poles, industrial minions manufactured in mind-boggling quantities.

During the nineteenth century, bridge builders seesawed between wood and iron. (The switch from iron to steel rails was also complicated; in the United States, it took twenty years and, like the choice between wood and iron, cannot be explained simply as a cost issue.[33]) Around mid-century, the New York and Erie Railroad decided to replace its few iron bridges with wooden ones after the collapse of a patented iron bridge type, which failed apparently because the design of specific members had been insufficient for the loads it had to carry.[34] Perhaps related was the construction of the Portage Bridge (figures 1.6, 5.21, 8.7). The wood came from 160 (also reported as 300) acres of pine forest owned by the contractor, approximately a mile away, where a steam-powered sawmill operated. In 1875, this reputedly largest wooden bridge in the world burned; surely the conflagration confirmed the inferiority of wood. Fire protection included a watchman day and night, a water reservoir, and pumps at various levels. But during the night, the watchman discovered that the fire had begun *under* the deck; moreover, "It was raining, and the top of the bridge was so wet that it seemed impossible for a spark from the engine to fire it." Arson was the likely cause; the Erie had long wanted to replace the bridge but could not move ahead because the structure had become such a popular visitor attraction, due in no small part to its own promotional efforts. Only eighty-two days after the fire the iron replacement bridge was complete. Had planning started before the fire?[35]

When the first Portage Bridge was new, an engineering journal based in Paris remarked, "It is not possible to do better in countries that are still new and have primitive materials, limited resources, and little income expected in coming years." The writer stated that in France and England iron, stone, and "cements" were replacing wood in large structures, but that "the general principle of American work," that is, to build quickly and cheaply with the expectation of replacing later—a strategy some consider to be a breakthrough of the digital age—should be emulated.[36] But the "wooden" Portage Bridge was far from primitive.

There is an oft-expressed but rarely explored idea that if a structure is around for a long time, it has proven its superior quality, rather than demonstrated that it was

Figure 1.6
Genesee Valley, as seen through the Portage Bridge, Portage, New York. The Erie Railway offered touristic experiences featuring awe-inspiring forests, waterways, and bridges. Tourists viewed American scenery framed by wooden parts that were avatars of interchangeability. The Miriam and Ira D. Wallach Division of Art, Prints and Photographs: Photography Collection, New York Public Library.

overdesigned (perhaps deliberately, as more material can enhance safety and durability), well maintained, or rebuilt. During the formative years of railroad development and iron manufacturing, limited capital and increasing rail loads inspired engineers to design relatively short-lived, less costly bridges. As Robert Fulton noted (and Ithiel Town cited), using wood meant that annual repair would be economical, compared to investing a large sum in a stone bridge.[37] Producers may have advertised metal as a more permanent material, but that isn't necessarily what was wanted. At the end of the nineteenth century, engineers Merriman and Jacoby wrote in their standard textbook that "the average life of iron railroad bridges is probably not far from twenty years, although under heavy traffic many are replaced after fewer years of service." (The Portage Bridge was torched after twenty-three years.) They asserted that metal was cheap enough to make frequent replacement economically viable, but for shorter spans, railroads were replacing iron bridges with stone arch.[38] At that time, it was reported that unprotected spruce and pine would last six and fourteen years, respectively; a covered bridge, kept ventilated and clean, would last forty to fifty years.[39] It all depended on context. If a wooden bridge had (an unplanned) short life span, it was likely due to factors such as inadequate design, incomplete understanding of forces in the members, defective workmanship, the type of wood chosen, an insufficient inspection and maintenance program, improper repair, or the increasing weight of trains—rather than because wood was flammable, prone to decay, or evolutionarily and morally inferior to stone and metal.[40] In places where iron production was important to the economy, and the new was often proclaimed a national achievement, wood, although always an important building material, was in a vulnerable position. And if a bridge was deemed "wooden," its failure could be attributed to the material. As an engineer pointed out in 1882 regarding a collapsed bridge in Tariffville, Connecticut (figure 1.7): "Though nominally a wooden bridge, like all such structures, it relied entirely upon iron rods to keep the woodwork together."[41] In other words, the wooden bridge was an iron bridge. In this case, the wood was not properly protected nor were adequate inspections conducted. But narratives shape material pasts and futures.

Those futures came about through events sometimes described as "local," a confusing term that often hides the interactions of extended networks.[42] By 1847, Squire Whipple, the pioneering iron bridge builder and early developer of practical methods of structural analysis, carefully compared wood and iron in bridges and concluded that, though it was not always the clear choice in every context, metal was the material of the future.[43] At mid-century, as the New York & Erie replaced metal bridges with wood, the Compagnie de l'Ouest in France replaced its wooden arch bridges with cast-iron arches, in order to protect against sabotage. (The Asnières oak and pine arch

Figure 1.7
Failure of iron tension rods in a supposedly wooden Howe bridge, Tariffville, Connecticut, 1878. Daniel S. Camp, Hartford, CT. Connecticut Museum of Culture and History.

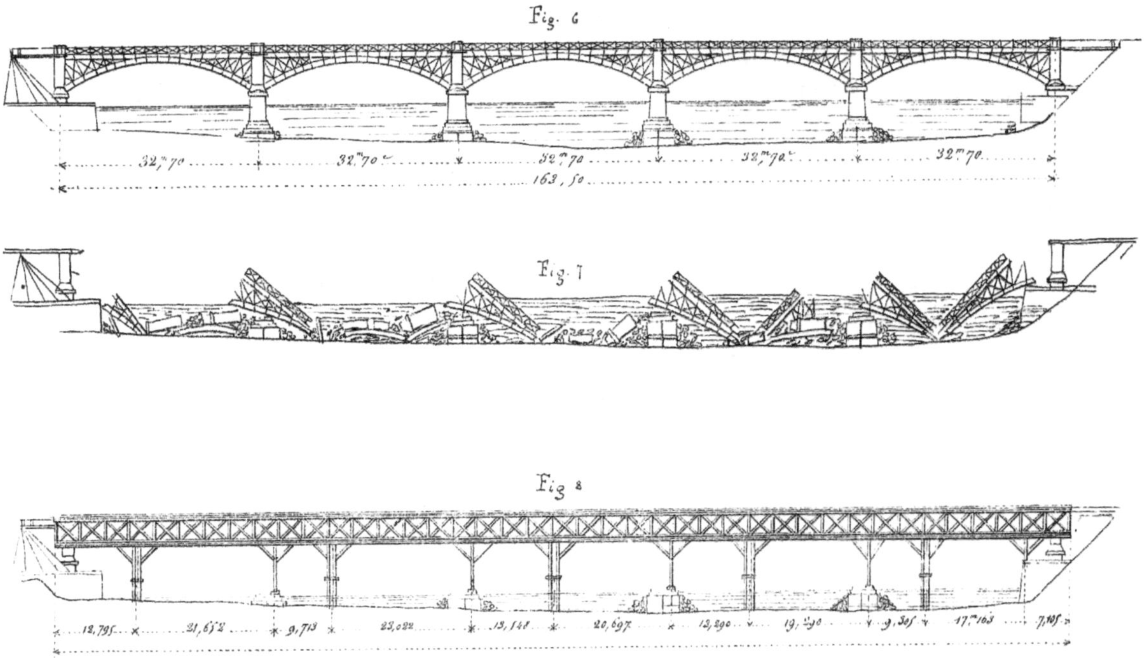

Figure 1.8
Arch bridge at Asnières (Karl Etzel, 1836), before and after its destruction in 1848. The design of the temporary replacement (Eugène Flachat, 1848) was based on the bridges of Ithiel Town and especially Stephen Long. It needed to be very rigid and to accommodate irregularly spaced piers; it was designed to serve as scaffolding for the installation of its replacement, an iron box girder bridge, in 1852. *Mémoires et compte rendu des travaux de la Société des ingénieurs civils* 1 (1848): pl. 3. University of Minnesota.

bridge, 1836 [figure 1.8], which carried three important rail lines to Paris, had required major reinforcing and maintenance efforts and was burned during political upheaval of 1848.)[44] These reversals were driven by a variety of factors. By 1854, though, German states had banned wood for railway use, and by 1863 both wood and cast iron had been "removed from the field of investigation[:] the former by negation, and the latter by direct condemnation."[45] Flammability and difficulty of manufacture and maintenance were apparent reasons, although the weight of these factors varied with local decision-makers. Designers could evaluate wood and iron in terms of manufacture, availability, transportation, labor, and maintenance and replacement costs. Their decisions were dependent on changing contexts, which meant that today's assumptions might prove inaccurate tomorrow. The Erie Railroad's turn to wood was not permanent, and in

1859, after convincing New York Central Railroad's management—via testing and the construction of two small bridges in Schenectady and Buffalo and a longer one in Batavia—that iron bridges were structurally and economically viable, a New York Central vice-president declared, "No more wooden bridges for this company." In the following decade, the repeated burning and building of wooden bridges during the US Civil War may appear to signal a deficiency, but wooden structures could be quickly built and more easily repaired and reinforced than iron (figures 4.11). This helps explain why some railroads in the United States continued to build wooden bridges into the twentieth century.[46]

Material did enable structural change; iron members with iron connections eventually were able to provide a rigidity, continuity, and durability that altered how bridges functioned. Still, it is a narrative commonplace to find "wooden bridges" and "iron bridges" presented as if they are natural rather than mind-crafted categories. Evolutionist hierarchies boosted metal supremacy, to the benefit of those pursuing power, wealth, or status.

MATERIAL DESIGNS

Evolutionist storytellers have for centuries reinforced iron as a gauge of progress. They employed technology as "the measure of men" and portrayed iron as the quintessential industrial-revolution material. Their calculation was rooted in the late eighteenth-century transition in the British economy's basis from wood and water to iron and coal.[47] Within this ferric landscape, wood was deemed appropriate for less complex, less civilized societies. The words of historian Carl Condit, whose publications on building technology were widely read, ring through the decades: "Wherever wood was plentiful and industrial techniques less advanced than in Western Europe, timber construction was bound to be the natural choice." He believed that wood framing belonged to a "vernacular tradition," that is, unscientific, less advanced. Wood's mythic nature—unlearned, craft-based, inflammable—helps explain why in the early twenty-first century the use of wooden members in high-rise buildings can still evoke surprise.[48]

Understandings of materials reflect engineering, evolutionist, economic, and nationalist perspectives. These shape how we see materials and how designers use them. Engineer John Roebling in 1860 proclaimed, "Iron has emphatically become *the material of the age*. Upon its proper use the future comfort and physical advancement of the human race will principally depend. It will yet be the harbinger of peace, as already it has given us the means of locomotion and of intelligent intercourse."[49] His rhetorical fervor aligned with the industrial-evolutionist reasoning of his time and is comparable to the language of today's promoters of digital technologies. Roebling believed that technology provided evidence of a superior civilization and that technological progress would benefit the world; his metal cable manufacturing company was helping make it happen ("intelligent intercourse" probably refers to the far-reaching impact of telegraph wires, which Roebling's company manufactured). Roebling, like Abraham Darby and William Fairbairn and Robert Stephenson, prominent engineers we'll meet in chapter 6, had a financial stake in iron. (This is not to say they were venal; a belief in a future they would profit from was integral to their entrepreneurial mindset.) Roebling's design for the Niagara railroad suspension bridge (figure 6.14), an American symbol whose image traveled around the world, depended on cables, though its deck initially made substantial use of wood and iron.[50] When the bridge opened in 1855, the *American Railroad Journal* proclaimed: "It must place the name of *Roebling* high among the greatest and best of those who have accomplished most for the advancement of their species."[51] This Prussian-born technologist remains an American hero to this day, thanks to his Brooklyn Bridge.[52] But historians who adopt material ideologies and biases from innovator-entrepreneurs such as Roebling and integrate them uncritically into storymaking create puzzling scenarios. They write things like "the iron truss came soon after the iron arch."[53] This affirmation, which links disparate structures and seems to exclude wood, is a celebration of progress rather than an insight into innovation history.

Wood and metal structures each required its own design method. With wood, the connection type determined the size of the entire member; with iron and steel, designers established the size of the members first. As an engineer explained in 1933, "If he designs a steel bridge while standing on his feet, he should stand on his head while designing a timber structure. In other words, the processes are reversed."[54] In metal structures, a much smaller material area was required for the connections, which were simpler, could take a variety of forces from various directions, and could be designed as an independent feature of the structure; plus, the shape of the member could be more precisely specified and manufactured. Metal enabled designers to create structures that physically were closer to one-dimensional depictions—that is, closer to the diagrams used in structural analysis. Lumber was superimposed and connected at the overlaps, while metal construction approached a single plane, with connection points where members met, usually at their ends.[55]

Designers had to think about wood and metal in different ways. Essential for successful design with wood was knowledge of the characteristics of wood types, which was shared by manufacturer and builder.[56] Because wood fails with forewarning, builders could learn through observation and repair. (A recent study reports that nineteenth-century wooden railway bridges had a safety advantage; they were not known to collapse while trains crossed them.[57]) As railways rejected wooden bridges, the importance of the type of knowledge and experience required to build with wood diminished. Iron can break with little or no warning; this may be due to metal type, how it is used, or to the quality of manufacture. Metal members were more of a black box, not knowable in the ways that wood was.[58] Bridge designers and builders trusted the metal maker to produce a material that met a testing standard, which in most cases was independent of the designer. The alienation from first-hand knowledge, along with the foregrounding of analytical tools for designing structures, became fundamental to the mass-construction of bridges. The designer could develop a structural idea in the abstract—and then seek materials that fit the design.[59]

The evolutionist ascent-of-iron narrative lowered the status of the carpenter as it elevated that of the engineer, who possessed a different kind of knowledge. No matter the material, however, intuition (that is, tacit knowledge and skill based on experience) remained basic to design. As historian Joachim Radkau explained, craftsmanship and a feeling for materials were still important, before iron and industrialized building and after, but "human skill was pushed to the edge of technologists' consciousness."[60]

The professionalization of engineering took place along with the development of structural metals. Civil and mechanical engineers became closely identified with new types of structures that employed metal; this enabled them to distinguish themselves from contractors—whose participation in iron construction was also essential.[61] Eminent structural engineer Corydon T. Purdy's assertion in 1895 that "it is only with the advent of steel that the engineer has become a necessity" transmitted heavy metal reverberations about progress: new materials require new people with new knowledge to replace those who came before. That year, in the same journal, engineer J. Parker Snow, while describing wooden lattice railway bridges he was maintaining, shared an "impression" that wooden bridges had become "obsolete." Already thirty years earlier, when an engineer mentioned "the lattice, long since abandoned as a wooden structure," he appeared to be confirming extinction; yet wooden lattices continued to be built, though in smaller numbers.[62] But metal was the material on which engineers, entrepreneurs, and historians were building professional status, careers, and wealth.

The idea that materials provide an evolutionary track for designers to follow, rather than being commodities designers manipulate, is related to the belief that each material has a form through which it can best be expressed. This notion endures through the oft-quoted bromide associated with architect Louis Sullivan, "Form follows function," which fits with his convictions about nature and evolution. As if function dictated a unique form, or each form had only one function! Sullivan's business partner, engineer Dankmar Adler, more astutely deciphered design: "Form follows historical precedent."[63] Or as an engineer in 1844 remarked regarding the succession of bridge types, "There is a fashion which rages for a certain time."[64] This is evident in the number of cable-stayed bridges built in recent years. While evolution clarifies a process of change in an animal species, biology cannot account for the myriad decisions that drive the design of individual objects or innovation over time.[65] There is always a menu of possibilities to choose from.

But evolutionism can circumscribe that choice. Anthropological experts traced cultural evolution through the West with the advent of iron as climax. In the late nineteenth century, John Wesley Powell, a founder of the field of anthropology, explained, "The age of savagery is the age of stone; the age of barbarism the age of clay; the age of civilization the age of iron." Ethnologist Otis T. Mason confirmed that "the civilized man passes his whole life in the midst of wheels and cranks and engines of iron." Metal embodied a majestic symbolic potency: its mythical strength and permanence provided proof of the durability and civilizational direction of nation and imperial empire.[66] So cultural narrators ignored the role of wood at a pivotal inventive moment—the reinvention of building—and likely remained ignorant of the moment because of wood's centrality. Lumber and enslaved people were considered primitive or preindustrial, on the other side of the divide, even if they played a giant role in the making of industrial capitalism and management.[67] Like the racial classifications often employed to define society, "iron bridge" and "wooden bridge" in historical accounts often are factitious labels that reveal evolutionary caste. They refer to imaginary homogeneous types rather than actual mixed heritage and composition.

The evolutionist-progressive narrative also functions as a political power tool. Entrepreneurs and engineers used it to rationalize intentions and minimize mistakes.[68] It enabled industrialists John D. Rockefeller and Andrew Carnegie to justify aggressive corporate tactics and the mistreatment of individuals; Carnegie claimed that inequality and concentration of wealth were "essential to the future progress of the race." Evolutionism supported the view that efforts to make society equitable were unnecessary and perhaps unnatural.[69] In China, evolutionism would become a replacement for traditional values and a tool for making massive change.[70] Just as biologists have applied evolution to all scales of life, from genes to species change over thousands of years, historian-evolutionists have turned their attention to all scales of invention and industry, ranging from "arrow into rocket" fantasias to seemingly small alterations. In defense against a lawsuit accusing them of theft, a group of dismayed engineers asked: "How can you patent something that is in the natural evolution of technology?"[71] Indeed, if designers are evolution's agents, they would not be responsible for illicit appropriation.

By the beginning of the twentieth century, corporations in the United States were less likely to publicly espouse survival of the fittest. The idea went underground and fertilized the evolutionist-progress narrative of technology that nurtures today's neoliberal thought. In the twenty-first century, evolutionist narratives can deflect attention from the inequity behind, for example, digital devices—gleaming avatars of progress made of metals whose manufacture and disposal depend on environmental harm and brutal working and living conditions in places Western consumers never see. This is a dark side of the evolution-of-materials tale. While evolutionism's inextricable ties to the openly nationalist and racist currents of the nineteenth and twentieth centuries are well known, less discussed are its support for contemporary corporate "innovation" and its impacts. The question is: Who are technological evolution and progress for?[72]

"Arguably, no folk theory of human nature has done more harm—or is more mistaken—than the 'survival of the fittest,'" asserts anthropologist Brian Hare and science writer Vanessa Woods.[73] The catchphrase justified and perhaps inspired a couple centuries of human and environmental destruction. It's so deeply ingrained that it's hard to extract from our understandings of history and society.[74] Struggle represents only one way of viewing the world, however. For humans and animals, cooperation may be the strongest outcome of evolution.[75] The ever-changing relationships of interdependent individuals create stories, materials, and communities. Are "American bridges" like processed American cheese, whose development runs through England and Switzerland? Are they like Gruyère, officially made only in Switzerland and France, though the United States, which imports more cheese by that name from the Netherlands and Germany, claims the name is "generic?" Are American bridges like French dressing, whose origins do not lie in France and whose contents the US government controlled for seventy-two years, until 2022, twenty-four years after the Association for Dressings & Sauces asked that it cease doing so?[76] Or do they have something in common with chocolate? Eighty percent of cocoa beans come from West Africa, although only 1 percent of chocolate is made there.[77] Exploring how social and political flows define and redefine manufacturing, innovation, and consumption can lead us to shift our understandings of national, local, and global. Today, Kinshasha and Paris have the same number of Francophones; 60 percent of French speakers live in Africa, where they are remaking the language.[78]

Biologist and computer scientist David Krakauer wrote, "Genes, minds and societies are all involved in various forms of construction. A better understanding of life requires that we abandon the view that organisms are account books recording in their behaviour past ages of the Earth and see them rather as builders engaged actively in the planet's construction."[79] Instead of regarding people as enactors of biological tropes and national destinies, *American Bridge* depicts creators working in a multiplicity of places through networks that range across construction sites, businesses, governments, factories, and universities, while building a diversity of futures.

2

ASSEMBLING THE PARTS

> There is also frequently a secret pride which urges many to conceal their speculative enquiries, rather than meet criticism, or not be thought the first in their favourite pursuit; ever anxious to claim the merit of invention, they cannot brook the idea of having their works dissected, and the minute parts attributed to the genius of other men. But in mechanics, I conceive, we should rather consider them improvements than inventions, unless improvement may be called invention, as the component parts of all new machines may be said to be old; but it is that nice discriminating judgment, which discovers that a particular arrangement will produce a new and desired effect, that stamps the merit. And this may perhaps, with propriety, be called either invention or improvement; which certainly exhibits that the artist has that penetration which is usually dignified with the term Genius. Therefore the mechanic should sit down among levers, screws, wedges, wheels, &c. like a poet among the letters of the alphabet, considering them as the exhibition of his thoughts; in which a new arrangement transmits a new idea to the world.
>
> —Robert Fulton, *Treatise on the Improvement of Canal Navigation*, 1796

Inventor Robert Fulton was confident that great change would come "when the true principle of building bridges of wood is discovered." Technology's poets, with planks and pegs, did indeed bridge that gap into the unknown, turning familiar elements into something transformational. Designers then took it and made it their own.[1] This chapter explores the contexts behind "a new arrangement" that conveyed "a new idea" about construction. We'll introduce Ithiel Town and reflect on the concepts, objects, and design strategies behind the development of his patent bridge.

Construction's primacy as an economic activity and the indispensability of transportation and communication networks for society and economy meant that bridge building would be in the forefront of industrialized practice.[2] Although Ithiel Town could not have foreseen the future of railways when he patented the lattice and opened an office in New York in 1820, at that moment the construction of the Erie Canal, longer than any previously built in Europe and North America and whose economic

Figure 2.1
Lattice bridge, Jordan River, Salt Lake City, Utah (built c. 1860, demolished 1908). Courtesy of Smithsonian Institution, National Museum of American History.

significance was widely understood, demonstrated the market for the mass construction of bridges.[3] The vast scale of "internal improvements," as public works projects were called, meant that inventors and entrepreneurs who enabled rapid and cost-effective construction could become rich. In the United States between 1814 and 1818, Town's home state of Connecticut granted concessions to fifty companies for fifty roads, 1,300 kilometers of which were completed by 1818. In France at that time, engineers rethinking road, canal, railway, and water supply strategies helped ground the concept of "network."[4]

While the scales of road- and bridgebuilding are different, the fundamental challenge was comparable. Town summed it up by paraphrasing a question Fulton had posed twenty-five years earlier: "By what construction or arrangement will the least quantity of materials, and cost of labor, erect a bridge of any practicable span or opening between piers or abutments, to be the strongest and most permanent, and to admit of the easiest repair?"[5] The challenge may sound straightforward but arriving at viable solutions was anything but. This was not about the design of one bridge but of a structure (and, by extension, infrastructure) composed of parts and connectors that manufacturers could produce and transport in large quantities and that builders could rapidly assemble. In 1823, Town built one of his earliest lattice bridges at the armory of New Haven–based cotton gin patentee and promoter of interchangeable parts, Eli Whitney. In 1820, Whitney had enthusiastically attested to the worthiness of the newly patented lattice based on an examination of a model: "The construction is such as to afford great facility in taking out any piece of timber in the Bridge, and replacing it with another." Town included in two of his promotional brochures Whitney's testimonial, which was endorsed by four prominent Yale University men. We might regard this with irony, as we now know that Whitney had expertise in promising, rather than delivering, interchangeable parts; in the lattice, replacement of individual planks, which were sandwiched and pinned at many points, would have been challenging.[6] As a structure, the bridge did, however, deliver on its modular promise.

"A NEW ARRANGEMENT"

> This improvement differs from other Bridges principally in the arrangement or distribution of the materials, in the quantity of materials in the size of the pieces and in the manner of securing or connecting the materials at the joints.
> —Ithiel Town's first bridge patent (1820)

Historians know little about Ithiel Town, but enough to consider him one of the most significant designer-builders of the early nineteenth century, a founder of the architectural profession as we understand it today, and a leader in the spread of classical revival architecture. Town built the Federal-style Center Church (probably partly based on architect Asher Benjamin's design, 1814), the Gothic-style Trinity Church (Town, 1814–1816), and the Greek-style State House (Town and Davis, 1831), all on New Haven Green, the central public square established in the seventeenth century adjacent to Yale University. With partner Alexander J. Davis, his projects included residential commissions and state capitols in Indianapolis (1835) and Raleigh (1840).[7] Town began his career in Boston and in 1812 established his base in New Haven, an intellectual capital anchored by Yale; both cities had public libraries by the seventeenth century, which may have given him access to engineering and architecture books as he began his career.[8]

From an early age, Town enjoyed playing with numbers and rearranging components and forms.[9] In his only known autobiographical statement, he recounted:

> I have had a great attachment for curious and uncommon books, ever since the age of eight years. At that time, in consequence of the death of my father, I commenced living with my uncle, and the girls of his family were in the habit of sending me to borrow a book, which contained an alphabet of letters, suitable for marking linen. In this book, which was entitled *The Young Man's Best Companion*, were various copyslips—mathematical diagrams—dials for different latitudes—mariner's compass, etc., neatly engraved on copper-plates. These diagrams I examined frequently, while on the way, and can yet recollect the strong and vivid impressions they made on my young and astonished mind, from the great ingenuity which I conceived it required to make and understand such very curious figures. Ever since, I have had a great propensity and love for mathematics, drawings, paintings, engravings, etc., and if this *book-affair* did not lead to it, I know not what did.[10]

A painted portrait of Town presents him as a learned man sitting among his books, with the lattice bridge displayed as his key achievement (figure 2.2).[11] He claimed that he was afflicted from an early age with a "disease" he called "book mania." Thomas Jefferson had made the same claim in 1789; Town eventually outdid the former president by assembling one of the largest art and architecture libraries in the United States and beyond. He also maintained in his New York City office a smaller professional library, which served as a place to display publications and artwork and to host gatherings. From June 1837, he lived next to Yale professor Benjamin Silliman in the villa-cum-library he designed. This edifice (figure 2.3), evidence of the wealth Town's bridge patent generated, housed about 11,500 books plus art prints and artifacts. It served as a worldwide web from which Town drew ideas as it linked him to a network of designers with whom he shared his collection. Davis apparently first came into contact with Town while studying Greek architecture in his library; together they popularized historic architectural styles.[12]

Town interacted with a constellation of prominent nineteenth-century figures. With painter-inventor Samuel F. B. Morse and painter-engraver Nathaniel Jocelyn, he traveled to Europe for the first time in 1829, after his career and reputation had been established. When Town saw his "first cathedral of large dimension" in Exeter, England, he exclaimed that "its effect on my mind can hardly be imagined."[13]

Figure 2.2
Frederick R. Spencer, *Portrait of Ithiel Town, Esq.*, 1840. The First Church of Christ in New Haven.

Early, Town may have worked under Asher Benjamin (1773–1845), author of popular builder's guides that offered constructional and stylistic advice.[14] During the 1810s, he collaborated on some of his earliest bridge building with Isaac Damon (1781–1862), a prolific contractor whose work would include a reported twenty-five lattice bridges.[15] Town's earliest documented lattice bridge was built near Salisbury, North Carolina, in 1818; through the 1830s, contractors, under his direction or licensing, built lattice bridges in states including South Carolina, Massachusetts, Connecticut (figure 2.4), Pennsylvania, Rhode Island, New Hampshire, New York, Indiana, Georgia, Virginia, New Jersey, Alabama, and Ohio. Architect Henry Austin, who began his career within Town's orbit, observed that Town "was fond of talking with first-class mechanics; was always inquiring from everybody, no matter who, from whom he

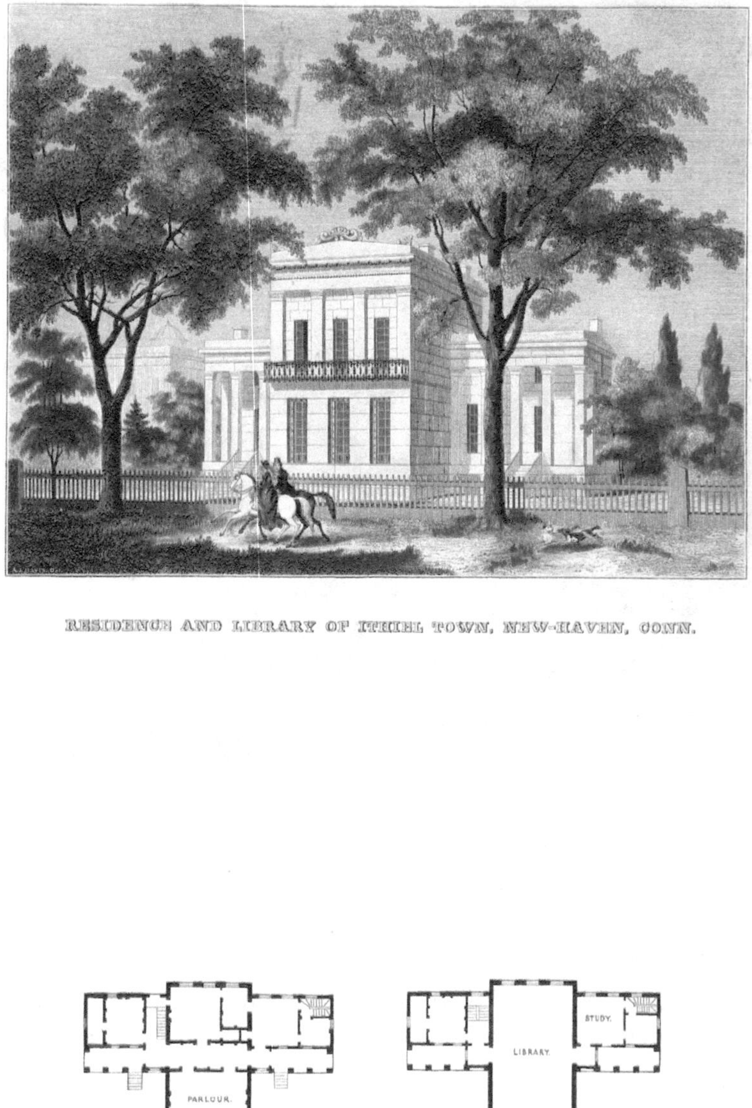

Figure 2.3
Ithiel Town's residence and library (1837) on Hillhouse Avenue, New Haven, was a six-minute walk from the New Haven Green. Hood Museum of Art, Dartmouth College: Purchased through the Julia L. Whittier Fund PR.949.107.

Figure 2.4
Early lattice bridge over the Farmington River, on Palisado Avenue near Congregational Church, Windsor, Connecticut (1833; washed away, 1854). John Warner Barber, *Connecticut Historical Collections* (J. W. Barber; A. Willard, 1836), 124. UC Berkeley Library, University of California.

could get information."[16] His inventive activity included a metal storefront, fireproof construction, a roof truss, and an octagonal schoolhouse. He may be responsible for the earliest US patent for hollow brick, a type which later appears to have become a staple of the building industry.[17] But it was the lattice design that had impact around the world.

When Ithiel Town asserted in his advertising brochure that bridge building should be based on "practice and experience" and implemented by "men scientifically and practically acquainted," who possessed "classic and well-established taste," he was surely describing himself.[18] As carpenter, designer, and bibliomaniac, Town blended hands-on with book-learnt knowledge of materials and structures. He brought together empirical know-how, design rules, scientific principles, and mathematical theories within his geometric and aesthetic modular mindset.[19] He announced the outcome as a "General System of Bridge-Building."[20] (The construction process, to be explored in chapter 5, was essential to the invention.) The lattice bridge reflected a holistic vision across scales of design and construction: multiplication of individual parts (planks and connectors); multiplication of subsystems (triangles, the most stable geometric shape) within each lattice frame (or beam); multiple lattice frames in each bridge; the replication of bridges in transportation networks. Modularity enabled opportunities for alterations and substitutions, which included the spacing and angle of the diagonals, types of materials (wood and iron, as announced on the brochures' covers), connection types, and the number of lattice frames employed in each bridge. Modularity enabled builders, designers, and manufacturers to imagine a diversity of possibilities. It remains an important feature of mass-buildable systems and is considered indispensable for the successful implementation of large, complex infrastructure projects.[21]

Town's 1820 patent, which he developed "after a number of years' attention to the theory and practice of this subject," described a bridge supported by horizontal structures (beams) that contained multiple diagonally intersecting planks (figure 2.5).[22] In simple terms, the bridge consisted of two lattice beams in a parallel position connected by a deck (floor) at the bottom and a roof at the top. Lattice planks were pinned tightly against each other at the intersections with wooden pegs or pins, also known as treenails or trunnels, through auger-bored holes. Pegs were used in shipbuilding to join the interior structure to the planks of the outer skin, which helped increase the rigidity of the vessel. The many connectors that kept the lattice together hindered loosening and slippage among members. Deck and roof structures served as lateral (sideways) bracing. If the deck was carried at mid-height or the roof served as the deck, additional lateral bracing in the open space under the deck enhanced rigidity (figure 2.11). The sides were partially or wholly clad with vertical boards; weather protection was essential.

The network of diagonally intersecting planks was to be made of "flat pieces of timber or other substances" (iron), laid parallel and intersecting each other at a "45 degree or any angle that may be necessary for a brace."[23] Town stated in his brochure of 1839: "the trusses are composed of planks 10 or 12 inches wide, three inches thick, and about 20 feet long, of white pine or spruce."[24] The closer the diagonals, the greater the "strength" of the bridge was supposed to be. Diagonals were pinned to the upper and lower chords (continuous horizontal members that Town called "string-pieces") and at each crossing. They were made of two thicknesses of plank, same dimensions as the diagonals. A 100-foot-long bridge that included two lattice beams required 2,592 holes and 912 treenails.[25] Such numbers achieved a key design goal: avoiding tensile connections (discussed below) that were difficult to craft and maintain, while taking

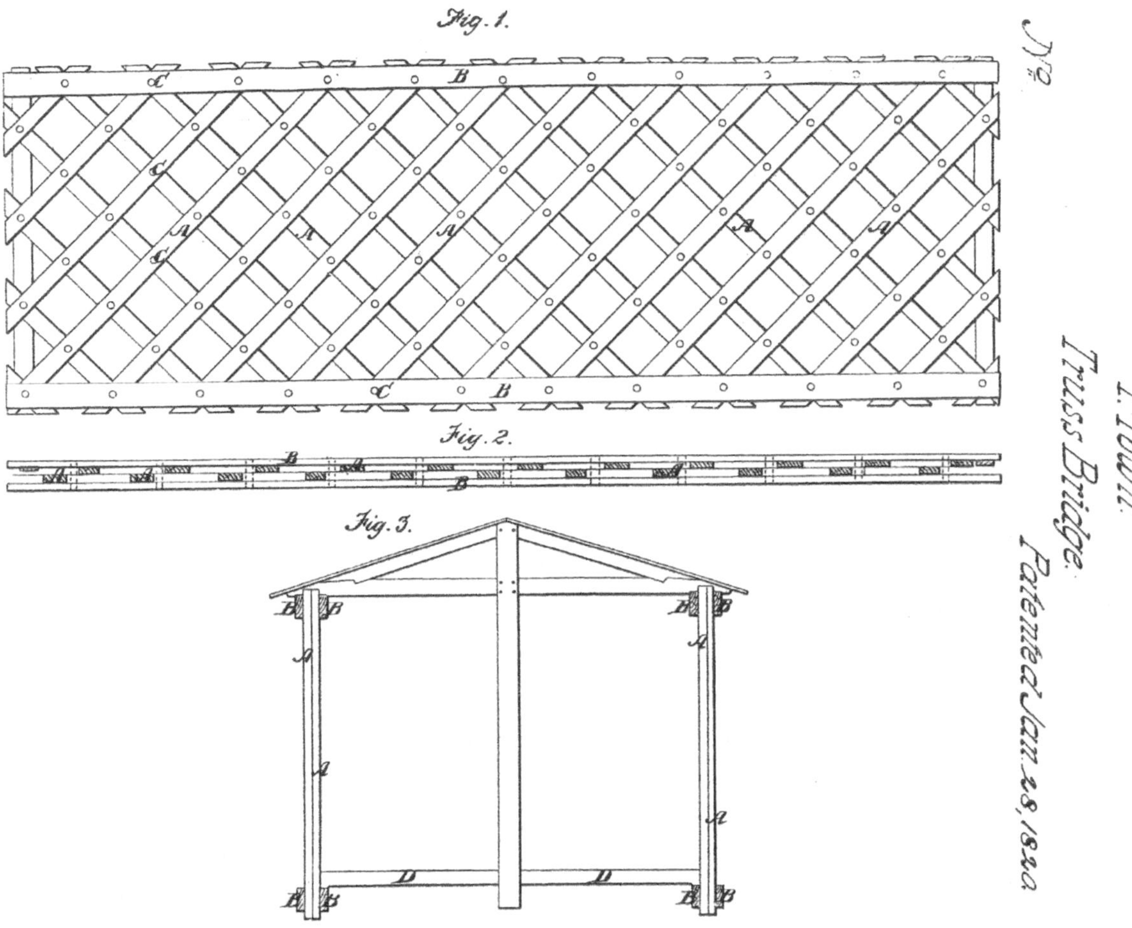

Figure 2.5
Ithiel Town's "Truss Bridge" patent 3,169X (January 28, 1820). National Archives and Records Administration.

advantage of the mechanization of wood fabrication methods. It divided the labor, as peg making, too, became a job that could be performed independently. The planks that made up the chords were staggered (a practice known as "breaking joints") for additional strength and durability; the jointless design of the bottom chord, which typically experiences the most tension, was novel (figure 2.6). As was holding a bridge together with pegs. For spans longer than 130 feet, Town recommended additional top and bottom chords to be placed, respectively, below and above the initial chords. He specified that the height of the lattice frames be one-tenth the span. Ordinarily, there

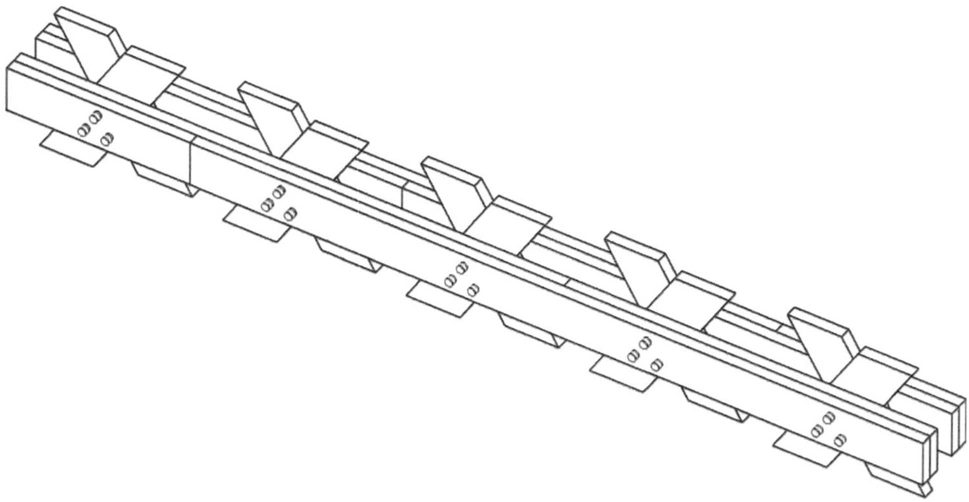

Figure 2.6
Chord detail. The bottom chord was assembled out of planks that met end-to-end, without a joint; these meeting points in adjoining layers were staggered through the length of the chord. Planks were pinned together at the intersection of lattice diagonals. Todd Craig Radford, "Town Lattice Truss" (PhD diss., MIT, 2010), 108. Courtesy of Todd Radford.

was a lattice at each side of the roadway; an intermediate lattice was recommended for wide bridges.[26]

Town's second patent (1835) doubled the lattices (figures 2.7–2.8). This was intended to prevent "trusses twisting, leaning side ways, or curving in the direction of their length." Damp environments can cause wood to shrink and swell, resulting in loosened connections and cracking; sagging under its own weight and deflecting under heavy loads could cause vertical cracking at the peg holes. These deficiencies spurred designers to improve the structure. (For Town, additional motivation for a second patent would have been the initial patent's expiration after fourteen years.)[27] Bridge performance depended on a number of factors, including the dimensions, angle, and spacing of diagonal members, the number and dimensions of treenails, and the length of the chord planks.[28]

Town identified the main supports of the bridge (that is, the lattice beams or frames) in several ways, including "trusses," "lattice-braces," "truss-braces," "lattices," "sides," and "side-trusses." He knew that wood was strongest and stiffest along its length (parallel to the grain); in his 1839 brochure, he stated that the members of the lattice would be loaded along their length (axially) in tension or compression. (Which members were under tension or compression at any moment depended on the location of the load moving across the bridge.)[29] With members pinned together at

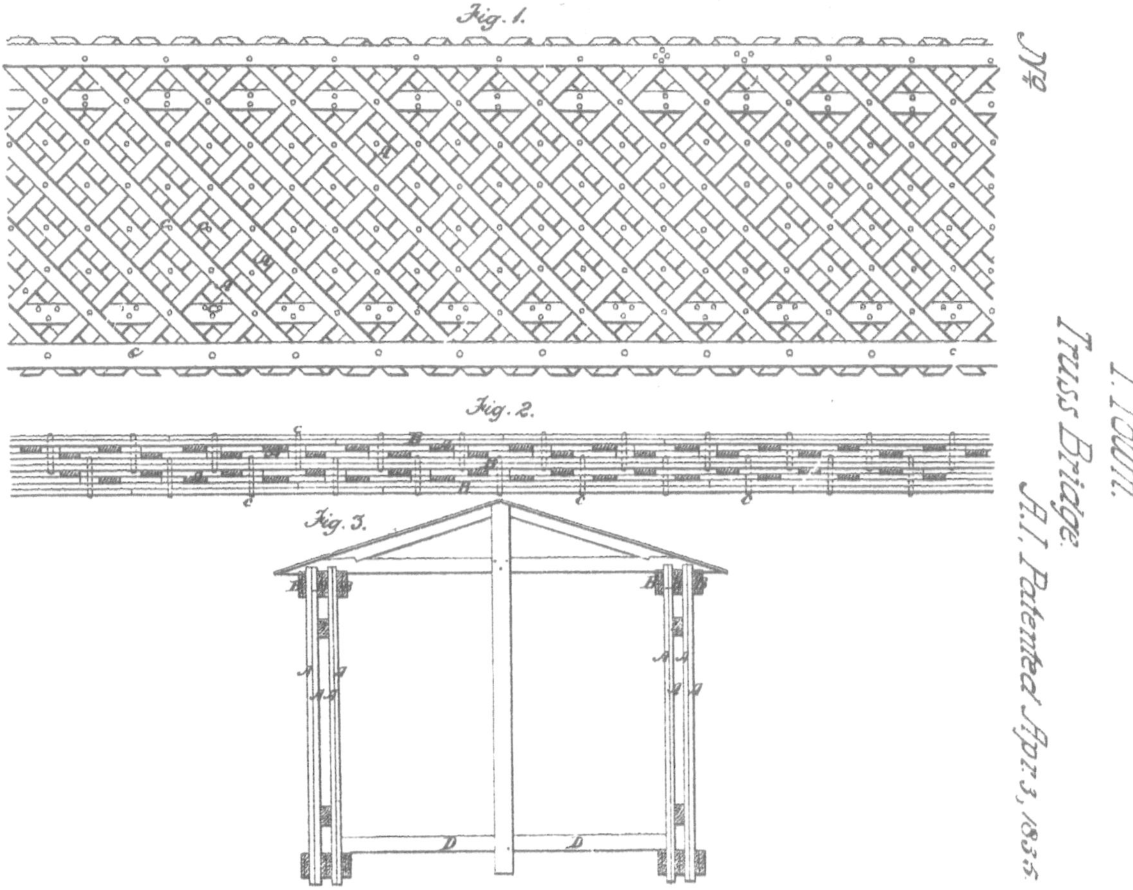

Figure 2.7
Ithiel Town's second "Truss Bridge" patent 8,743X (April 3, 1835). National Archives and Records Administration.

every intersection—which makes precise analysis of the forces in each member quite complex—the original lattice was easier to evaluate as a beam, which works in bending, than as a truss (figure 2.9).[30] Town stated in his patent that the lower chord should be "of sufficient *tension strength*" and that the upper chord should be "capable of bearing a sufficient *thrust strain*" (i.e., compressive stress). This indicates that he regarded the lattice as a beam resting freely on two points; when loaded, the largest stresses occur in the bottom and top portions of the structure, stretched at the bottom and pushed together at the top.[31]

The lattice beam, assembled out of relatively small, uniform parts, was intended to rigidly span between supports, which might be piers standing in a river or valley, or

DESCRIPTION OF THE PLATES, WITH DEFINITIONS OF TECHNICAL TERMS, &c.

Numbers 1, 2, and 3 exhibit all the parts of a bridge on the last improvement, with suspension posts, &c., suited to a span of 170 to 220 feet, according to the size and number of its parts, which may be so varied as to suit the width of span required.

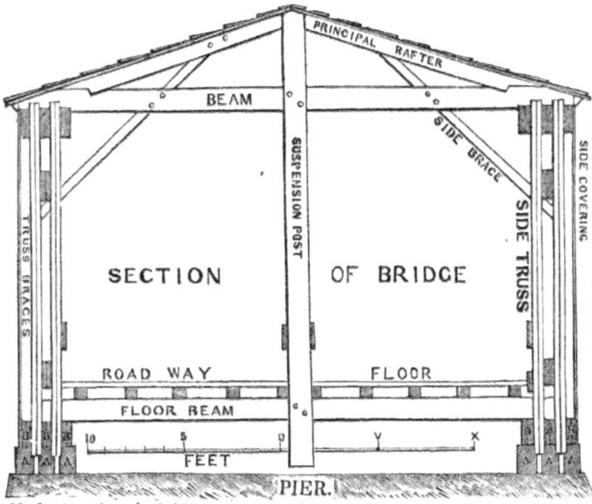

No. 1. A section of a bridge, with the roof, suspension posts, and all the parts, upon the pier, with a scale of feet.

No. 2. The side elevation of a truss, to the same scale. The height of a truss may be greater or less, by any number of half diamonds, or by a change of the angle of the truss-braces.

Definitions, Explanations, and General Remarks.—1st. A truss is that combination (No. 2) of materials, which, with one or more other similar ones, constitute the whole vertical support of the bridge. They are placed in a vertical position on the piers, and are kept so by the addition of bottom beams, top beams, and horizontal and side braces, as shown in No. 1.

These few parts may be termed the skeleton of the bridge, and constitute the whole strength and support of the superstructure; to which a floor is added, for the conveyance of travel, produce, goods, &c., and generally a roof and side covering, to save the skeleton and floor from that decay which is incident to its exposure to the weather.

2d. Truss-braces, or diagonal braces, are those crossing each other, and forming the truss into diamonds. There are two series or sets of such truss-braces, separated by middle string-pieces six or seven inches thick. (See figs. 1 and 2.) Each series of truss-braces may be termed a lattice; hence it is, by some, termed the double lattice bridge. The diamond-truss bridge, double diamond-truss, or double diamond bridge, would be more appropriate and better understood.

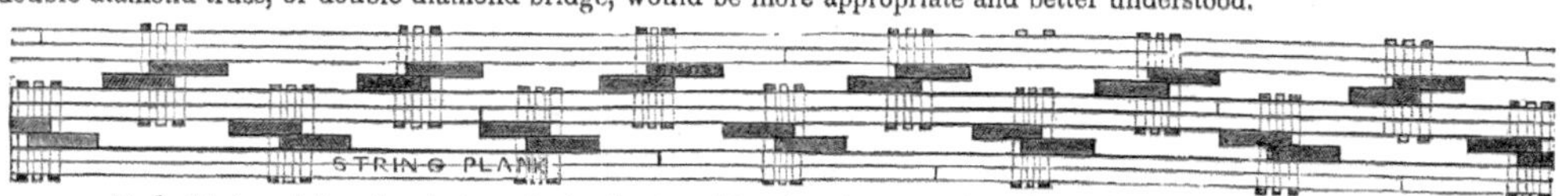

No. 3. The top or bottom view of a truss, showing the edges of the string-pieces, tree-nails, and ends of the truss-braces, to a double scale.

3d. String-pieces are all those horizontal parts of a truss, at top and bottom, or intermediate, which run the whole length, and are secured to the truss-braces by three or four tree-nails, at each intersection of them, over which the string-pieces pass.

4th. Intermediate string-pieces are those which pass over the lines of intersections of the truss-braces, between the extreme top and bottom string-pieces. Middle ones are here added, and the others may be added when more strength is required.

5th. Centre, or middle string-pieces, are all those which run between the two series of truss-braces, whether at top, at bottom, or intermediate; intermediate middle ones only, are here represented.

6th. String-plank are those pieces of which all the string-pieces are composed, and are half the thickness of the string-pieces, and are, in this case, 27 feet long. They break joints in their centres.

7th. Floor-beams are those which rest upon the trusses, and support the flooring of the bridge. The size of them should be in suitable proportion to the distance between their bearings upon the trusses, or suspension-posts and trusses, when the former are introduced; great depth in proportion to their thickness is important.

8th. Suspension-posts (see No. 1) are pieces 12 or 14 inches by 4 or 5 inches, firmly secured by tree-nails, locking, &c. to the top and bottom beams, and to each pair of principal rafters. They are introduced instead of a middle truss, for the support of the floor beams of a covered bridge, when the floor is on the bottom beams, and when the width of the bridge is so great that floor beams of sufficient strength cannot be procured.

9th. Floor-joists are pieces running lengthways of the bridge, and resting upon the top of the floor beams. They should be $4\frac{1}{2}$ to 6 inches square, according to the distance apart of the floor beams.

10th. Floor-planks are the pieces resting on the floor-joists, of various widths, from 6 to 12 inches, and from $2\frac{1}{2}$ to 4 inches thick, as may be required. In some instances, two thicknesses of planks are preferred; in which case, they cross each other at right angles or obliquely, and may be of less thickness, or such as their use may require.

4

Figure 2.8

Lattice bridge, section and elevation of beam ("side truss"), and section of chord ("string piece"). Ithiel Town, *Description of Ithiel Town's Improvement in the Principle, Construction, and Practical Execution of Bridges* (1839), 13. Courtesy of Smithsonian Libraries and Archives.

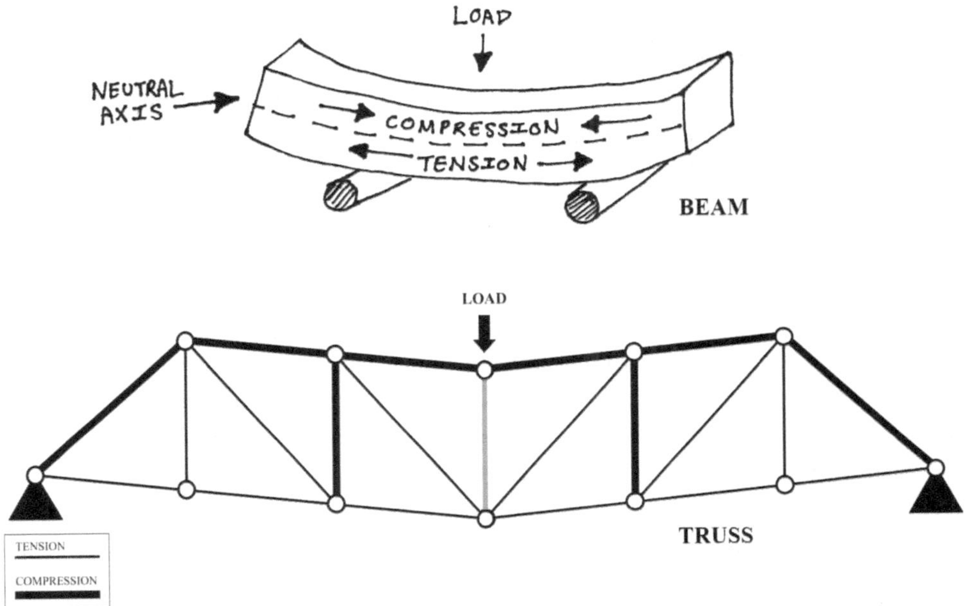

Figure 2.9
Beam and truss: theoretical (ideal) models. A load applied to the center of a beam results in bending, with shortening (compression) in the top portion and lengthening (tension) in the bottom portion; these stresses increase the further they are from the neutral axis, where stresses vanish. A load applied to the center of a truss whose members are pinned (that is, free to rotate relative to each other) puts individual members into either tension or compression, instead of bending the entire structure. In actual conditions, the members are not free to rotate and develop bending and shear.

the walls of a building. The lattices, as primary supporting elements of a long span, conceived with neither vertical (as trusses usually were) nor arched members, were unusual, if not unique.[32] Astonishing was the use of pegs as connectors. Architect-engineer Charles Zimpel, who worked in the United States for an extended period, wrote that as American bridges were "built with such simple connections, only the fact that these bridges have actually endured steadfast for years convinces us that such structures really are feasible."[33] Simple, uniform parts became an invitation to innovators to imagine a variety of materials, modes of connection, and structural configurations.

Town, like the average carpenter, knew that a triangle remains rigid even if its sides are only pinned together at their ends—that is, if they are free to rotate. (The ability to achieve structural rigidity without rigid connections helps builders conserve on materials, labor, and construction time.) The lattice's numerous, overlapping triangles enabled Town to devise a structure without connections in tension (which can pull apart) and to avoid compressive loads (pushing forces) that can cause buckling (bowing or deforming sideways).[34] The multiplicity of pinned intersections divided the effective length of each plank into short pieces, making them less likely to buckle. In other words, rather than risking concentrated stresses in a few axial connections that might pull apart, he divided the stress among many pinned connections. We'll come back to this idea.

A single long structural element such as a log or plank may bend excessively when there is a load on it (figure 2.10). So, before the advent of "American bridges," when a builder wanted to span a significant distance, a common solution was a trussed beam or arch, or a series of these.[35] ("Arch bridge" includes a spectrum of types that functioned in a variety of ways.[36]) A truss employed triangles, the most efficient way to ensure rigidity. For example, carpenters in seventeenth-century Britain built king and queen post trusses to support roofs; these were standard at the beginning of the nineteenth century (figure 2.23).[37] The traditional masonry arch bridge, a structure with relatively small openings beneath it, enabled builders to achieve rigidity through mass; thick and heavy piers resisted the sideways push (thrust). But construction of masonry arch bridges was usually costly, and they were built upon temporary bridge-like structures called centering. These held the blocks or bricks in place until construction was complete and the bridge could support itself. Wooden arch bridges were lighter in weight but likely to be less rigid; they, too, required complex construction methods. In some arch bridges, the roadway rose in the middle, causing inconvenience.

Bridge builders were acutely aware of the limitations of the standard bridge designs. Bavarian builder and entrepreneur Carl Friedrich von Wiebeking (1762–1842), who worked in Germany and France, summed it up in 1812:

> To find a method of building wooden frame bridges as solidly as stone bridges, whose existence would be assured for centuries and whose wide arch openings would facilitate navigation, provide against flooding, and join the elegance of its form to its solidity; this is the problem whose solution is of the greatest interest to all civilized states.[38]

Wiebeking's solution, a laminated timber arch (developed 1807–1809), was criticized because of a lack of rigidity and the tendency of wooden members to decay or be crushed. The preparation of timber was a complex undertaking involving "scarfing, notching, wedging, drilling and fitting" before erection.[39] Much of the body of the structure neared the waterway, which made it vulnerable to moisture damage and more likely to obstruct and be harmed by floods, ice, and watercraft.[40]

Compared to an arch, a horizontal structure such as the lattice offered significant advantages, which Town detailed when pitching his bridge to potential clients.[41] It

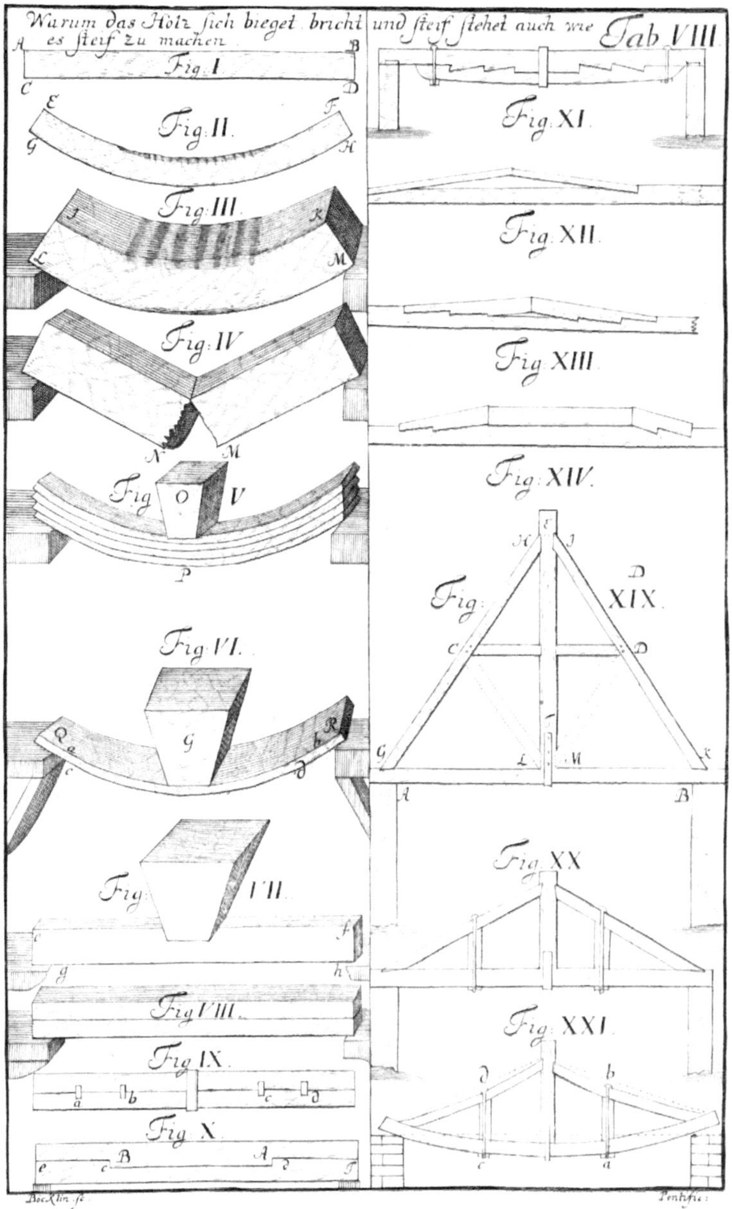

Figure 2.10

Beam and truss. Technologist Jacob Leupold showed a beam bending (and failing) under a load, along with methods for enhancing rigidity: adding material or creating trusses. A single plank is adequate for carrying small loads over a short span; over a wide opening, it would bend and wobble. Increasing the depth (height) of a plank increases resistance to bending; this can be achieved by turning the plank sideways or adding material. The longer the plank, however, the more material required to keep it rigid; eventually the beam can't carry much besides its own weight. A solution is to truss the beam, shown in the right column. Jacob Leupold, *Theatrum Pontificiale, oder Schau-Platz der Brücken und Brücken-Baues* (Zunkel, 1726): pl. 8. The Getty Research Institute.

was rigid and relatively light. It could be built independently of the piers and subsequently placed on top of them; this independence provided freedom regarding the location and number of piers. Its horizontality meant that it was less exposed to moisture, rot, and ice; if the top rather than the interior of the bridge was used as the roadway, piers could be shorter, enabling cost savings (figure 2.11). Loads were in theory sent straight downward into piers and abutments. This meant that piers could be less massive than those of arch bridges, offering savings in materials, labor, and cost. The lattice enabled longer spans, which required fewer piers, meaning lower costs and less waterway obstruction. In service, destruction of one span might minimally impact the rest, unlike an arch bridge, where one damaged arch span could affect the entire structure.[42]

BUILDING ON THE VERGE OF INDUSTRIALIZATION

Historians sometimes fetishize the object at the center of their story, elevating its significance.[43] In this case, contemporary evidence indicates that the lattice made a world-changing contribution. It demonstrated the viability of a long, rapidly buildable beam made of simple uniform parts. While a beam, like a spoon or screw, seems so basic that it's hard to imagine someone inventing it, the success of the lattice and its enduring impact as a model of mass-producibility and dependability makes it an industrial icon on par with Ford's Model T. Below, we'll consider lumber, connection, and structure, and then ponder how innovators including Town tinkered with these elements to create something new.

LUMBER

We've already outed "wooden bridges"; they were iron, too. The lattice bridge was therefore a great rarity, a wooden bridge actually made of wood. But "wood" and "iron" are catchall categories encompassing a diversity of materials: some types are more easily worked than others; some are less prone to decay or corrosion; some are more resistant to impact; some are more flexible; some have greater strength in tension (e.g., pine and wrought iron); others are more appropriate for compression (e.g., oak and cast iron). A tree can be felled, hewn, and integrated into a structure; this method is viable when builders have the expertise, skill, strength, laborers, and time. This is not adequate when thousands of structures have to be erected quickly and meet a uniform level of reliability.[44] Instead, wood, extracted from a forest as ore is from a mine, is transformed at the sawmill into products of standard dimensions and sold to builders. Storytellers sometimes claim that iron has "no natural form"—as if stone and wood

14 A DESCRIPTION OF ITHIEL TOWN'S IMPROVEMENT

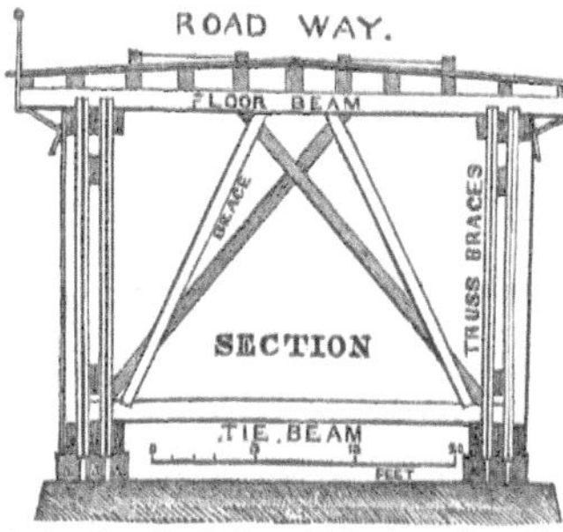

No. 4. A section of a bridge, with the floor, side railing, &c. on the top of the trusses. The floor may easily be made to turn off the water. With a scale; which, with side boarding, would secure it from the weather effectually.

11th. Horizontal braces at the top are either framed in diagonally, between the top beams, or consist of long planks of suitable dimensions, spiked and tree-nailed to the top of the beams, so as to give the most secure support, to keep the trusses in a straight line at top. This is a very important support, and should be done in the most secure manner, in every respect.

12th. Principal rafters are those which stand upon the top beams, and are thoroughly secured to them at their feet. They may be 10 to 12 inches deep, and 4 to 5 inches thick, according to their length.

13th. Small rafters are those between the principal ones, to nail the roof-boarding to, for the purpose of shingling.

14th. Floor-braces, or bottom-braces, are pieces about 4½ by 6½ inches, as their length may require, framed in and keyed, between the bottom beams, to keep the bottom part of the bridge secure against any motion or vibration sideways, by wind, travel, or transportation.

15th. Side-braces (see No. 1) are pieces about 4½ by 5½ inches, connected with the side-truss, the top beam, and the principal rafters, for the purpose of preventing the side-trusses from leaning, or inclining up or down stream. In the section No. 4, the side-braces are differently secured, and may cross each other or not, as shown in the section. By crossing, greater strength is obtained.

16th. Side-walks are sometimes made for foot travel, and may be in the inside or on the outside. They are parted off by a railing 3 or 4 feet high, and if on the outside, may be made very ornamental, as seen in No. 5.

No. 5. A perspective view of a roofed bridge, with side-walks on the outside, protected by colonnades and railings, finished in a very rich style of decoration.

17th. Tree-nails of two inches diameter, made of white oak or other hard wood, are used for securing the truss-braces and string-pieces together, as seen in Nos. 2 and 3. They should be exactly fitted to the augers used, so as when seasoned to drive tight, and make solid work. Tree-nails may be made different ways, but the best and most economical is, to saw them out square from plank, with a circular saw, and then turn them with a small lathe, attached to some water or other machinery. They should be unseasoned, to be easily made, but must afterwards be well seasoned before driven into the work; they will season quick, or may be kiln-dried. Tallow, or oil, &c. may be rubbed upon them, to make them drive more easily, if necessary.

18th. Abutments are the two supports of a bridge or aqueduct, at the ends, and are erected against and connected with the banks, on each side of the stream or ravine. They may be constructed in various ways, the same as the piers, if any, with which they are connected. The abutments for a bridge, on this improvement, requires no other strength or solidity than a pier of equal exposure, there being no lateral or horizontal pressure, to affect either the abutments or piers, and therefore a perpendicular support only is required, sufficient to bear up the weight of the bridge, and that which may pass over it. One or more draws, for the passage of shipping, &c., may consequently be anywhere placed or constructed, without producing the least injury to this kind of bridge, as its strength and safety do not depend, in the least, on being connected throughout; except very partially, as respects the action of severe gales of wind, in which case, a structure in one entire connected mass, is of course somewhat less liable to be acted upon.

Figure 2.11
Lattice bridge cross section and elevation. The roadway might run between or atop the lattice beams, as shown here. The modularity of classical elements reinforced the structure's industrial nature. Ithiel Town, *Description of Ithiel Town's Improvement in the Principle, Construction, and Practical Execution of Bridges* (1839), 14. Courtesy of Smithsonian Libraries and Archives.

required less attention to design, labor, processing, and shaping than iron.[45] In any case, factors that determine wood's successful use include the part of the log lumber it is cut from, how it is dried (and its resulting moisture content), and the direction of the grain.

Historian Werner Sombart believed that the end of what he termed the "wooden age" occurred in Europe during the eighteenth century. According to Sombart, European capitalism and civilization—threatened by a dearth of trees needed for structures and fuel—were subsequently saved by coal and iron.[46] Historian Brooke Hindle wrote of a wooden age that continued in the United States as it waned in Europe; wood, abundant and "cheap" in heavily forested North America, was readily available for the axing. But these widely adopted perspectives may have been exaggerated—or were imprecise, at least. The availability of wood varied according to region and season. Moreover, wood scarcity was a function of political and social circumstances rather than a mere question of tree supply.[47]

Wood was an essential product in the United States and elsewhere, but the ease with which it could be transported was critical to its growing consumption in nineteenth-century industrialization.[48] The continental United States originally contained an immense forestland, approximately 668 million acres, with four-fifths of the commercial quality forests located east of the Great Plains.[49] Manufacturers transformed trees into products including machinery parts, barrels, pitch, and fuel. There is a paradoxical aspect, however, to stories of the wooden wealth of the United States: there was both abundance and scarcity. An adequate supply of wood was not consistently available; in 1841 Scottish engineer David Stevenson warned, "should . . . the clearing of land be conducted in the same indiscriminate manner as hitherto, another hundred years may see the United States a *treeless* country."[50]

Already in 1745, Benjamin Franklin had noted the difficulty in obtaining wood for fuel, which historian Michael Williams reports "had to be brought nearly 100 miles to the large coastal towns, and at considerable cost."[51] Fifty years later, William Strickland, an English visitor to the United States, reported that large lumber had to be transported 200 miles from the Albany area for the construction of two bridges over the Hackensack and Passaic rivers, observing "every one complains of the scarcity and increasing price."[52] By 1802, "live oak [a superior type of wood] is rarely met with north of the Carolinas: that used in the Boston ship-yards is brought from Georgia; a distance of more than a thousand miles."[53] An instance of Christianity-inspired harmony reveals the challenges. During the War of 1812, when Ithiel Town was working on his first prominent project, Center Church, he wrote to Royal Navy Captain Thomas Hardy, whose vessels were blockading New Haven; as a result, Hardy permitted the wood to be delivered so that the church could be built.[54] It was reported in

South Carolina in 1823 that, because the forests had been cut for agriculture, there was no timber available to repair wood-paved roads.[55] Still, old-growth forests in coastal South Carolina, in the region where the earliest lattice bridges were built, contained trees whose wood was of extraordinary quality; enslaved labor in the South helped establish and operate mills that supplied lumber and wood products to shipbuilders in Maine, for example. The demolition of forests in the Northeast and Midwest led industrialists to denude the forests of the South for their building needs during the nineteenth century.[56] For the 1838–1840 building of lattice bridges on the Raleigh and Gaston Railroad in North Carolina, with its copious pine forests, transportation and contractor challenges were so stressful for the engineer in charge that he ordered lumber from a supplier in Maryland. Scarcity was worsened by the fact that as much as 20 percent of the material was routinely wasted in the milling process.[57]

So the "wooden age" of the United States was not as woody as sometimes suggested. Wood's availability for building varied according to place and period, and prices reflected it. In the 1830s, lumber in the United States was on average more expensive than on the European continent.[58] Charles Zimpel reported in 1840 that prices were the same or more expensive than prices in Prussia and that, in the United States, transport of lumber accounted for one-fifth to one-sixth of its price; it has been estimated that it could account for 50 percent of the price.[59] Town likely contended with lumber issues including a dearth of long logs.[60] It was challenging to transport logs to mill and lumber from mill to customer; in the middle of the century, large structural members were still often hand-hewn. Moreover, mills were not devoted solely to the production of planks.[61]

The difficulty and cost of obtaining suitable wooden building material helps explain why inventors in North America focused on devising structural systems that required smaller parts. They were inspired by scarcity rather than abundance. From a builder's perspective, moreover, smaller parts made mass-scale construction easier to achieve. It's not surprising that the use of uniform, relatively short pieces was a fixation of structural innovators. Painter and inventor Charles W. Peale, who in 1797 may have been the first bridge patentee in the United States, declared:

> The novelty of building with boards or planks only in works, which before this time have called for the largest timber, is a change so great, that experience only can convince the timid and doubtful mind, of the powerful effect of this combination in producing a strength to bear burdens of great weight, far beyond what has been thought possible, to be acquired by such small and light materials.[62]

Town was trained during a lumber industry boom in a region that was home to leaders in wood-processing technologies. He constructed his first lattice bridges when milling technologies (including saws, machinery, and power sources) were a focus of innovation. In Massachusetts, between 1634 and 1695, at least seventy-two water-driven saw mills were built; by 1800, there were 250. In 1810, there were 2,106 sawmills in Pennsylvania. One of Town's earliest bridges, the Clarendon Bridge over the Cape Fear River in Fayetteville, North Carolina, was built 1819–1820; the Cape Fear Steam Sawmill Company was established in 1819, but already in 1776 there were about fifty sawmills along the river.[63] Circular saws began to be used in mills by the early 1820s and within a decade logs were cut with gang saws (parallel blades mounted on a frame). This did not mean that precision was quickly achieved; an inspection of a lattice bridge built in 1829 indicates that carpenters completed milled lumber with planes and finishing tools before construction.[64]

Today, when most structures are composed of standard materials and parts available to the trade and consumers, it may be difficult to appreciate the achievement of an ideal so long out of reach. But this is why the lattice bridge, designed to be assembled of smaller, simple, uniform components, and take advantage of their increasing availability, became an archetype in industrializing regions.[65]

CONNECTION

Even before sawmills made lumber widely available, builders had designed structures that consisted of uniform parts. Builders could more readily achieve uniformity without the precision of standardization because tolerances in large structures were not as small as in, for example, firearms.[66] An advantage of lumber is that it enabled designs consisting of a seemingly limitless reproduction of relatively lightweight easy-to-handle parts, but this was also a weakness, because many members means many connectors. In bridges, these must remain rigid over time if the structure is to function as intended.[67] Let's take a quick look at how the reinventors of building took on connection challenges.

Subject to a variety of stresses and exposed to the weather, connections are among the most vulnerable parts of a bridge. An effective connection cannot be pulled apart, crushed, or become excessively tight or loose over time. It should not invite rot; it might be adjustable during maintenance.[68] A connection might appear merely to be a detail, but in some designs the failure of a detail can lead to collapse.

Traditional mortise-and-tenon connections (figure 2.12) required skilled, costly, time-consuming labor to fabricate, often at the construction site; this type of joinery might loosen or break when subjected to frequent alternating pulling and pushing, changes in weather, or deterioration. Wood shrinkage, which occurs when timber is freshly cut and not allowed to dry over an extended period, can weaken connections

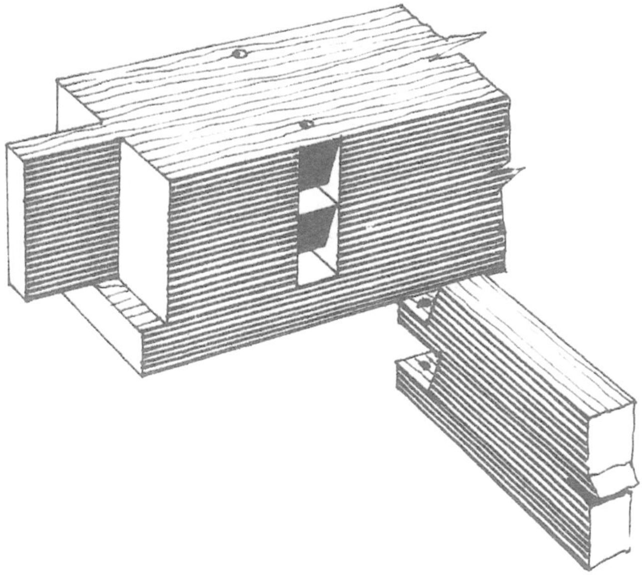

Figure 2.12
Sixteenth-century joint, Essex, England. Mortise-and-tenon joints consisted of a hole or recess that received a projection at the end of another member. A peg sometimes secured the connection. Joint types included notched, keyed, dovetailed, and scarfed. Cecil A. Hewett, *English Historic Carpentry* (Phillimore, 1980), 282. The History Press.

and even cause them to fail.[69] When a mortise-and-tenon connection is in tension (pulling), the stresses, which can be large, are concentrated in a small area of material—the tenon. This reduced amount of wood had to transmit loads and maintain rigidity at a vulnerable point where parts came together. So the avoidance of parts being subjected to pulling, with a risk of being pulled apart, was a prime challenge to carpenters. It explains why metal hardware was widely employed to connect wooden members.[70]

In one of the earliest publications reporting on the Town bridge, the author characterized it as "à la Philibert De l'Orme."[71] The structures of architect and engineer de l'Orme (1510–1570) were widely known in the United States during the early nineteenth century. In his book *Nouvelles inventions pour bien bastir et à petit fraiz* (New Inventions for Building Well and at Low Cost, 1561), he described a system for constructing vaults, domes, floors, and bridges (figure 2.13).[72] Each arch was made of two courses of small, curved pieces of wood (approximately 130 × 25 × 3 centimeters) held together by wooden pegs. Purlins (horizontal members) passed through the middle of these connections and were wedged on either side of each rib. For large domes, a second set of purlins was added to the exterior. De l'Orme's dome did not find success in its

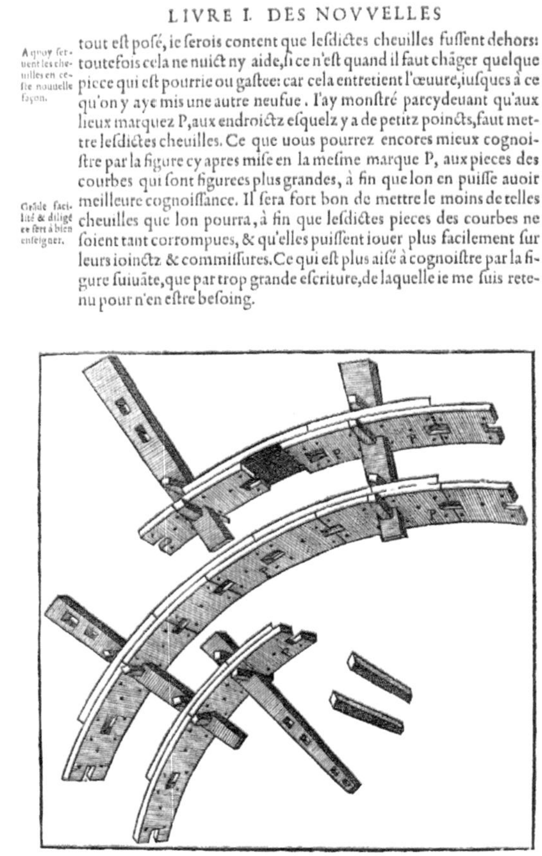

LIVRE I. DES NOVVELLES

A quoy seruent les cheuilles en ceste nouuelle façon.

tout eſt poſé, ie ſerois content que leſdictes cheuilles fuſſent dehors: toutefois cela ne nuict ny aide, ſi ce n'eſt quand il faut chãger quelque piece qui eſt pourrie ou gaſtee: car cela entretient l'œuure, iuſques à ce qu'on y aye mis une autre neufue. I'ay monſtré parcydeuant qu'aux lieux marquez P, aux endroictz eſquelz y a de petitz poincts, faut mettre leſdictes cheuilles. Ce que uous pourrez encores mieux cognoiſtre par la figure cy apres miſe en la meſme marque P, aux pieces des courbes qui ſont figurees plus grandes, à fin que lon en puiſſe auoir meilleure cognoiſſance. Il ſera fort bon de mettre le moins de telles cheuilles que lon pourra, à fin que leſdictes pieces des courbes ne ſoient tant corrompues, & qu'elles puiſſent iouer plus facilement ſur leurs ioinctz & commiſſures. Ce qui eſt plus aiſé à cognoiſtre par la figure ſuiuãte, que par trop grande eſcriture, de laquelle ie me ſuis retenu pour n'en eſtre beſoing.

Grãde facilité & diligẽce ſert à bien enſeigner.

Figure 2.13
De l'Orme's domed roof structure. Philibert de l'Orme, *Nouvelles inventions pour bien bastir et à petits fraiz* (Morel, 1561), book 1, chapter 7. The Getty Research Institute.

own time, but Thomas Jefferson built a version at Monticello and coerced architect Benjamin Latrobe into using it in Washington, D.C.'s Hall of Representatives (1805).[73]

De l'Orme's designs shared fundamental characteristics with the lattice bridge. It consisted of a few types of parts assembled with many simple connections. It required no metal hardware and was intended as a replacement for trussed-beam roof structures with complex mortise-and-tenon connections. It was lightweight and produced only compression in the walls that supported it; that is, walls did not have to withstand forces that pushed sideways. De l'Orme emphasized, as Town would, the advantages of using smaller, simpler pieces of wood: they were easier to find, transport, assemble, and inspect.[74] But wedged systems such as de l'Orme's can loosen over time.

What might have encouraged Town to realize that, instead of wedges, simple pegs pounded into tight holes could hold materials tightly together in the face of moisture, dirt, heat, vibration, heavy loads, and intense stress? Here's one possibility. In 1811 and 1812, inventors exploring connection solutions, working in Massachusetts and Connecticut, not far from New Haven, filed several patents for methods for uniting the sole and upper of a shoe with a series of wee pegs. Over the next half century, pegs were included in at least three-quarters of US shoe production. *Scientific American* later proclaimed that the peg "worked perhaps as great a revolution in a most important branch of industry as was ever effected by a single device." As in the lattice, pegs cut costs by reducing the need for skilled labor; previously, shoemakers sewed parts together. Although the pegged shoe was more rigid, it was significantly cheaper.[75]

Why did Town avoid metal connectors? As we've noted, iron parts such as nails, bolts, and straps had often been an essential part of the mix. But structural iron connectors were challenging. For centuries, iron was expensive and unreliable. And it rusted. Furthermore, wood and iron expand and contract at different rates when exposed to changes in temperature. This caused loosened connections or damage to bridge members.[76] Town addressed the iron deficiency by simplifying and multiplying connections and leaving iron out.[77] This altered how the structure held itself together. Traditional mortise-and-tenon connections enabled forces to pass through their length, that is, axially; but treenails were nonaxial connectors that aimed to produce a monolithic bond among parts. The pegging of layers of wood is similar to riveting, which had been employed for centuries to fasten the parts of metal objects; Town was likely familiar with it. At mid-nineteenth century, the lattice design was a widely known precedent when the riveting of large iron structural members became more common (figure 7.10).[78] The likely adaptation of a metal connection technique, riveting, to wood, and its subsequent re-adaptation to metal, represents inventive extrapolation and intermaterial leaps that are standard in design and can lead to extraordinary outcomes, and that repudiate evolutionist explanations and material hierarchies.

STRUCTURE

Lattices were long known as much for decorative as for functional qualities. The pleasingly complex, regular pattern enhanced landscape architecture and window design. Lattices appeared in paintings as early as the fifteenth century (figure 2.14); by the seventeenth, they served as fences, screen walls, and plant supports.[79] Latticework stiffened medieval window panes and furniture. It was sometimes used in half-timbering, a building system employed in Europe and the United States. Town's church-building work in New Haven would have focused his attention on the qualities of the lattice; the stained-glass windows of Trinity Church (Town, 1814–1816) were "all glazed with diamond

Figure 2.14
Medieval latticework. This portrait features the carpenter Joseph, whose wife was Jesus's mother. One panel of the *Annunciation Triptych* (Merode Altarpiece), workshop of Robert Campin (ca. 1427–1432). Image copyright © The Metropolitan Museum of Art. Image source: Art Resource, New York.

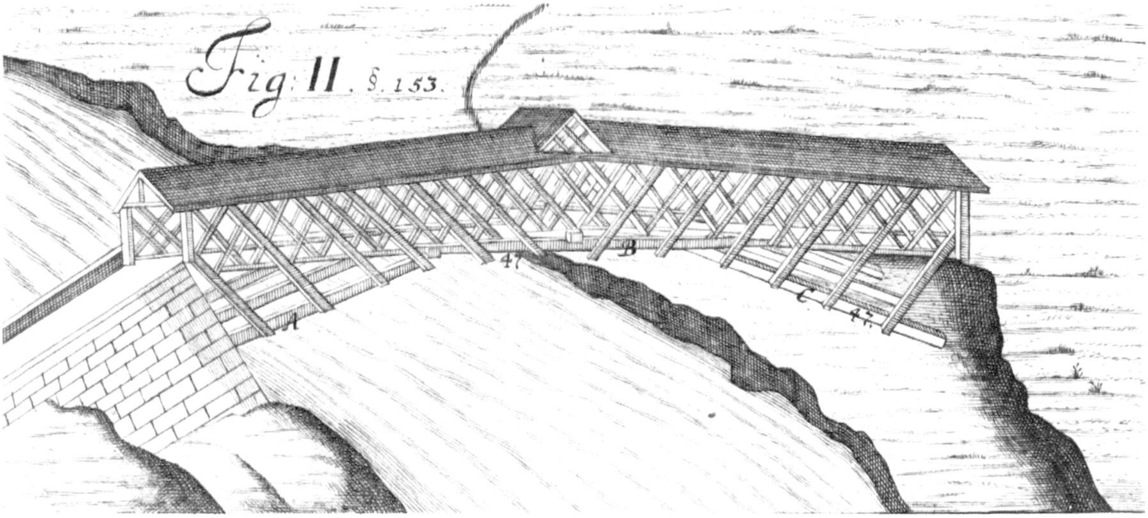

Figure 2.15
In this concept for a lattice-arch bridge, iron hardware connected the wooden members. Leupold, *Theatrum Pontificiale* (Zunkel, 1726), 79–80, table 13, figure 2. The Getty Research Institute.

panes" including one considered "not equaled in the United States." Town at one point suggested calling his invention a "diamond-truss bridge."[80]

Before Town's intervention, lattice-like components rarely appeared in bridge design.[81] A "model or sketch" of a structural idea that merged arch and lattice was depicted in one of the earliest bridge compendia in the German language (figure 2.15).[82] The most widely diffused images of a structural latticework—although not part of the main support of the bridge—was the "diamond-shaped" floor support of the bridge over the Rhine at Schaffhausen in Switzerland (figure 2.16), built (1756–1758) by the Grubenmann brothers, led by Hans Ulrich (1709–1783). This long-span, roofed, heavy timber structure, which mingled elements of arch, truss, and suspension bridges, was considered a technological marvel in the late eighteenth and early nineteenth centuries. Illustrations and scale models of Grubenmann bridges circulated internationally and were referenced when bridge structures were discussed, though the latticework floor was not the focus.[83] The bridge was known in the United States; the *Weekly Raleigh* [NC] *Register*, for example, reported in 1803 that a new bridge over the Connecticut River was "an improvement upon the very wonderful bridge built at Wittengen, in Switzerland, by Mr. Ulch Grubenham . . . who also built the celebrated bridge over the Rhine at Schaffhausen."[84] It is likely that Town saw descriptions, images, or models of the bridge; he knew Robert Fulton's *Treatise*, which discussed it. Reproducing the

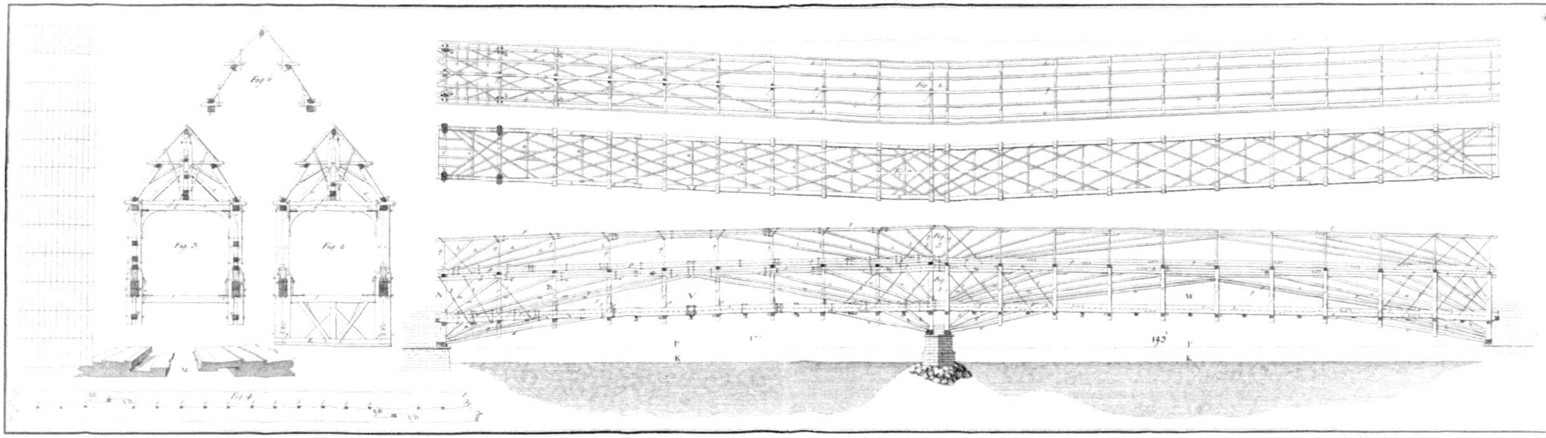

Figure 2.16
Plan, section, and elevation of Schaffhausen bridge over the Rhine, Switzerland (Hans Ulrich Grubenmann, built 1756–1758, destroyed 1799). Plan view at center shows lattice-like cross-bracing in the floor. Christoph Jetzler, ***Beschreibung der hölzernen Brükke über den Rhein in*** *Schaffhausen* (Steiner, 1778); print, 1776. ETH-Bibliothek Zürich, Rare Books Collection.

structure likely would've required the experience of the Grubenmanns themselves; plus, images that circulated contained variations that would have challenged builders.[85] Still, the Schaffhausen bridge, an amalgam that was challenging to classify, analyze, and build, inspired creators through their (mis)understandings of its design and function.

The Grubenmanns did not develop a system designed for mass construction, but their structures sparked construction-focused minds. Their superposing and overlaying of (box) beam and arch likely touched Town via the work of Massachusetts bridge builder Theodore Burr (1771–1822), whose popular design combined arch and truss in a simplified, systematized manner (figure 2.17). Images and descriptions of Burr bridges (patented 1806 and 1817), built in the United States at least through the 1850s, were widely published.[86] Burr's structure stiffened the arch and distributed the load so that the arch did not receive excessive load at any single point. This might be called an arch bridge, but whether the horizontal or arched component served as the main support depended on the specific design, construction sequence (i.e., which component took the load first), and changes to the structure over time.[87] According to engineering theorist Karl Culmann: "With iron consistency, [Burr] stood by his arch at all times, using, however, every newly discovered frame, without ever devising a new way of distributing the load on his arch."[88] Culmann perceived that the bridge might have helped builders imagine and test a variety of options. Reportedly, some of Town's earliest bridge building projects involved Burr bridges; he believed that the arch was the "main support."[89] Perhaps it gave him the opportunity to test the lattice design.

From the 1790s, James Finley (1768–1828) was assembling another type of long, horizontal structure out of small parts. He patented in 1808 a suspension bridge with an independent horizonal roadway stiffened by its floor structure and cross-bracing on both sides (figure 2.18). Suspended chains carried loads to the piers. It was a landmark achievement because the roadway, rather than following the curve of the chains, was level and hung from the chains. Per historian Eda Kranakis, the bridge, which incorporated uniformity of parts, represented "a basic new structural type." But the bridge, which required tall piers, substantial use of iron, and trust in a novel system, fell completely out of favor by the 1820s, replaced by bridges whose designers, such as Burr and Town, might have been aware of its deck design.[90]

Technologists were also developing structures that sailed on water rather than above it. While a bridge sags under its load, the reverse may take place in a ship; it may arch upward ("hogging") when the water does not adequately support its relatively thin ends. So ship designers focused on structures that could resist the instability and battering of the sea. Between 1806 and 1811, British naval architect Robert Seppings (1767–1840) developed a system of diagonal bracing (figure 2.19). As he explained, a frame in which

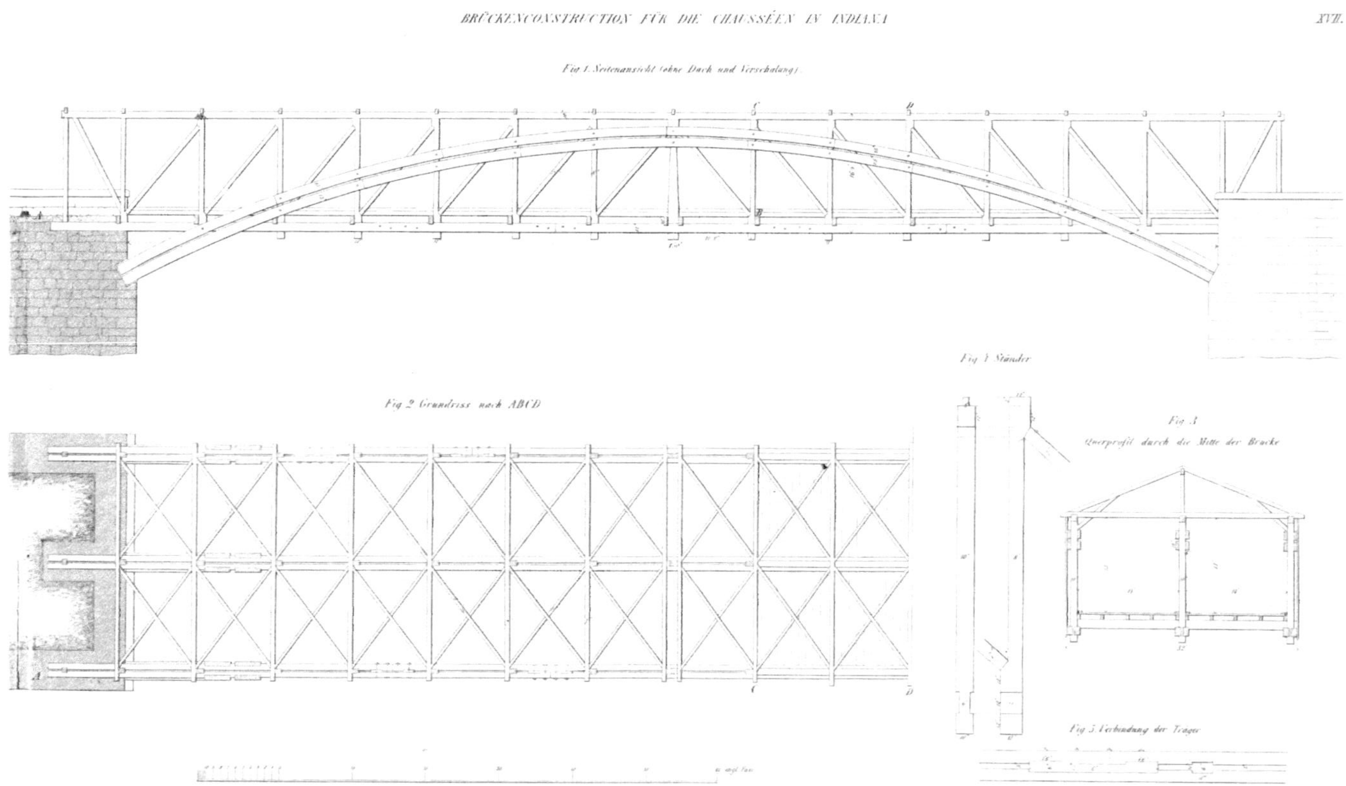

Figure 2.17
Inventor-builder Theodore Burr combined arch and truss. Franz Anton Gerstner, *Die innern Communicationen der vereinigten Staaten von Nordamerika*, vol. 2 (Förster, 1842–1843), pl. 27. Courtesy of Smithsonian Libraries and Archives.

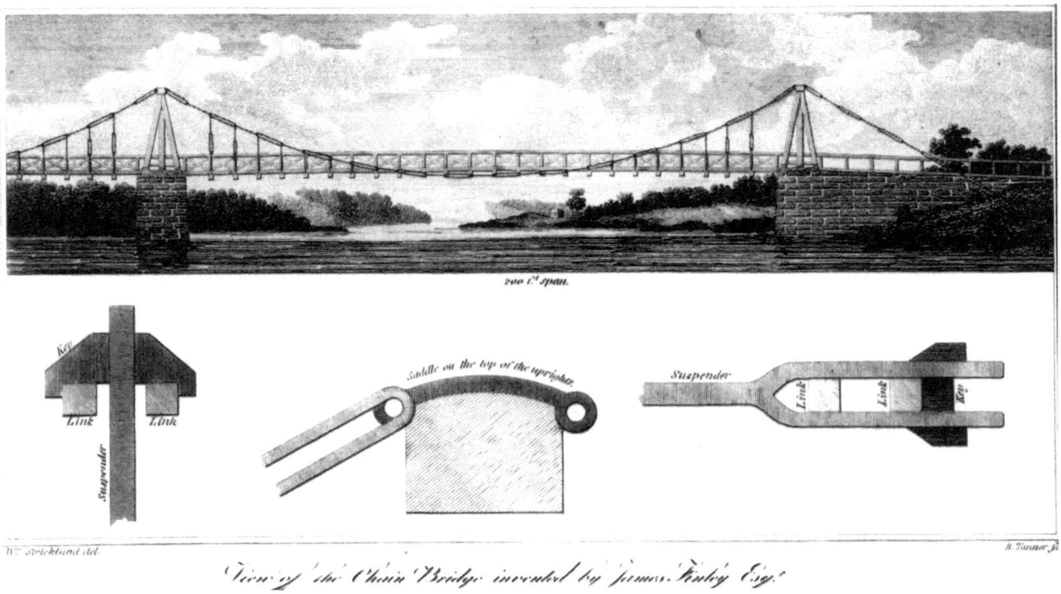

Figure 2.18
Chain suspension bridge (James Finley, tested 1801, patented 1808). A level bridge deck was the transformative feature of this design. *Port Folio* 3 (1810): 441. Internet Archive.

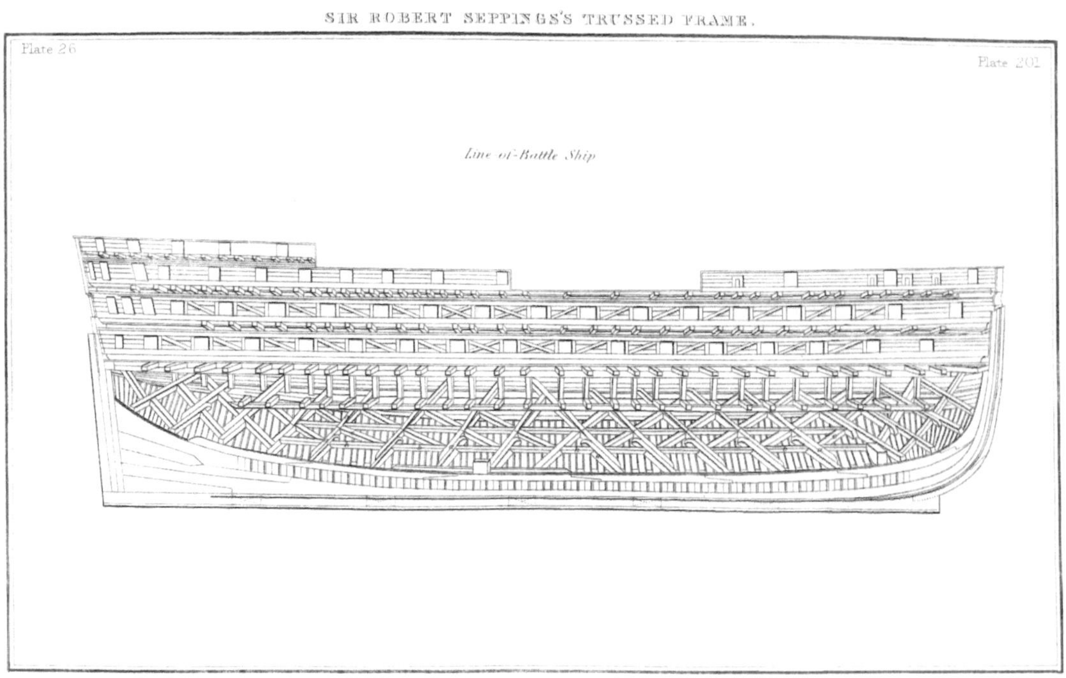

Figure 2.19
Seppings's lattice-like ship frame (1814). John Fincham, *History of Naval Architecture* (Whittaker, 1851), 200. San Francisco Maritime National Historical Park.

290 *Mr.* SEPPINGS *on a new principle of constructing*

and securing them with iron pins in the form of a square; which on the least pressure may be made to change its form to the rhombus; but let another piece be fixed to it diagonally, and the figure of the frame will be found immoveable. Place a bar in the middle parallel to two of the sides, and secure it firmly by iron pins, still the figure will easily be moved by the hand, like a parallel ruler, and assume the rhomboidal shape of A or B; but apply to the frame what the

carpenters term the brace in a common field gate as Z, and the figure will remain, as before, immoveable.

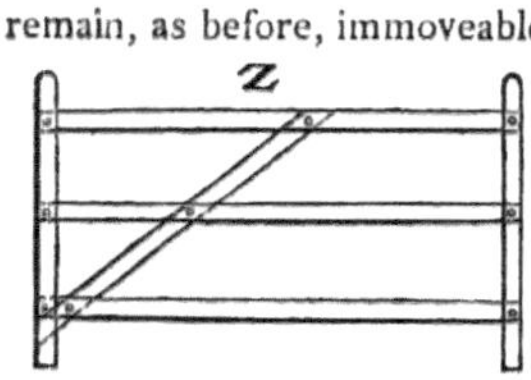

If this brace or diagonal piece is not fixed to it, the outer part of the gate (or that part most distant from the hinges) will have a constant tendency downwards, until at length it will reach the ground.

Let fig. 1 and 2 represent two frames of wood composed of parts strongly connected by bolts or iron pins.

Fig. 1. Will represent the principle on which the present system of ship building is conducted.

Figure 2.20
Seppings's structural design for ships. With a common field gate, Seppings demonstrated the rigidity of a triangle, that is, the effectiveness of diagonal bracing. Robert Seppings, "On a New Principle of Constructing His Majesty's Ships of War," *Philosophical Transactions of the Royal Society of London* 104 (1814): 290. Princeton University.

members are at right angles (i.e., having no diagonal members or triangles) "is well known to the meanest mechanic, to be the weakest"—this was the "present system of ship building." Seppings explained his "new principle," whose rigidity- and strength-enhancing diagonal members would be subject to axial forces (tension and compression), by referring to bracing on a gate (figure 2.20).[91] Seppings revealed, when the originality of his design was challenged, that his inspiration had been "the celebrated bridge at Schaffhausen."[92]

Seppings explained his design strategy:

> [I have] constantly kept in view these two leading axioms, "That the strength of a fabric consists not so much in the quantity of the materials of which it its composed, as in the disposition, the connections, and the security of its several parts." And, "that the strength of a ship . . . can never exceed that of its weakest parts," and consequently, "that partial strength produces general weakness."[93]

He was emphasizing the relationship between local weakness and overall strength, that is, how a detail might weaken or endanger an entire structure. For designers such as Seppings and Town, this was a salient issue. How could an apparent liability, many parts and connections, become an advantage?

Town understood that "the greater number of joints . . . the more strength and less strain upon each joint when weight or power is applied." A multiplicity of connections, working together, would mean that each had less work to do. According to Town:

> By dividing the strain or stress into so many parts . . . what falls upon any one part or joint is easily sustained by it without either the mode of securing the joints, or the strength of the material being insufficient.[94]

The lattice, as a unit, enabled its numerous parts and connections to carry loads through multiple paths: partial strength contributed to overall strength. This power of interdependence undergirds bridges, collaborative teams of people, and computer networks. In the lattice, forces appear to follow the structure, that is, they go where there is support to carry them: "Forces can detour around trouble just as traffic does in a grid street system like that of Manhattan."[95] Town was confident that this phenomenon distinguished the lattice bridge from other structures: "Such is the advantage of this mode in this one respect alone, that no substitute in other modes, that can possibly be introduced, will ever equal it; this is equal to mathematical truth, and cannot be erroneous."[96]

It was a key insight. A structure or system that is dependent on one path is no stronger than its weakest point. As an engineer noted at the end of the nineteenth century, "The strength of the weakest member of the weakest bridge on the line of a railroad is a measure of the [railroad's] capacity."[97] The question of how local weakness might be integrated into overall strength was essential to lattice bridge design, in which stress is spread among a multitude of connections, so each has less load to carry and each is comparably strong. One or several may fail, but the whole functions as others take on the load. Today this is called "redundancy." This is likely why a report of a catastrophic failure of a lattice bridge has yet to be discovered; a structure with this degree of interdependence fails gradually, which becomes obvious before it collapses. This

Figure 2.21
New Haven State House (Ithiel Town, 1827–1831), New Haven, Connecticut. The architecture exemplifies Town's modular strategy and nation-forming mindset. Postcard, collection of the author.

explains why an engineer inserted the lattice into an early twentieth-century debate about bridge design and public safety, as we'll see in chapter 7.

INVENTIVE DESIGN STRATEGIES

Ithiel Town's design work sprang from a modular, geometric perspective that integrated appearance and function and reduced structures to an aggregation of uniform, duplicable parts. In his buildings and bridges, he assembled a set of components, whether columns and pediments or planks and pegs (figure 2.21). Regularity, repetition, and symmetry, conspicuous in the handbooks of Town's reported teacher Asher Benjamin, characterize his designs. Modular thinking had also guided classicism-loving innovators such as Philibert de l'Orme and Thomas Jefferson, who was well acquainted with Greek architectural elements. "Grecian" to such designers meant geometric rigor, simplicity,

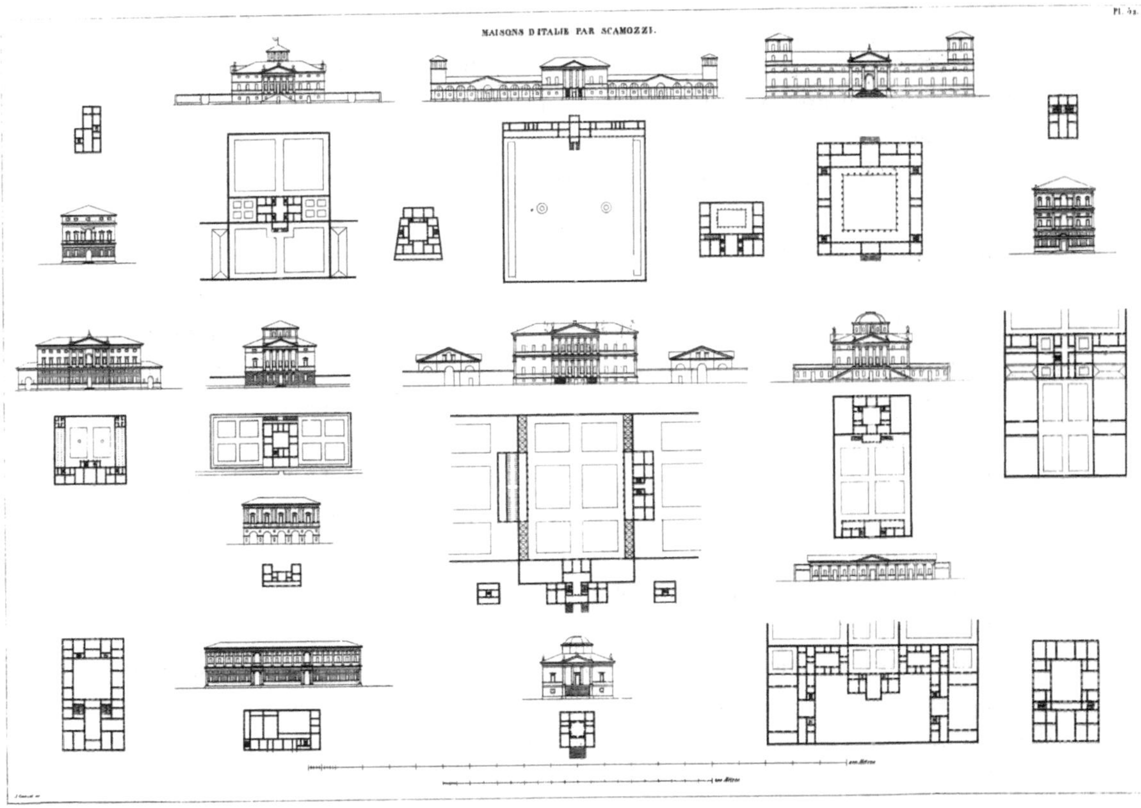

Figure 2.22
Durand's teachings exemplified geometric modularity, underpinned by ideas about economy and utility—the industrial ethos. Building plans, from Jean-Nicolas-Louis Durand, *Recueil et parallèle des édifices de tout genre, anciens et modernes* (1801): pl. 52. Reprint, Nördlingen Uhl, 1986.

and economy; it was a sign of good taste and a prescription for America's future.[98] Town, along with partner Alexander Davis, was familiar with a theory of design founded on the systematization of the design process as espoused by influential École Polytechnique professor Jean-Nicolas-Louis Durand (1760–1834). Durand's publications, which laid out principles for student architects, were among the most widely known design guides from the beginning of the nineteenth century (figure 2.22); their perspective shaped the teaching of architecture and engineering at West Point. Durand's graphic approach resembled both trade catalog and botanical taxonomy; it laid out classical typologies without reference to the theory behind them. Within a grid on a sheet of paper, architectural designers could focus on utility, space, and structure, employing a Greco-Roman kit of parts. The Hellenic ethos was a current rippling through the minds of elite makers of technology, industry, and nation in the United States and Europe.[99]

With a retrospective gaze across two centuries, we've been considering ideas and objects as emerging fundamentals of industrial strategies. Careful as we've been, it's risky to make assumptions about how designers two hundred years ago thought as they brought solutions into play. Histories of engineering have revealed little about how inventors such as Ithiel Town developed new structural systems and construction methods.[100] As we've noted, Town manipulated (overlapping) hands-on, theoretical, and industrial knowledge. It was embedded in things such as windows, ships, shoes, roof trusses, bridges, and classical architecture, along with unstable factors such as the availability of labor and materials. It was not obvious beforehand what the results would be, or if others would buy into them.

Analysts have explained the design process through three modes: (1) accretion, that is, adding one member at a time to produce a structure; (2) combination of small constructs to form a whole, to produce what historian Tom F. Peters has called "overlapping overlays"; and (3) systemic conception, that is, considering the bridge as a structure in which each member has a specific function yet is subservient to the whole. In the first, the designer does not have in mind a finished structural system; in the second, he or she considers and reconsiders how to assemble component structures; in the third, the designer from the outset has a vision of the outcome as a unified whole and works to achieve it.[101] The modes are not mutually exclusive.

Engineer Hans H. Hauri explored the accretionary method in his discussion of Grubenmann's Schaffhausen bridge (figure 2.16): "These structures were built up systematically from the same basic element: the inclined strut. The static behavior of a strut of this type was well-known to the carpenters of the [last third of the eighteenth century]."[102] The strut angle determined loadbearing capacity, which the builder could determine experimentally. Hauri continued: "This loadbearing element was then applied, using only a few variations, both above and below the deck, either in the form of a supporting brace, a trussed beam or a cantilevered strut. By the addition and superposition of many such elements, structural systems could be developed for large bridges."[103] Using a relatively small member, the strut (which we might anachronistically understand as a standard part), designers developed trussed beams, arch-like structures, and components. Additive technique enabled structures to have a longer span, carry a greater load, or be more rigid, without the builder having to find or fabricate large individual members (figures 2.23 and 2.24).[104] As Fulton observed, the small steps of accretionary design can be transformative.

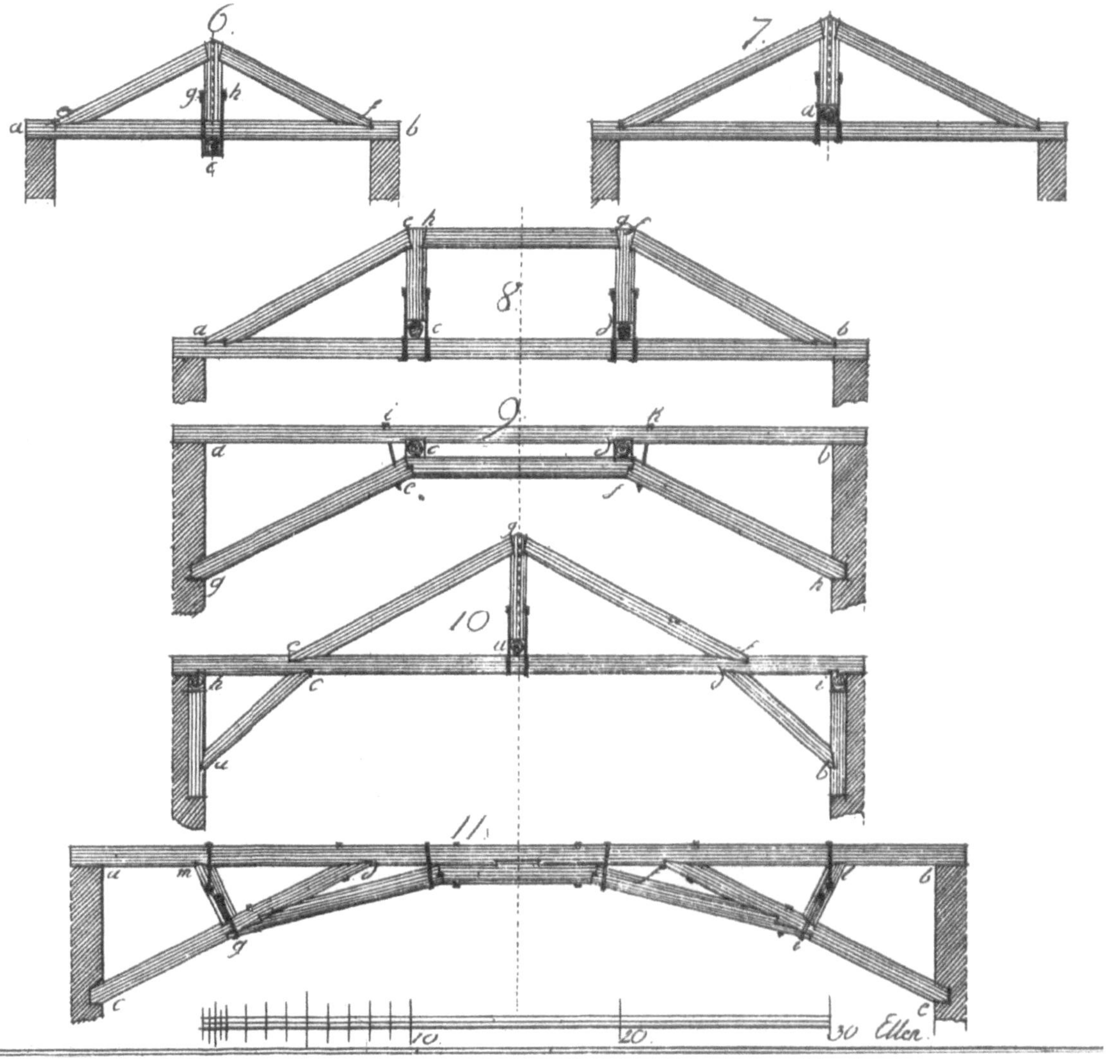

Figure 2.23
Building a trussed beam with inclined struts. A basic form of truss is the king post (7 above, single *Hängewerk*, in German). Rather than build a column underneath at the bridge's midpoint, two inclined beams, fixed at the ends of the bridge, meet at an angle and hold a vertical member from which the midpoint hangs. To increase the span, it is necessary to heighten the apex with longer struts; after a certain point it is more practical to employ alternative designs, such as a queen post truss (8 above, double *Hängewerk*) with shorter struts. Christian Gottlob Reuss, *Anweisung zur Zimmermannskunst*, 3rd ed. (Breitkopf, 1764), pl. 13. Deutsches Museum Library.

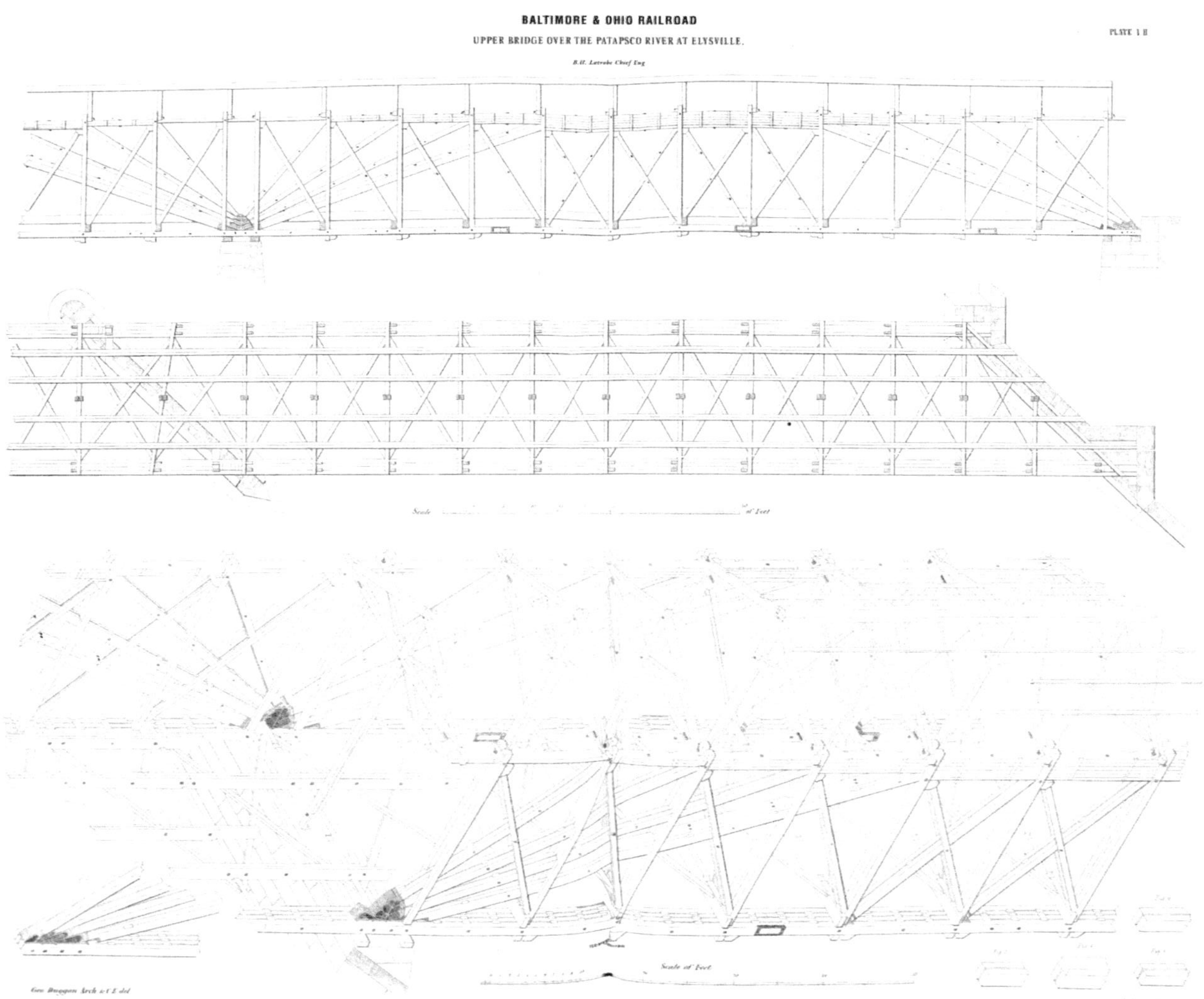

Figure 2.24
Upper Bridge over the Patapsco River at Elysville (Benjamin H. Latrobe, 1838–1839; demolished 1853). George Duggan, *Specimens of the Stone, Iron and Wood Bridges, Viaducts, Tunnels, Culverts, &c. of the United States Rail Roads* (Appleton, 1850), pl. 1.2. Courtesy of Smithsonian Libraries and Archives.

Designers who did not have the theoretical tools to analyze the type and intensity of the forces in all parts were still able to assemble resilient structures that they knew from testing or experience would function.[105] Additive and overlay strategies do appear to be behind the Grubenmanns' creative practice and designers who took inspiration from them. The complexity of their structures could be said to have made them safer.[106] That is, parts that may appear to be theoretically unnecessary or redundant provide for the security of pedestrians and passengers by compensating for the possibility of failure at any point. If one detail was inadequately designed, fabricated, installed, or maintained, or if a heavy vehicle or violent storm struck the bridge, the so-called redundant members could hold it in place, saving economies and lives.

Town integrated uniform, overlapping members into a relatively simple, rigorous system.[107] The lattice, an enfilade of interwoven struts, hovered in a space where "accretion" becomes "system," and "traditional" becomes "industrial." Words that engineers today use to describe trusses, such as "regular," "uniform," and "rational," seem to fit the structure.[108] Town remained true to his vision; he did not introduce curved members or elements that would have altered the pattern, though others would.

Town brought a new language to construction. Before he filed his patent, there was no "lattice bridge" in any language. The terms *Gitter* in German and *treillis* in French, which previously referred to structures such as fences, were applied to the lattice and sometimes to other bridge types.[109] The German word *Fachwerk*, considered the equivalent of "truss," which itself designates an large diversity of structures, was apparently first employed by engineering theorist Karl Culmann, who had studied bridges in the US (see chapter 4); the word till then had designated half-timbering. As "lattice bridge" and "American bridge" imprinted themselves onto designers' consciousness, their connotations extended beyond the mere description of an object. The category reflected a way of viewing the potential of structures and the infrastructure of the world.

Technological labels, like all words, function within a web of associations. These categories guide our thinking and communication as they nurture the illusion of stability. "Engineer," for example, might seem established and neutral, but a German speaker once proposed deleting the word from his mother tongue because of its French origin.[110] The lattice bridge's apparent clarity as a system belied functional ambiguities and suggested to technology's poets possibilities that were inventive catnip.

3

BUILDING FUTURE STATES

By the 1830s, it was claimed that builders in the United States had constructed canals and railroads equal in length to all of those in Europe.[1] So it's hardly surprising that engineers, industrialists, historians, scientists, social observers, novelists, and economists scrutinized the natural and built landscapes of the US. They made their observations public through "a rapid and ceaseless succession" of books and articles.[2] One critic, noting the number of reports, claimed in 1838 that "the people of the United States are like persons surrounded by mirrors," reflecting ideas everywhere.[3] Capitalism drove the explosive expansion of publishing, whose output sailed through global channels of transportation, providing information and inspiration. Countries that had the greatest industrial growth during the nineteenth century more than doubled their literacy rates during those years.[4] Printed words and images boosted people's ability, locally and globally, to envision a future through national-technological narratives that made "America" a gauge for industrializing societies in various spots on the planet.

US business and political leaders responsible for states' physical and administrative infrastructures felt that less costly and quicker methods of doing business would drive innovation.[5] Technological solutions to social challenges was the dream. Observers who spotlighted US infrastructure believed that expanding the "American" mode of building into other lands would unify the world. (Bygone dreams of the World Wide Web might be found in the burial ground containing the remains of nineteenth-century aspirations linking communication networks, free trade, and world peace.)[6] American bridges were a model for a revamp: how to build a New World and restructure the Old.

The previous chapter surveyed the technological menu that nurtured Ithiel Town's industrial ingenuity. Here, we'll take a quick look at the dynamics of that environment, the variety and mobility of texts and images, along with the relationships they reveal, as people embedded the lattice in industrial-political conversations and builders inserted the lattice into the landscape. This patent bridge, a product of and stimulus for creative exchange, was a climax rather than a conclusion. Here, we'll stalk the lattice concept through a series of published moments while introducing prominent participants and their perspectives on nation and bridge building.

Town's tremendous library and the connections he made as book lover and entrepreneur enhanced the capacity of his inventive networks (figure 3.1). Innovators then as now avidly consumed information in order to achieve goals. In the 1820s, even in the most rural parts of a rural United States, literacy and books were community transformers. In 1825, lawyer-politician Daniel Webster noted "a community of opinions and knowledge amongst men in different nations existing in a degree heretofore unknown. . . . Every breeze wafts intelligence from country to country; every wave rolls it."[7] The front page of an 1825 issue of the *Western Carolinian*, which included an update on one of Town's earliest lattice bridges on page two, included a lengthy article on the design of roads, with reference to England and Sweden. Next to it was an item entitled "Marriage in China."[8] A comparison to the internet is apt and may be understated because, beyond the multinational precipitation of information from every discipline, people in the nineteenth century were also *learning* to read and experience the jolts of pleasure provided by the regular consumption of factoids, pictures, and multiple perspectives—abilities that in the latter twentieth century were already ingrained in consumers of digitally transmitted narratives.

The internet provided the foundation for the creation of the Web, the information system that enables instantaneous global data connectivity; in the 1800s, the rapid building of transportation infrastructures turbocharged oral and printed intercourse. Even during the earliest decades of the nineteenth century in the United States, before the establishment of big industry, upgraded roadways, and expanded trade, civil engineers and mechanics formed social networks that ran east and west of the continent's Atlantic coast.[9] Water was the main connector; until mid-century, for example, publishing companies' success in the United States was dependent on the ease and cost of transportation of publications via ocean and river.[10]

Engineering journals that form the foundation of this book constituted significant channels of communication. At the same time, popular publications brought news to a broad audience. To name three whose images appear in *American Bridge*: *L'Illustration* (Paris, founded 1843) might send half its print run abroad; pictures from *Illustrated London News* (founded 1842) and *Illustrirte Zeitung* (Leipzig, founded 1843) would have been seen "all over the world." They sold their images to other newspapers, too. Among elite audiences and beyond, these publications helped create professional and national identities as they encouraged shared perspectives that cut across national borders.[11] Pictures were as important as the texts that teemed around them.

Technologists operating within and across regional and state boundaries discussed, debated, and developed a common language about novel engineering structures—although

Figure 3.1
Nathaniel Jocelyn, *Ithiel Town*, 1826. For Town, books represented learning, inspiration, communication, business, and a means of connecting with others. National Academy of Design, New York. © National Academy of Design, New York / Gift of George Dudley.

consensus on how to build and understand them, and how to finance and manage them, was not universal or uniform within any set of boundaries. Similarly, once the concept of "nation" became real in the Americas, it served as a model that could be pirated (Benedict Anderson's term); it would be further debated and developed around the world as it displaced dynastic and religious power structures.[12] The idea of a lattice structure as a means of support, once crystallized and portrayed and built, became a model for appropriation, analysis, and transformation. The idea of "America," brought to life through a way of building, sparked people's imaginations in combustible capitalist, industrialist, nationalist domains. The reinvention of construction was enmeshed in the construction of nations; each was about implementing a conceptual and physical vision. The resulting global patchwork of nationalized spaces belies the web of transnational (and prenational) relationships that defy nationalist myths about innovation, which hide connections across borders and differences within them.

Inventors and engineers in North America did not have unique access to a distinct American knowledge. They were in some respects visitors in their own lands. In order to supply themselves with a stock of ideas, they, like visitors from other places, traveled and observed. In Europe, the name for a place that was, and still is, under negotiation, American citizens met Europeans who were traveling, too, in order to learn of the latest technological "improvements." Transatlantic travel became frequent during the second quarter of the nineteenth century; by 1840 there was regular steamship service. Travel enabled people to provide and glean information that could not be gotten from books. Although *American Bridge* relies on publications, person-to-person communication was as critical then as it is today.[13]

While conversations, visits, lectures, and correspondence were means of exchanging knowledge, evidence of who said or wrote what to whom and what people actually knew often cannot be recovered; conveniently, the proof-impoverished strategies of evolutionist and nationalist storytellers often render this evidence optional.[14] But long before the term "industrial research lab" existed, groups of civil, mechanical, and theoretical engineers, along with entrepreneurs and managers, worked with one another on the development of railroads, structures, and construction processes. This activity occurred through border-crossing social networks and within local, corporate settings. For example, the Baltimore and Ohio and Saint Germain railroad companies were hotbeds of engineering that set standards and whose engineers would play key roles in a variety of projects.[15]

Engineering organizations whose names anchored them to specific places (for example, "America" or "Bavaria" or "London") used meetings and publications as a glue to bring together professionals and promote exchange. These organizations constructed and sustained the significance of the places in their names. The idea that a professional organization or academic institution, any more than a nation, represents a unified viewpoint, however, is likely an illusion. But decades or centuries later, under the shadows of celebratory narratives and busts of great men, it can be difficult to decode the divisions still simmering on the pages of old journals. It might feel natural to tell a story about "Germany" or the "United States," but, even if one makes the overweening assumption of national uniformity and continuity, how does one disentangle national from individual? For example, owners, promoters, and investors in bridges, railways, and newspapers had personal and financial ties that encouraged them to exaggerate success and hide failure.[16] Whose stories were they telling?

Information arrived in diverse forms ranging from news briefs to lavishly illustrated volumes. The account by the duke of Saxon-Weimar-Eisenach (today part of Thuringia in Germany), Karl Bernhard, of his 1825–1826 journey through North America, for example, published in German- and English-language editions, noted, sometimes with disapproval, the many bridges he crossed. He shared "local" information widely. For example, he reported seeing a lattice bridge near Fayetteville, North Carolina, which he called a "gegatterten Hängewerk" ("hanging latticework" in the translation).[17] Bernhard, who met with luminaries including President John Quincy Adams and Thomas Jefferson, reported seeing a model of the bridge at the Patent Office in Washington, DC.[18] Models of lattice bridges were on display in several cities; in the eighteenth and nineteenth centuries models were widely used for promotion and testing and instruction at technical institutions.[19]

Observers described, analyzed, and classified tools, machines, techniques, systems, and people. They told stories whose purpose was to prevent or promote, in the environments in which they worked, conditions and achievements they perceived elsewhere. The United States, foreign and familiar, seemed bold, hardworking, quick, and practical to observers from places also blazing new national and industrial paths. American people and practices enthralled and repelled observers. Were the denizens of the New World a different species? In the opinion of engineer Guillaume Lambert, who at mid-century embarked from Belgium on a technological tour of the United States:

> For the newly arrived European in New York, the audacity of the American goes as far as recklessness, while on his side, the latter accuses us of pushing caution to the point of timidity. These judgments are equally unfounded, because they take people, rather than their situation, too much into account. I'm led to believe that on Union soil most Europeans would soon be Americans, and vice-versa.[20]

National boundaries intensify the curiosity and angst that foster cross-border collaboration and competition.[21] The idea that the mindset of an author living in France would be preordained by political borders, or that information shared there would be more resonant locally than elsewhere, are unproven assumptions. "Foreign" design, a frequent subject in technological journals, was relevant to local and global audiences and consumed by both. General interest newspapers kept readers up to date on innovations near and far. When the newspaper *Moniteur Universel* (Paris), reported on December 3, 1803, that "the bridge over the Schuylkill on the other side of Philadelphia, is well under way, and will be completely finished during the coming summer," we can assume the progress of Palmer's Permanent Bridge (figure 3.2 and 3.5) was of interest beyond the borders of Pennsylvania.[22] Technical journals borrowed information from each other (there were no international copyright laws); they published translations and reviews and critiques and excerpts and bibliographies and announcements. Some publications, such as the *American Railroad Journal* (New York, established 1832) and the *Journal des chemins de fer* (Paris, established 1842) apparently exchanged copies of each issue.[23] Nations exchanged books: in 1846, for example, the United States government, as well as the governments of individual states, gave hundreds of books to France as gifts; France reciprocated.[24]

General technical journals existed, but specialized periodicals were a novelty in the early nineteenth century. The editor of *American Railroad Journal* began the third issue in 1832:

> As to the title of this periodical, we have been not a little amused at the comments it has excited. "The Rail-road Journal!" "Phoebus what a name!" "I should as soon think," cried a gentleman in our hearing, "of a Patent-Furnace or Cooking-Stove-Journal! A newspaper devoted to Rail-roads? You might as well have an 'Aqueduct Chronicle,' or a 'Turnpike Commentator,' as a *Rail-road Journal!*" "Certainly," echoes another; "and 'the Steamboat-Aegis,' or 'the Steam-Bath-Locomotive,' or 'the Steam-Scouring-Visiter,' would be a far more attractive title!" So they might—so they might, Gentlemen; and you might add, that 'the Automaton-Working-Man' would be a more engaging title still to those who sit with their arms folded quietly at home, and when the whole world is awake and bustling about them, not only put their hands to no work or enterprize, but close their ears to the din of business, and shut out all sounds that would remind them of strenuous exertion.[25]

While the whole world was not engaged in railway building in 1832, a subset was envisioning the possibility, or soon would be, supported by journals and books that nurtured technological specializations and expectations. On their pages, the lattice bridge was part of what nineteenth-century philosopher Gabriel Tarde described as a "current of inventions."[26] The lattice, which seemed to first-time observers to be without precedent, became an object for designers and builders to reject or adapt at sites stretching from America to Europe to Russia. Because of its apparent novelty, its industrial efficiency, its structural ambiguity, and its role as an analytic and design catalyst—as well as its association with America, the nation of the future—the lattice left a printed trail that enables us to begin to reconstruct the transnational creation of knowledge.

Figure 3.2
William Russell Birch, Timothy Palmer's Permanent Bridge (1805) over the Schuylkill, ca. 1820. The systematized structure, along with cladding (visible in the portrait in the foreground) that provided essential weather protection, attracted wide attention. Library Company of Philadelphia.

WORD AND IMAGE

Ithiel Town launched the news of a structure intended for mass construction in four illustrated promotional brochures, published in 1821, 1825, 1831, and 1839 (figure 3.3); he also placed newspaper advertisements.[27] As an innovative designer amassing one of the world's largest collections of art and architecture books and prints, he was familiar with the mobility and potency of images. As a successful engineer and architect, he traveled through the northeast and southern United States and had the political connections to get

A
DESCRIPTION
OF
ITHIEL TOWN'S IMPROVEMENT
IN THE
PRINCIPLE, CONSTRUCTION, AND PRACTICAL EXECUTION
OF
BRIDGES,
FOR
ROADS, RAILROADS, AND AQUEDUCTS,
WHETHER BUILT ENTIRELY OF WOOD, OR OF CAST OR WROUGHT IRON:
BEING INTENDED AS A GENERAL SYSTEM OF CONSTRUCTING BRIDGES OVER
RIVERS, CREEKS, HARBORS, &c.,
WHERE A CONVEYANCE IS REQUIRED: EITHER WITH WIDE OPENINGS BETWEEN THE PIERS, OF FROM TWO HUNDRED TO FIVE HUNDRED FEET; OR WITH OPENINGS OF LESS EXTENT: WHETHER THE PIERS AND ABUTMENTS ARE BUILT OF STONE, FRAMED TIMBER UPON MUD SILLS, OR PILES DRIVEN INTO THE BED OF THE RIVER; OR WHETHER THE BRIDGE BE SECURELY COVERED FROM THE WEATHER, PARTIALLY SECURED, AS IT MAY BE, OR ENTIRELY EXPOSED.

Economy of expense, great strength, durability, and a level road-way, united in a general system of building Bridges, Aqueducts, &c., suited to any part of the country, which may be constructed of such materials as any part of the country affords, are what is here attempted. If these objects have been accomplished, as it is confidently believed they have been, they are important to the country.

ALSO,
CRITICAL REMARKS
ON THE
GENERAL PRINCIPLES AND PRACTICAL EXECUTION OF OTHER MODES OF CONSTRUCTION USED IN THIS IMPORTANT BRANCH OF IMPROVEMENT,
WHICH NOW ENGROSSES SO MUCH ATTENTION AND CAPITAL, THROUGHOUT NOT ONLY THIS EXTENSIVE COUNTRY, BUT MOST OTHERS.
WITH
PRACTICAL AND SCIENTIFIC REMARKS AND OBSERVATIONS,
ON THE STRENGTH, &c. OF SUITABLE MATERIALS, EXPOSURE TO WINDS, CAUSES OF DECAY, AND OTHER GENERAL DEFECTS IN THE ARCHED BRIDGES, AS WELL AS OTHER MODES, CONSTRUCTED OF WOOD.

BY ITHIEL TOWN,
ARCHITECT AND BRIDGE ENGINEER.

NEW-YORK:
PUBLISHED BY THE AUTHOR.
HITCHCOCK & STAFFORD, PRINTERS, NEW HAVEN.
1839.

Figure 3.3
Town's final brochure, title page. Ithiel Town, *A Description of Ithiel Town's Improvement in the Principle, Construction, and Practical Execution of Bridges* (1839). Courtesy of Smithsonian Libraries and Archives.

himself state capitol building, as well as bridge, commissions. Town twice toured Europe, though if and how he promoted his bridge there is yet to be discovered.[28] He apparently enjoyed a warm relationship with British technical publisher John Weale.[29] Architect Edward Shaw's popular handbook *Civil Architecture* contained a description of Town's bridge; it appeared in eleven editions from 1830 until 1876.[30] Illustrated descriptions of the two lattice bridge patents appeared in the publication of Town's future New Haven neighbor, well-known Yale professor, scientist, and widely traveled lecturer Benjamin Silliman's *American Journal of Science*.[31] Volume three, published in 1821, contained an article that duplicated Town's brochure; this volume could be found by April 21 of that year in the library of the University of Edinburgh—a city that was an intellectual capital of Europe.[32] After Town registered his 1835 bridge patent, a brief notice appeared in journals promoting innovation and industry, including the *Bulletin de la société d'encouragement pour l'industrie nationale* (Paris), *Allgemeine Polytechnische Zeitung* (Nuremberg), and the *London Journal of Arts and Sciences*. The *Mississippi Free Trader* ran an item that described the bridge and judged it "well worthy of attention."[33]

The circulation of the lattice idea and image is bound to the growth in technical publishing that took place during the nineteenth century. In the first known in situ portrait of a lattice bridge, created by Nathaniel Jocelyn for Town's 1821 brochure and reproduced in some of the publications we'll explore below, wild vegetation and rocky hills framed a vista featuring a manor house with manicured greenery (figure 3.4).[34] A three-span covered bridge, incongruously dropped into the scene, cuts across the foreground and seems to end abruptly in a rocky hillside, but likely leads to the proprietor's residence. This image portrays industrial mastery of the landscape—part of ongoing continental conquest made manifest through evolutionist markers of "civilization": agriculture, architecture, and civil engineering.

The tableau was a come-hither to civic managers, engineers, and builders to look into the future. The leftmost bridge span, unclad, reveals the lattice. Although it is the image's raison d'être, a leafy branch half obscures the crisscrossing members, as if Town wanted to cover the invention's nakedness in its Edenic setting. Exposed was a structure well suited to graphic reproduction. "The mode of construction is so simple and plain to inspection, as to require little explanation," Town stated.[35] Its geometry did not express a hierarchy typical of bridges that depended on a key structural element such as an arch. Indeed, in comparison to existing frame systems, the lattice possessed a curveless, repetitive precision, a brain-pleasing orderliness and symmetry that made it one of the earliest engineering pin-ups of the railway age. In the landscape, however, as observers noted, "It is not possible to conceive any thing more ungraceful than these huge, snail-like housings appear" and "the architectural effect . . . is what might be

Figure 3.4
Nathaniel Jocelyn, *Town's Bridge*, 1821. A lattice bridge stretches across a genesis portrait of civilization and industrialization. *Description of Ithiel Town's Improvement in the Construction of Wood and Iron Bridges* (1821), frontispiece. Courtesy of Dartmouth Library.

expected from an enormous square box and, whether boarded in or not, may be safely put down as a minimum."[36]

The reproducibility of the American-bridge structural image highlights its industrial character. Previously, novel bridges had been associated with specific geographic places and particular designer-builders: for example, Grubenmann's bridge in Schaffhausen (figure 2.16), and Timothy Palmer's Permanent Bridge (figure 3.2) and Louis Wernwag's Colossus, both in Philadelphia (all three depicted in figure 3.5). Jocelyn, however, portrays the lattice in an unidentified setting. It would look more or less the same wherever builders might erect it; they would specify the length of spans and cladding style. Soon, imagery in engineering texts would mostly concentrate on one-dimensional, decontextualized representations of bridges.[37] Town interpreted the implacable geometry of the lattice as a kind of truth; for some it was a foreshadow of a future state. An engineering diagram, like the shape of a country reproduced endlessly on maps, can become a national symbol.[38] The lattice, and what were known in English as truss bridges, became icons of invention and industry, designed to be replicated in any context, anywhere in the world. Behind the lattice stood an emerging transnational figure: the professional engineer.

"*AUS AMERIKA* GLORIOUS NEWS!"

Lattice bridges erected during the 1820s and 1830s attracted increasing attention. "American bridges": this is what observers called structures patented by Burr, Town, Long, and Howe.[39] The latter two inventors we'll introduce in the next chapter. Below, quickly, we'll boldface nine people who shared what they saw in a new country and a new structure.

An extended visit to the United States shaped the future-focused writings of **Henri de Saint-Simon** (1760–1825), who noted, regarding his experience during the Revolutionary War: "It was in America, while fighting for the cause of industrial liberty, that I conceived the first desire to see this plant from another world flower in my country." Saint-Simon's fantastical portrayal of a New World released from the strictures of European society described a path forward for the advancement of civilization. His expression, "industrial liberty," highlights his awareness of the contours of the new freedom: the United States offered capitalist opportunity for select groups of individuals.[40] "America" was a lodestar for the Saint-Simonians, a political-economic-philosophical group that attracted pivotal early-nineteenth-century engineers and thinkers.

As infrastructure is inseparable from the nation-state, the construction of transportation, communication, and defense networks was often integral to discussions about the form of new nations. Saint-Simonians believed that transportation infrastructure, achievable through access to capital supplied by the banking system, would bring the world together while raising its financial and industrial capacity. They regarded the construction of "internal communications" in the United States as more than just a remarkable achievement; it was a manifestation of a social and political system they wanted to emulate. Former Saint-Simonians would later be instrumental in creating much of France's transportation infrastructure.[41]

Friedrich List (1789–1846) began an 1836 report on a Baltimore and Ohio Railroad locomotive test in his *Eisenbahn-Journal und National-Magazin* with the German-English mash-up, "*Aus Amerika* Glorious News!"[42] During the first half of the nineteenth century, this exuberant attitude was typical of reports by industrializing engineers who visited the United States seeking experience and inspiration. List, an enormously influential journalist, economist, entrepreneur, and politician, was from the early 1820s a promoter of industry and railways in the United States, France, and several German states; he lived in the US from 1825 to 1830, participated there in the establishment of one of the earliest railways (and was likely familiar with lattice bridges through his association with engineer Moncure Robinson), helped shape US tariff policy, was a prominent interpreter of life in the new states, and was an advocate for policies that resulted in the creation of Germany.[43] List regarded the railway as a national, cultural, and economic connector: this is an enduring idea, although transportation networks (for example, in European lands today) may be just as likely to confirm or create borders as to unify.[44]

The publications of engineering expert Major-General **Howard Douglas** (1776–1861), who served Britain at various locations in the world (he was governor of New Brunswick from 1823 to 1831), enables us to witness a first encounter with the lattice. In his 1816 publication *Essay on the Principles and Construction of Military Bridges*, Douglas (whose book also appeared in French and German translations) discussed basic concepts underlying triangulated bridge design. He described the combining of trusses to form a longer and stronger structure that could be analyzed as a beam. His additive approach appears to be on the verge of describing what we'd today recognize as a truss.[45]

In the second, expanded edition of *Military Bridges* (1832), Douglas announced a discovery:

> It is ever of great importance, in military operations, that the truss principles, by which small pieces of timber may be made to serve for great spans, should be well understood by officers employed in the bridge department. In this view the following method of forming truss frames of common planks, is a valuable addition to the resources of the military engineer. Vast bridges have been erected in this manner in the United States, and such may be constructed on any scale, for military communications, either as temporary or permanent bridges.[46]

With this, Douglas began a description of the lattice bridge. The accompanying image (figure 3.5) verifies his familiarity with Town's brochures, although Douglas never states the source of his information and only off-handedly mentions "Mr. Town."[47] Many engineers involved in early railways came from the military; Douglas's observation that the lattice bridge would be "highly useful in military operations" confirms that the army's constructive prerequisites—cheap and quick, with uniform, transportable parts—aligned with the mindset of railway constructors in the United States.[48]

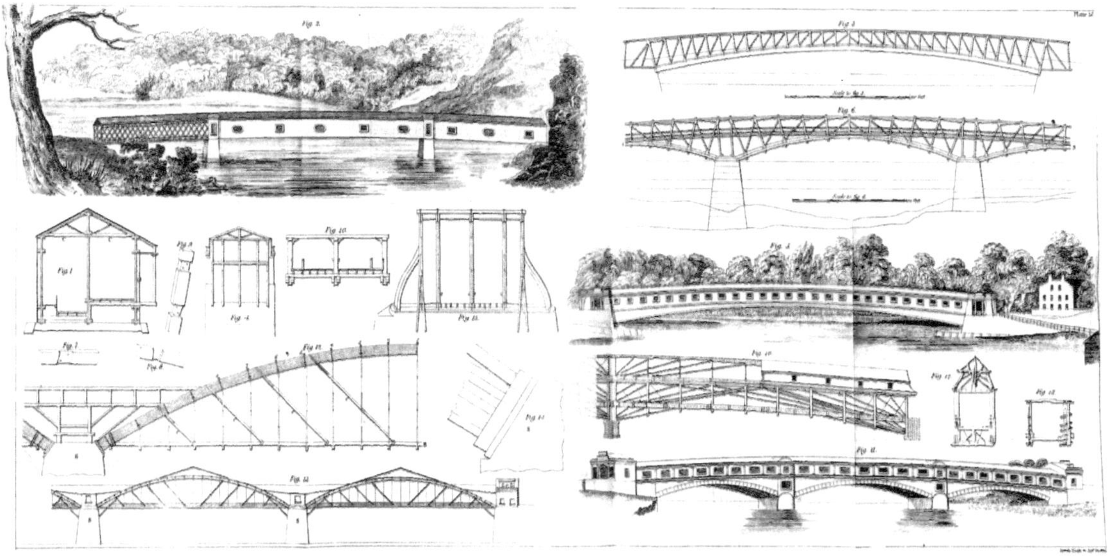

Figure 3.5
US bridges (and one Swiss). Howard Douglas's rendition of the lattice bridge (top left), adapted from Town's brochure, removes the leafy obstruction. Middle right (landscape portrait), and top right, uncovered: Colossus Bridge (Louis Wernwag, 1811–1812) over the Schuylkill, near Philadelphia. Between these images and at the bottom right is the Permanent Bridge, uncovered and covered (Timothy Palmer, 1805) over the Schuylkill, Philadelphia. Directly above the bottom right image is the Schaffhausen Bridge (Hans Ulrich Grubenmann, 1756–1758) over the Rhine. Lower left: bridge over the Delaware at Trenton (Theodore Burr,1804–1806); not a typical Burr bridge. Howard Douglas, *Essay on the Principles and Construction of Military Bridges*, 2nd ed. (Boone, 1832), pl. 12. Courtesy of Sutro Library, California State Library.

Engineer-architect-artist Guillaume Tell Poussin (1794–1876) became France's ambassador to the United States and worked for a time after his 1814 arrival in the US for renowned architect Benjamin Henry Latrobe on the restoration of the Capitol Building in Washington, DC. He provided a prominent account of the Town bridge. Within the US Army Corps of Engineers, between 1817 and 1832, much of that time as an assistant topographical engineer to Simon Bernard (former adjutant to Napoleon), Poussin participated in the construction of public works in North America before returning to France, where he became minister of war. Poussin traveled through the South with Bernard and worked in North Carolina, where Town developed his earliest lattice bridges.[49] His report on civil engineering in the United States, published in 1834 in France and in a German edition a couple of years later, contained an appendix devoted to the Town bridge (figure 3.6).[50] The description is mostly an excerpt of

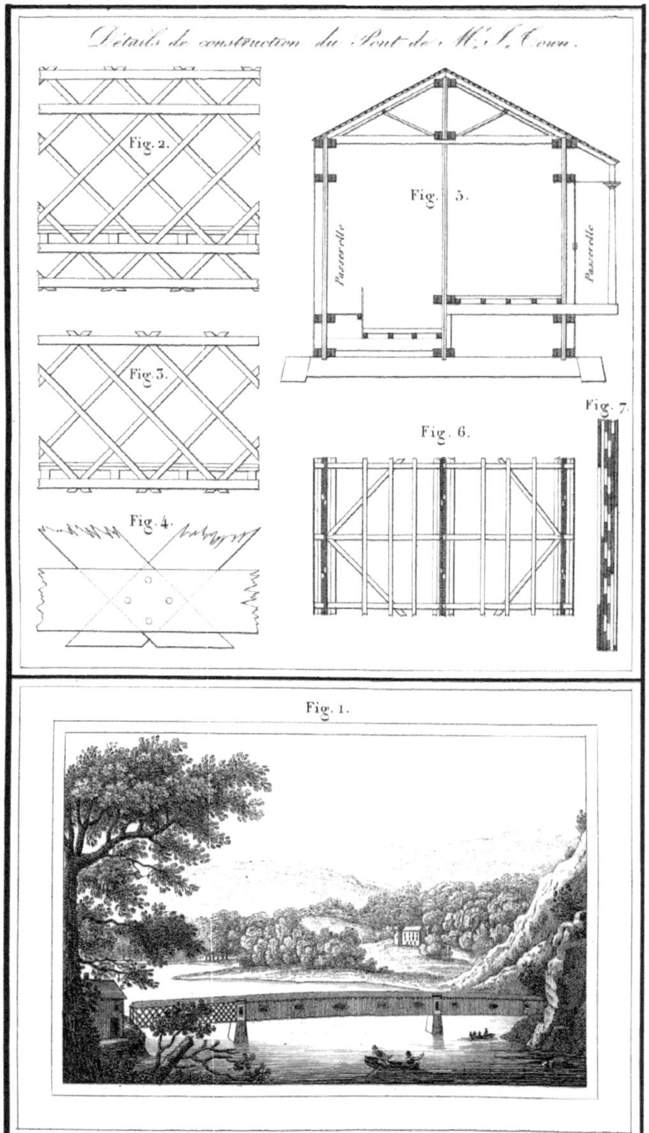

Figure 3.6
Lattice bridge. Poussin replicated the images that appeared in Town's 1821 brochure. G[uillau]me Tell Poussin, *Travaux d'améliorations intérieures projetés ou exécutés par le gouvernement général des États-Unis d'Amérique, de 1824 à 1831* (Anselin et Carilian-Goeury, 1834/1836), pl. 10. David Rumsey Map Collection, David Rumsey Map Center, Stanford Libraries.

Town's first brochure; Poussin redrew the illustrations.[51] In his 1836 book on American railways, Poussin's interest in "internal improvements" was embedded in a belief about transformation that surfaces in a tribute he pays to George Washington: "It was this truly great man who first understood that routes of communication were the best way to spread liberty and to elevate the human race, by improving its fate, to the highest degree of civilization."[52]

David Stevenson's (1815–1886) *Sketch of the Civil Engineering of North America*, published in London in 1838, included descriptions of navigation, canals, roads, and railways; it was excerpted and widely read in French- and German-language periodicals.[53] Stevenson (1815–1886), a twenty-two-year old Scottish marine engineer, took (according to a contemporary reviewer) "advantage of a short interval of professional leisure . . . for the express purpose of examining for himself the much talked of, but little known, engineering works of America."[54] During his three-month trip, Stevenson found the bridges "in general . . . specimens of good carpentry, and not unfrequently of good engineering."[55] (Stevenson was less restrained when addressing the Royal Scottish Society of Arts; he declared: "The finest specimens of timber-bridge architecture that I have met with are to be found in that country."[56]) He provided illustrations and written descriptions of structures, including Burr, Town, and Long bridges. He met Yale professor Benjamin Silliman in New Haven and received a drawing of a lattice bridge from prominent railroad engineer Moncure Robinson.[57] Town included excerpts from the *Sketch* in his brochure of 1839 and noted that Stevenson "did not see but a small number of those [lattice bridges] that are well constructed.[58] Indeed, Town's double lattice (patent of 1835) was not presented in Stevenson's book.[59]

"Isn't it true that along the great roads, canals and rivers, ideas circulate at the same time as goods, and that every commercial traveler is a kind of missionary?" asked prominent economist **Michel Chevalier** (1806–1879), who was an 1829 engineering graduate from the École Nationale des Mines and a Saint-Simonian.[60] Chevalier traveled through the United States from 1833 to 1835 on a mission for the French government to investigate canals and railroads; he believed that the railroad's ability to connect people would make it a global organizer and transformer. Chevalier's *Lettres sur l'Amerique du Nord* (1836), his first major publication on the United States, is considered a complement to, and in some ways more perceptive than, Tocqueville's *Democracy in America* (1835, 1840).[61] One of his younger brothers, Emile, was assistant engineer on the construction of the Philadelphia and Reading Railroad, which was likely a factor in his friendship with Moncure Robinson, who supervised the construction of numerous lattice bridges on that pioneering line and others (figure 3.7).[62]

Figure 3.7
Charles Fenderich, Railroad Bridge over the Wissahickon, Near Manayunk (Philadelphia), c. 1834. This rare early image shows a lattice bridge (Moncure Robinson[?], completed 1833; demolished 1844) on the Philadelphia, Germantown and Norristown Railroad, the first in Pennsylvania accessible to the public. Michel Chevalier would likely have known of or visited the structure. See figure 4.1. Collection of the Historical Society of Pennsylvania.

Chevalier's monumental publication would have captured the attention of engineers who had missed earlier descriptions of Town's bridge.[63] It elevated the lattice to an icon. In his exhaustive work *Histoire et description des voies de communication aux Etats-Unis (History and Description of the Routes of Communication of the United States)*, published in 1840–1841, Chevalier provided minutely detailed descriptions of various lattice bridges in a special "Appendix on the Construction of Bridges in America"; the bridges of Burr and Long were also described and illustrated in this enormous volume.[64] The text contained detailed descriptions of US geography, as well as the construction and management of canals and railroads, with comparisons to conditions in Europe. Chevalier's accounts of lattice bridges document the name (in French and English), quantity, and dimensions of each part, along with construction costs (figure 3.8).[65]

PONTS EN TREILLIS DE RICHMOND. 575

DÉTAIL DES BOIS DU PONT DE RICHMOND.

DÉSIGNATION DES PIÈCES.	QUANTITÉ DE PIEDS COURANTS ou nombre de pièces.	ÉQUARRISSAGE des pièces en pouces anglais.	NOMBRE DE MÈTRES COURANTS ou nombre des pièces avec toutes leurs dimensions en mesures françaises.	NOMBRE de mètres cubes.
Garde-fous. Main courante. . . . *oo* (1)	5,800 pieds courants. . . .	2× 8. . . .	1,769^m, × 0^m,051 × 0^m,203.	mèt.cub. 18 314
—— Couronnement de la main courante. . . . . *o'o'* (2)	5,800 —	5× 5. . . .	1,769^m × 0^m,127 × 0^m,127.	28 532
—— montants verticaux. *xx* (3)	720 pièces 5 pieds de long.	5× 6 1/2(*)	720 pièc. de 1^m,52 de long, 0^m,127×0^m,165	22 933
—— traverses. . . . . . *z'z'* (4)	1,440 — 8 pi. 1/2 de long.	2× 5. . . .	1,440 pièc. de 2^m,59 de long, 0^m,051×0^m,127	24 157
Rails de garde du côté intérieur de chaque voie. . . . . . . *vv* (5)	5,800 pieds courants. . .	5× 8. . . .	1,769^m × 0^m,127 × 0^m,203.	45 607
—— du côté extérieur. . *v'v'* (6)	5,800 — . . .	5×10. . . .	1,769^m × 0^m,127 × 0^m,254.	57 064
Rails proprement dits du côté intérieur de la voie. *rr* (7)	5,800 — . . .	3× 5. . . .	1,769^m × 0^m,076 × 0^m,127.	17 074
— du côté extérieur. *r'r'* (8)	5,800 — . . .	5× 5. . . .	1,769^m × 0^m,127 × 0^m,127.	28 532
Planchers. . . . . *tt, t't'* (9)	67,200 pieds de superficie.	2×12. . . .	20,496^m × 0^m,051 × 0^m,305.	318 815
Longrines qui support[t] le plancher *tt* *uu* (10)	2,900 pieds courants. . . .	4×12. . . .	884^m,50× 0^m,102 × 0^m,305.	27 517
— — *u'u'* (11)	5,800 —	4×10 1/4. .	1,769^m × 0^m,102 × 0^m,260.	46 914
— — *hh* (12)	5,800 —	5×11 1/2. .	1,769^m × 0^m,127 × 0^m,292.	65 602
— — *h'h'* (13)	5,800 —	5× 9. . . .	1,769^m × 0^m,127 × 0^m,229.	51 448
— — *x'x'* (14)	5,500 — en pièces de 7 pieds 7/12 de long.	7× 7 (**). .	1,677^m,50× 0^m,178 × 0^m,178	53 150
Traverses supérieures. . . . . . . *gg* (15).	360 pièces 22pi.1/2de long	10×14. . . .	360 pièces 6^m,86 × 0^m,254 × 0^m,356.	223 311
Contre-vents des traverses supér. . *ll* (16)	720 — 15 —	5× 6. . . .	720 — 4^m,57 × 0^m,127 × 0^m,152.	63 518
Cordons. . . . *c,c',c''; d,d',d''; e,e',e''* (17)	2,850 — 36 —	3×12. . . .	2,850 — 10^m,98 × 0^m,076 × 0^m,305.	725 372
Moises des treillis. . . *aa,a'a';bb,b'b'* (18)	5,700 — 21 1/2 —	3×11. . . .	5,700 — 6^m,56 × 0^m,076 × 0^m,280.	795 702
Contre-fiches. . . . . *pp* (19)	360 — 18 1/3 —	6× 6. . . .	360 — 5^m,59 × 0^m,152 × 0^m,152.	46 494
Traverses inférieures. . . . . . *mm* (20)	180 — 17 1/2 —	10×14. . . .	180 — 5^m,34 × 0^m,254 × 0^m,356.	86 916
Contre-vents des traverses infér. *nn* (21)	360 — 20 —	5× 8. . . .	360 — 6^m,10 × 0^m,127 × 0^m,203.	56 615
Chapeaux des piles qui soutiennent la charpente. . . . . *ii* (22)	80 — 20 —	18×18. . . .	80 — 6^m,10 × 0^m,457 × 0^m,457.	101 918
Pièces qui maintiennent le bordage latéral. . . . . *s's'* (23)	5,800 pieds courants. . . .	3× 6. . . .	1,769^m × 0^m,076 × 0^m,152.	20 435
Bordage latéral. . . . . *ss* (24)	95,000 — de superficie. .	3/4×12. . . .	28,975^m × 0^m,0190× 0^m,305.	167 910
			TOTAL.	mèt.cub 3,093 850

(*) Nous avons pris ici la dimension moyenne ; à l'une de leurs extrémités ces montants ont 5 pouces sur 5 ; à l'autre ils ont 5 pouces sur 8.

(**) Nous avons pris ici la moyenne des pièces. Il y en a de 7 pouces sur 8, et de 7 pouces sur 6.

(1) *Ribband for hand-railing. . . oo.* — (2) *Cap for hand-railing. . o'o'.* — (3) *Posts for hand-railing. . . xx.* — (4) *Braces for hand-railing. . . z'z'.* — (5) *Guard rails. . . . vv.* (6) *Guard rails. . . . . v'v'.* — (7) *Bearings rails. . . . . rr.* — (8) *Bearings rails. . . . . r'r'.* — (9) *Flooring planks. . . . . tt, t't'.* — (10) *Flooring joists. . . . . uu.* — (11) *Flooring joists. . . . . u'u'.* — (12) *Flooring joists. . . . . hh.* — (13) *Flooring joists. . . . . h'h'.* — (14) *Flooring joists. . . . . x'x'.* — (15) *Top girders. . . . . . gg.* — (16) *Top braces . . . . . ll.* — (17) *Chords. . . c,c',c''; d,d',d''; e,e',e''.* — (18) *Lattices. . . . aa,a'a'; bb,b'b'.* — (19) *Vertical braces. . . . pp.* — (20) *Bottom girders. . . mm.* — (21) *Bottom braces. . . nn.* — (22) *Support timbers. . . ii.* — (23) *Pieces to nail on weather-boarding. . . s's'.* — (24) *Weather boarding. . . ss.*

Figure 3.8

Chevalier enumerated the parts of the lattice bridge over the James River, Richmond, Virginia (Moncure Robinson, 1836–1838; destroyed April 2, 1865). Michel Chevalier, *Histoire et description des voies de communication aux Etats-Unis et des travaux d'art qui en dépendent* (Charles Gosselin, 1840–1841), 2: 575. Courtesy of Smithsonian Libraries and Archives.

Moncure Robinson (1802–1891) spent two years in Europe from 1825, attended lectures at the Sorbonne in Paris, traveled to the Netherlands, Wales, and England, and in the US became a force in railroad construction and management and a supplier of information about the lattice. He shared information with Poussin, Stevenson, and Chevalier, and met with Pavel Petrovich Melnikov and Nikolai Osipovich Kraft; they were sent by Czar Nicholas I, who asked Robinson to be engineer for the construction of the St. Peterburg to Moscow Railroad. He turned down the offer.[66] Robinson wrote that Chevalier's *Histoire*, "which has never been translated in English . . . made the internal improvements of the United States, at that time, better and more accurately known to Europeans than they were to ourselves."[67] Indeed, there was no US publication of the period that provided anything comparable. Chevalier's passion for infrastructure and the US government's involvement in driving its development, which he encouraged France to emulate, meant that constructional innovators and nation builders in Europe and beyond would pay attention to lattice bridges and the managerial strategies behind the transportation systems in which the bridges played an essential role.[68]

Franz Anton von Gerstner (1796–1840), Austrian mathematics professor, engineer, entrepreneur, and son of prominent Bohemian engineer and theorist Franz Joseph von Gerstner (1756–1832), was a builder of the earliest railroads in Austria and Russia. He traveled to England several times in the 1820s and went to the United States in 1838; the *American Railroad Journal* announced his arrival on its front page and encouraged exchange. A series of his letters describing his travels appeared in newspapers, including the *Preussische Staatszeitung*; they were published as a book in 1839 and serialized in the Pennsylvania-based *Journal of the Franklin Institute* in 1841.[69] (We'll explore his views in chapter 5.) At the time of his visit, Gerstner was one of the most well-traveled railroad engineers in the world. He stated emphatically that engineers could learn far more from American practice than British: US railways were not cheaply and poorly built; rather, money was spent where it was necessary.[70]

Gerstner died of pneumonia in 1840 (four years before Ithiel Town passed away) in Philadelphia after more than a year of inspecting US canals and railroads. His book *Die innern Communicationen der Vereinigten Staaten von Nordamerika (The Internal Communications of the United States of North America)* was published posthumously (1842–1843); the huge quantity of data found in this two-volume work chronicles the management and construction of railroad lines of the United States.[71] Lattice bridges often surface (figure 3.9).[72] Gerstner detailed the structure and cost of the lattice (1837) over the Harlem Inlet in Manhattan on the New York and Harlem Railroad, which "belongs to the strongest and most perfect of this kind in America."[73]

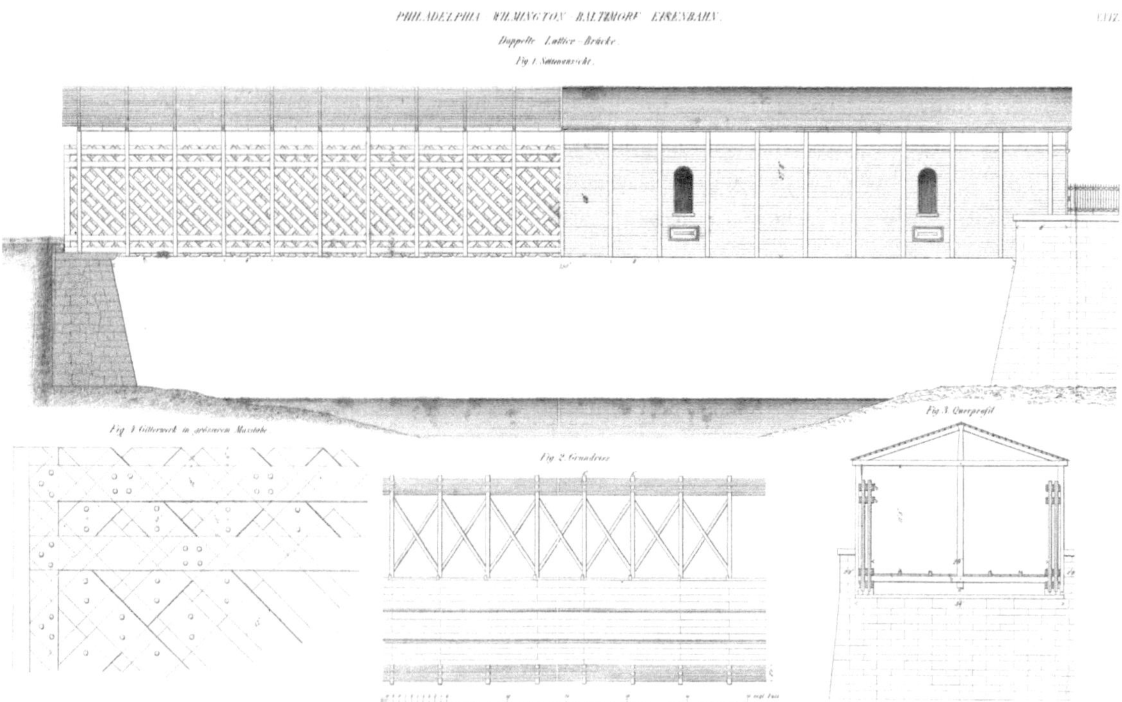

Figure 3.9
Lattice bridge, Philadelphia-Wilmington Railroad. Gerstner, *Die innern Communicationen* (1842–43), pl. 26.

The publications mentioned above circulated among technologists; the authors were familiar with each other's work.[74] Excerpts of Stevenson's *Sketch* appeared in various European journals. Poussin's, Chevalier's, and Stevenson's findings all appeared in the newly established *Annales des ponts et chaussées*, the official organ of the renowned engineering school; Henri-Charles Emmery (1789–1842), one of its founders and director of municipal engineering in Paris, who summarized Poussin's and Chevalier's reports in the journal, concluded: "We cannot sufficiently call engineers' attention to this object; the system is more economical than suspension bridges and would therefore permit the replacement of a certain number of ferries with wooden bridges known as 'American.'"[75] In 1844, Town's bridge appeared in the new, popular journal *L'Illustration*, published in Paris.[76] The accounts did not provide information adequate for construction, but the missing know-how inflamed engineers' curiosity. For example, prominent Swiss army officer and engineer Guillaume Henri Dufour, in his review of Poussin's 1836 book, gave a summary description of the characteristics and advantages of the lattice bridge; he then recommended that readers refer to the book for further details, "with the regret that Mr. Poussin did not take more care in the description of such an interesting structure; its novelty and the potential uses to which it can be put make it worthy of a separate page in the atlas."[77]

Some reports' accuracy might be questioned; information was sometimes secondhand, and observers were not necessarily competent to analyze what they saw. Stevenson, for example, was no bridge expert; his was a quick trip. Poussin, because of his surveying work, must have been quite familiar with the structures, yet his description is not very detailed. Chevalier traveled in the United States from 1833 to 1835, but some of the Town bridges he so painstakingly describes were built as late as 1840.[78]

In that year, architect-engineer **Charles F. Zimpel** (1801–1879) published a book in Vienna on railway engineering in North America, England, and other countries, in which he dealt with "practical" aspects only. He had worked for eight years in the United States on railroads and was impressed with civil engineering there. Regarding bridge building, Zimpel found "American systems" superior to the German ones. After his return to Europe, Zimpel built a lattice bridge on the Berlin to Frankfurt railway.[79] In 1845, an enthusiastic but balanced article by an unnamed writer in the *Allgemeine Bauzeitung* confirmed that in the United States, wood construction had been taken "to a high level."[80]

FROM PRINT TO BUILT

There was no uniformly shared national knowledge, as William Hosking (1800–1861), professor of the arts and construction at King's College, London, inadvertently makes clear in his compendium *Essay and Treatise on the Practice and Architecture of Bridges* (1843), which contained an account of the lattice bridge. "We are not aware of any situation in this country or in Europe where a bridge on Mr. Town's principle has been erected," Hosking wrote.[81]

But engineers had constructed and publicized several and were about to build more. In 1838 and 1841, French state engineer Joseph Hyacinthe Garella (1807–1866), grandson, son, and brother of state engineers, built two provisional lattice bridges over the Rhône in Lyon (figure 3.10); at that time Garella had not visited England or the United States.[82] (Town had visited Lyon in January, 1830.)[83] And an "American bridge of Mr. Town, described by Mr. Poussin" was built in France, under the guidance of de Sainte-Claire, who used it as a starting point for a reworked design; this was for a temporary structure after a flood in 1841 at Vaudreuil on Royal Road no. 182 from Mantes to Rouen. He selected the "American system" (as it was named in the header of his report) because it

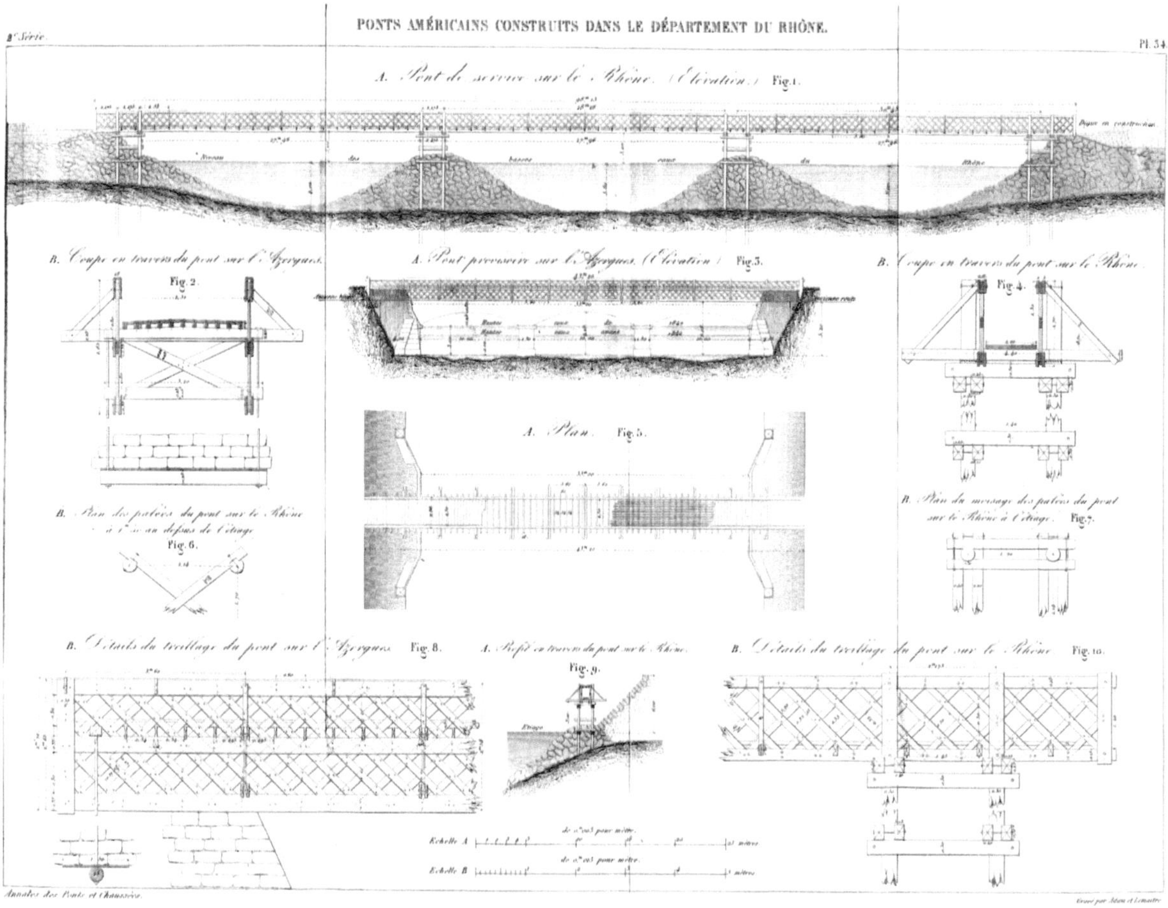

Figure 3.10
American Bridges Built in the Deparment of the Rhône. These provisional lattice spans replaced arch bridges without requiring the reconstruction of piers in the waterway. Joseph Hyacinthe Garella, "Note sur deux ponts provisoires en charpente construits dans le système américain de M. Town," *Annales des ponts et chaussées* 2/2 (1842): pl. 34. UC Berkeley Library, University of California.

produced no sideways thrust (as arches do) and required no intermediate supports.[84] These two engineers, who confirmed the viability and economy of the lattice as bridge (and beam), would have had official permission to build the bridges; their reports appeared in the state engineering organ *Annales des Ponts and Chaussées*. Between 1837 and 1840, engineer William Scarth Moorsom directed the construction of "at least a dozen" lattice bridges on the Birmingham and Gloucester Railway in England; one of these was illustrated in Samuel Charles Brees's *Railway Practice* (1840).[85] Moorsom, active in Ireland and England, had explored Nova Scotia while in the military during the 1820s.[86] In 1841, a Town bridge was built over the Yashchera in Russia. In 1841–1842, Carl Ghega (see chapter 4) built a Town bridge on the Austrian state railway, between Vienna and Glognitz. In 1842, Zimpel built the above-mentioned Town railroad bridge on the Berlin-Frankfort line. The same year, engineer Jan F. W. Conrad, technical director of the Amsterdam-Rotterdam Railroad, who became one of the most prominent engineers in the Netherlands during the nineteenth century, was responsible for the construction of a lattice bridge over the Trekvaart (Leyden Canal) at Vogelenzang in the Netherlands.[87] In Ireland, between 1843 and 1845, John Macneill (1793–1880), of the Dublin-Drogheda Railroad, who earlier had been one of Thomas Telford's assistants, built three lattice bridges of iron; this work was reported in the United States.[88] Yet Hosking was "not aware" of any of these. Experts in Britain did not possess British awareness, a state of knowledge that might be contrasted with American awareness. Overlapping and opposing viewpoints thrived within ever-shifting transtechnological, transnational spaces, with knowledge in various states of accuracy and completeness percolating through many minds. These minds developed techniques intended to generate profits, build infrastructure, and empower the industrializing, civilizing nations' mission to take humanity to a networked future state founded on diverse ideas about technology and liberty.

4

TRANSFORMING STRUCTURES

These were tumultuous times within the cross-border spaces of structural innovation. Designers converting the bridge into an industrial product grappled with the hands-on, theoretical, industrial, and social aspects of building. The introduction of analytical methods that did not require advanced knowledge of mathematics meant that designers could apply to their work a mental model of forces, along with experimentally verified knowledge about materials and structures.[1] Swirling around this were political and emotional ideas that pushed and pulled innovators propelling constructive transformations.

By the 1840s, news of Ithiel Town's invention had wended its way through the minds of bridge-preoccupied technologists in North America, Europe, and Russia. By the 1850s, the lattice was a public figure on the Continent, recognizable over the Kinzig, Dirschau, Vistula, Rhine, and Seine rivers. Engineers who duplicated, adapted, subverted, and transformed "American" ideas unleashed a diversity of structures and opinions. Technologists interacting transnationally and acting locally produced outcomes that contributed to the (re)construction of national and colonial landscapes.

Fascination with the lattice bridge revolved around modularity of parts and connections, construction methodology, and understanding how it all functioned as a whole: it could be viewed as a unit (that is, a beam) or as a truss (an assemblage of parts). The long, level box eventually became a familiar form in the landscape (figure 4.1). The early lattice bridges did not always function as desired, but malfunction inspired innovators. The so-called American bridges offered a design vocabulary, a lingua franca that spoke directly to the industrial yearnings of designer-builders. Some regarded the lattice with excitement, others with apprehension.

Inventions of the pontifical Americans—Burr, Town, Long, and Howe—became archetypes found in engineering textbooks, bridge compendia, and encyclopedias (figure 4.2). The new structures were learning opportunities for professionals and students. The fourth edition (1839) of the civil engineering textbook of Joseph-Mathieu Sganzin, an early professor at the École Polytechnique, adapted Poussin's description and illustrations of the "American bridges of Mr. Town," including a list of seven of its advantages.[2] By 1843, the lattice was one of the bridges that students in the Royal Bavarian

Figure 4.1
William L. Breton, *Bridge over Wissahickon Creek, near Manayunk* (Philadelphia, 1833; demolished 1844), ca. 1835. It must have been startling, in the early years, to catch sight of these high, level, long-span structures. See figure 3.7. Breton Watercolor Collection, The Athenaeum of Philadelphia.

Polytechnical School in Munich were assigned to draw.[3] Students at the École Nationale des Ponts et Chaussées, who went on technological missions to various European countries, brought back written descriptions and drawings of lattice bridges.[4] During the early 1850s, when lattices were counted among Europe's most striking engineering achievements, future tall-structure engineer-entrepreneurs Gustave Eiffel (1832–1923), and William Le Baron Jenney (1832–1907) would have learned about American bridges in the courses of Professor Louis Charles Mary (1791–1870), who for thirty years from 1833 taught at the École Centrale in Paris, a launching pad for industrial engineering. Mary, a former student of Durand (see chapter 2), relayed and enhanced the latter's modular teachings and design methodologies for upcoming generations. Mary's students, whose

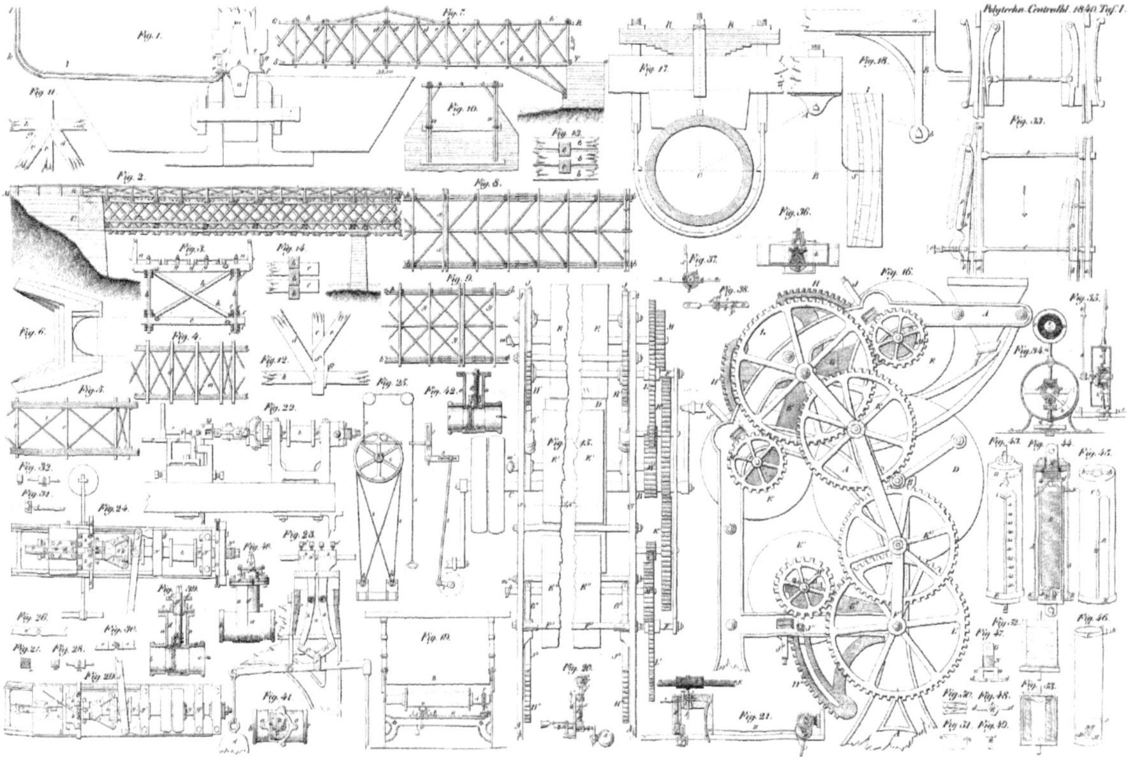

Figure 4.2
Town and Long bridges appear with mechanical devices in an engineering journal's illustration plate. *Polytechnisches Centralblatt* (Leipzig) 6 (1840), pl. 1. Deutsches Museum Library.

learning included site visits, focused intently on structure and the primacy of the construction process as the basis of engineering and architecture.[5]

Learning is integral to inventing. American bridges were an inspiration for the analysis, tinkering, repair, and revision behind the new way of building.[6] But innovators wanted a more precise structural understanding to take it further. This chapter investigates moments in the lattice's role as a muse for designers, builders, and analysts who participated in the development of structures classed under the capacious English rubric "truss." Innovators were not trying to invent the truss; they were developing bridge types intended to meet pressing needs in diverse environments. Here, we'll explore the lattice bridge becoming what we'd today call an open-source code for exchange, collaboration, research, publication, prototyping, and modification.[7]

LATTICE, TRUSS

Let's begin with timber merchant George Smart, who patented in England one of the earliest lattice variations. The *Register of Arts and Sciences* (London) pronounced it "the best bridge ever invented" (figure 4.3). Outside of test models, there is no report of one having been built, and subsequent designers did not cite it as inspiration. Mathematician Olinthus Gregory noted that, rather than a beam-like structure, the bridge was "in fact, a *suspension* bridge, and would require brace or tie chains at each pier." We mention it here because, over the years, its principal function has been as nationalist tool: as late as 2013, it has fueled British nationalist storytellers' false claims of lattice priority. As Town noted in his 1825 brochure, the Smart bridge patent was registered "two and a half years after mine."[8]

In 1830, military man, explorer, railroad engineer, and entrepreneur Stephen H. Long (1784–1863) delivered the lattice's earliest significant offspring when he registered the first of a series of bridge patents, which were depicted in US and European publications including a 100-page book published in German in 1840 (figure 4.4).[9] A key element of Long's bridges were the wedges that put the cross-braces into compression (in today's terms, prestressing them); this was intended to keep the structure rigid and make tension connections for the diagonal members unnecessary. It was also intended to prevent loosening of the structure as the wood shrank.[10] It is uncertain how durable these structures were in service.[11]

While emphasizing how special the wedged connections in his bridge were, Long in a footnote revealed the anxiety of influence:

> It is admitted that this principle is rendered operative in the lattice bridges invented by Mr. Town, but it should be borne in mind, that his bridge frames, are not truss frames, inasmuch as they are completely destitute of keys, wedges or trussing of any kind, and are trenailed, rather than truss frames.[12]

Novelty was a key requirement for a patent, so Long, who had an understanding of the forces in the individual members of a bridge, needed to differentiate his invention. (He was familiar with engineer and physicist Claude-Louis Navier's analytical model of the truss as a beam, which enabled the dimensioning of upper and lower chords.)[13] Placing his design in the "truss" category and contrasting it with Town's beam helped him do so. Still, Long's analysis is not unfounded, as the lattice diagonals, pinned at all intersections, did not possess basic characteristics of what is usually called a truss, a structure whose individual members are not subject to bending, only to tensile and compressive forces. In any case, Long would confirm again that the lattice was his bugbear and inspiration. On January 23, 1836, he received a patent for "certain improvements in the

REGISTER

OF

THE ARTS AND SCIENCES.

No. 28.] SATURDAY, OCTOBER 23, 1824. [Price 3*d.*

CONTENTS :

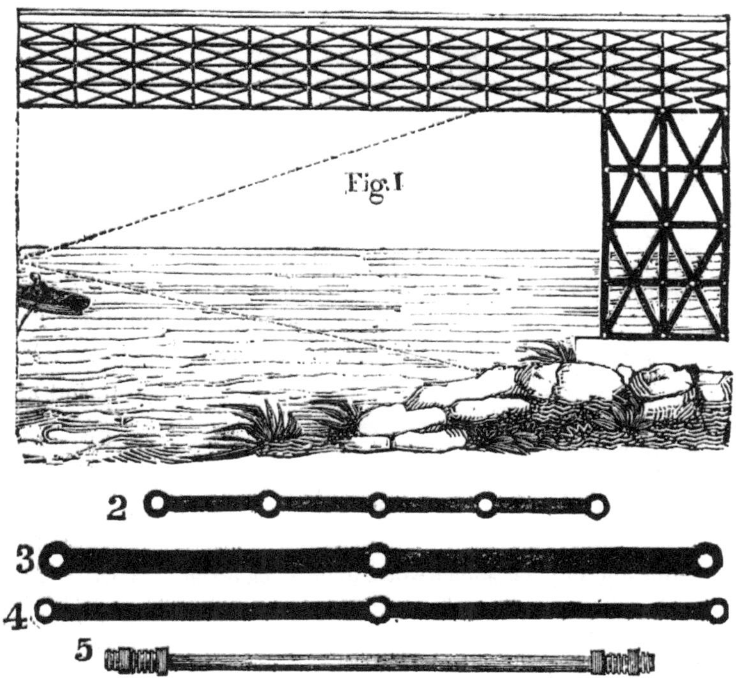

MR. GEORGE SMART'S PATENT IRON BRIDGE.

In consequence of the account which we published in the 23rd number of this Work, of Mr. Smart's beautiful invention of the Bow and String Rafter, our attention has been directed to another invention

VOL. II. E

Figure 4.3
Smart's patent iron bridge consisted of four types of wrought-iron bars. *Register of Arts and Sciences* 2 (1825): 49. The New York Public Library.

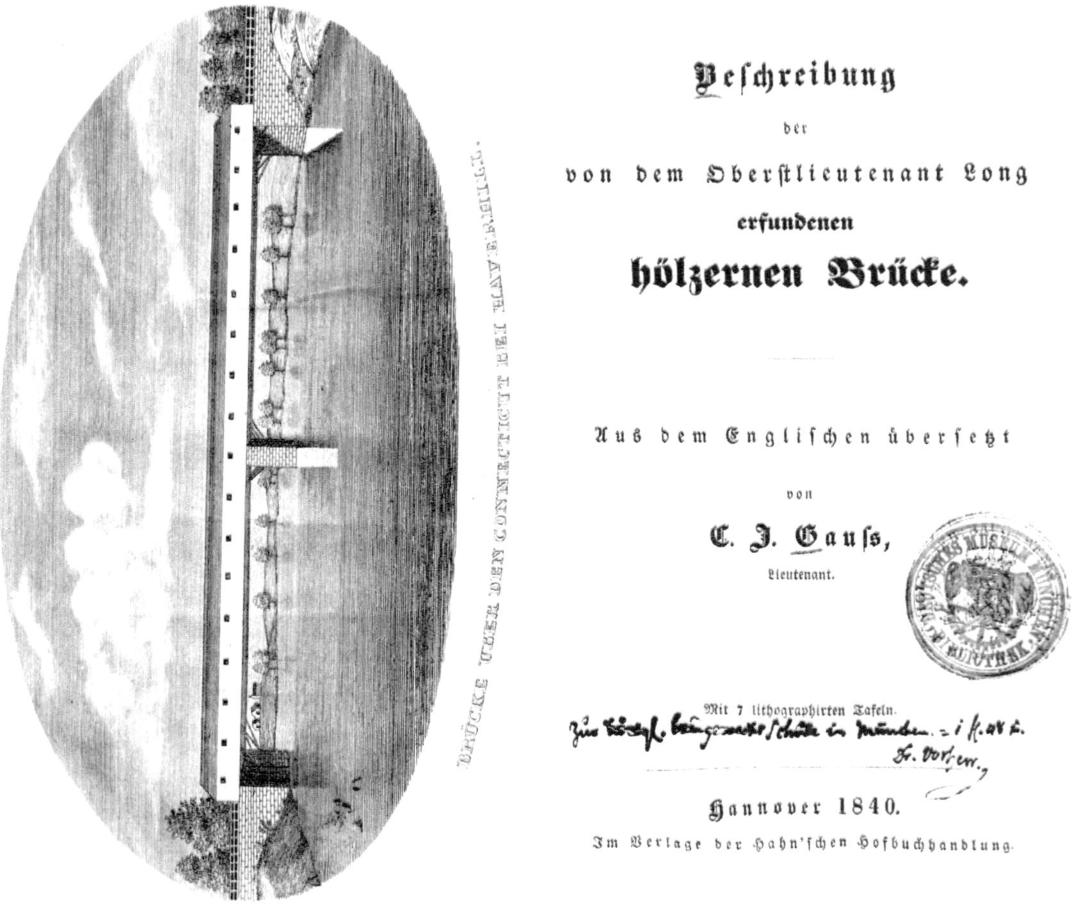

Figure 4.4
Frontispiece and title page of publication devoted to the Long bridge. Stephen H. Long, *Beschreibung der von dem Oberstlieutenant Long erfundenden hölzernen Brücke* (Hahn'schen Hofbuchhandlung, 1840). Deutsches Museum Library.

construction of wooden or Frame Bridges"; Long called these improvements "lattice bracing" (figure 4.5). The invention was "new and original," he claimed, because "*Lateral* and *horizontal* stiffness, in contradistinction to *vertical* and *transverse* inflexibility, is the object of this invention."[14] This lattice, in a horizontal position, would support the floor and provide lateral (sideways) stiffness—comparable to the (unpatented) Grubenmann design (figure 2.16). Although Town had recommended such a use, specifying it enabled Long to try to profit from a lattice. Whether or not he actually did, Long's

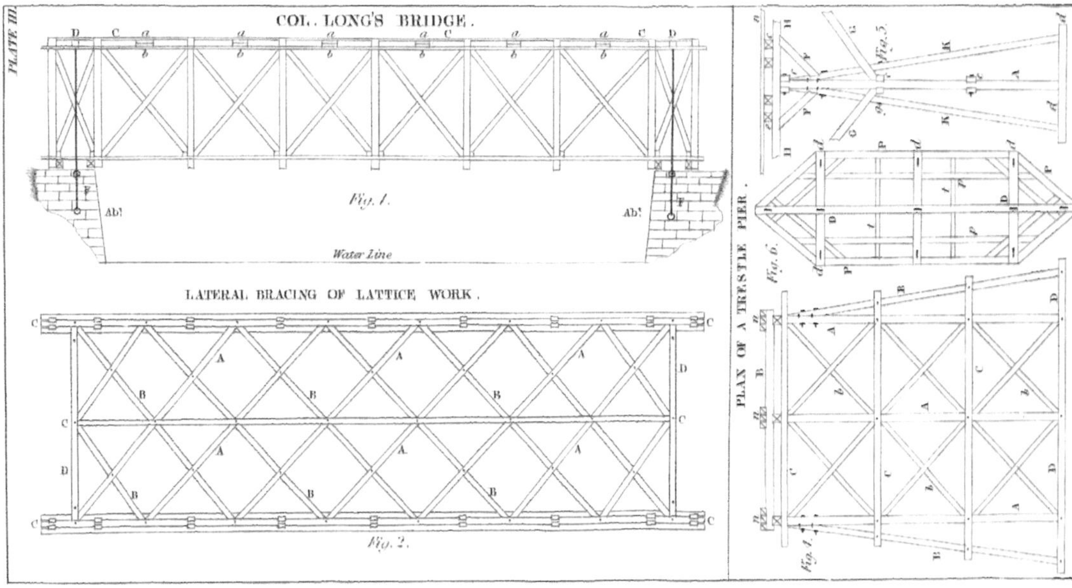

Figure 4.5
Long bridge lattice cross-bracing and pier design. Stephen H. Long, *Description of Col. Long's Bridges, together with a Series of Directions to Bridge Builders* (Brown, 1836), pl. 3. Courtesy of Smithsonian Libraries and Archives.

patent bridges appear not to have been built over a long period. (A Howe bridge, which we'll explore below, melded Long's and Town's ideas, while omitting the challenging-to-maintain wedges.) Long's astute application of theory to design and the power of print, however, made his ideas well-known and amplified the potential of American bridges in engineers' minds.[15]

West Point graduate, railroad engineer, and manager Herman Haupt (1817–1905) was another lattice-inspired technologist. He acknowledged its "beautiful simplicity, light appearance, and especially its economy," while noting that its original design required improvement.[16] In his first publication, a pamphlet on bridge analysis, this American engineer, who would become "Lincoln's railway man" during the Civil War, admitted that he, at that point, had received his information on lattice bridges from colleagues.[17] (This is just as analysts in other countries had. As we've noted, citizens do not have exclusive access to national knowledge any more than they are insulated from learnings hailing from the other side of a national border.) "Theory, observation, and experiment" indicated that "half of the inclined pieces are of no use as counterbraces [compression members] and badly answer the purpose of ties [tension members]," and

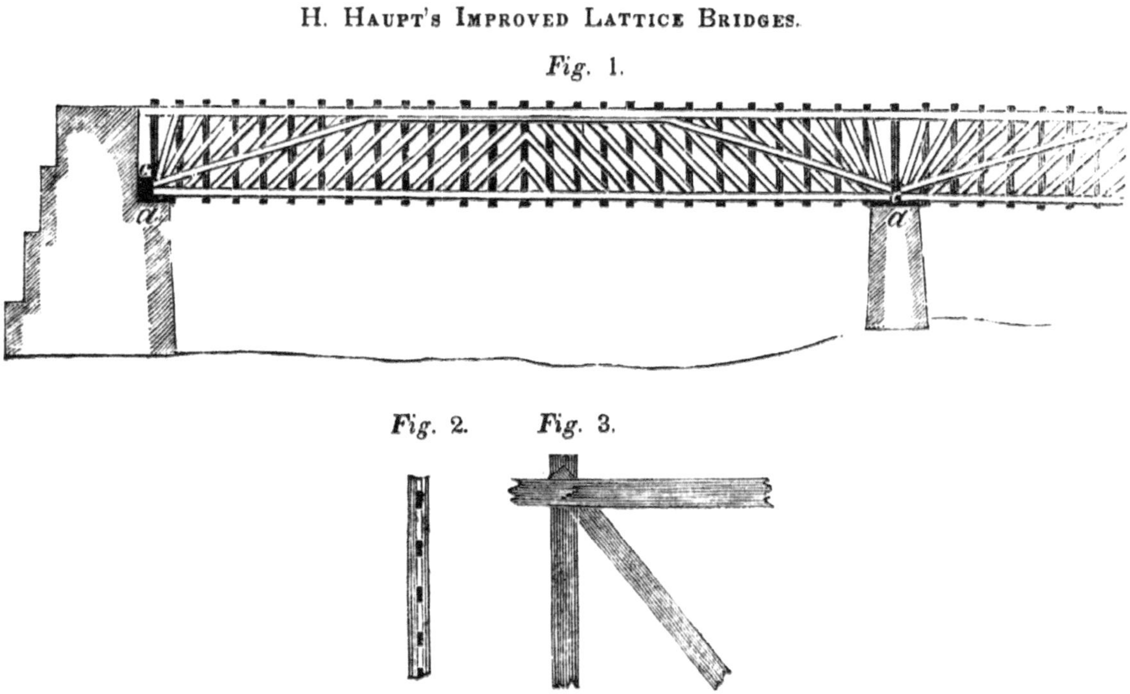

Figure 4.6
Herman Haupt's Improved Lattice Bridge (Patent 1445, December 27, 1839). The structure contains an arch-like overlay. Herman Haupt, "To the Editors," *American Railroad Journal* 11 (1840): 199. University of Illinois at Urbana-Champaign.

that the resulting stresses caused these planks to split at the pegs and put more stress on the lower intersections of the lattice. His solution was a design that included a series of vertical members and short struts (diagonal members) with an arch-like overlay (figure 4.6).[18] Haupt apparently contacted many engineers in order to discover a method of determining the forces in the individual members of a structure such as the lattice, in order to maximize "strength and stiffness," and found that it did not exist.[19] So he devised a method, which he published in 1841 and further developed to enable the calculation of forces in the members of a lattice bridge.[20]

Engineers' improvisational blending of knowledge was taking shape in a variety of patents.[21] In 1840, civil engineer Ithamar A. Beard devised an "improvement," which he presented in an informal report in the *Journal of the Franklin Institute*. Like Haupt, he added closely spaced vertical members—a solution that would be applied, for example, to two mid-century iron bridges over the Nogat and Vistula rivers (see chapter 6). In Beard's bridge, "the principal intersections are firmly secured with bolts," which, he claimed, reduced the number of connectors needed, made them more rigid, and enabled tightening during maintenance.[22] Smaller, uniform parts were an advantage, but *too* many parts were less so. During the period in which Beard was erecting a lattice structure in Brunswick, Maine, engineer Joseph Hyacinthe Garella (figure 3.10) built two provisional "ponts américains" near Lyon, France. These also included bolts and vertical members.[23] In their publications, Beard and Garella provided basic structural information, including sizes of the members, types of connections, and cost. Both observed the bridges in service and measured how much they deflected when loaded. Garella presented a paper in the *Annales des ponts et chaussées* that included an analysis of the strength of the lattice structures, which he considered as a hollow solid, per Navier's beam analogy.[24]

In 1840, the year that news of Haupt's and Beard's lattices (neither of which seemed to enjoy success) appeared in print, Massachusetts-based bridge designer, builder, and millwright William Howe (1803–1852) obtained the first of three bridge patents.[25] Howe took a different approach: the first patent's principal improvement was the addition to a lattice of vertical members that were wedged (Long's method) in order to "strain up" (that is, prestress) the structure.[26] This increased rigidity reduced vibration and enhanced camber; although wedges remained effective only as long as they remained tight. (Camber, curving the bridge slightly upward in the center, lessened how much the bridge would deflect or sag under its own weight when loaded and over time.[27] Achieving camber meant varying slightly and precisely the lengths of chord members—which meant that chord parts were not interchangeable.) In his second patent, granted the same year (figure 4.7), Howe replaced the challenging-to-maintain wedged wooden vertical members with vertical iron rods ran that could be put into tension by tightening a nut on the threaded end of each rod. This squeezed together the upper and lower chords, putting the wooden diagonals into compression.[28] The prestressing enabled Howe to omit tensile wood connections in the diagonals and do so without mortise-and-tenon joints or treenails; instead, the diagonals were seated in cast iron blocks.

Howe made a "principle" (rigidity achieved through a prestressed, adjustable system, as Long had done) available to builders through a kit of simplified members (even if all were not interchangeable) and connectors that enabled rapid construction. It also substituted a standard iron member for wood that performed in a way wood could not.[29] The design allowed the entirety of each metal rod to be used in tension and the full length of wooden members to be used in compression, which exploited each material's key qualities. It included a capability that the lattice bridge lacked: adjustability. During maintenance (which was essential), tension in the rods could be adjusted to keep the diagonals adequately compressed, maintain camber, and provide for wood shrinkage or creep.[30]

As an engineer in 1844 wrote, "The wood work of [the Howe] bridge is a sort of compound of the bridges of Town (lattice) and Col. Long."[31] The bridge addressed the

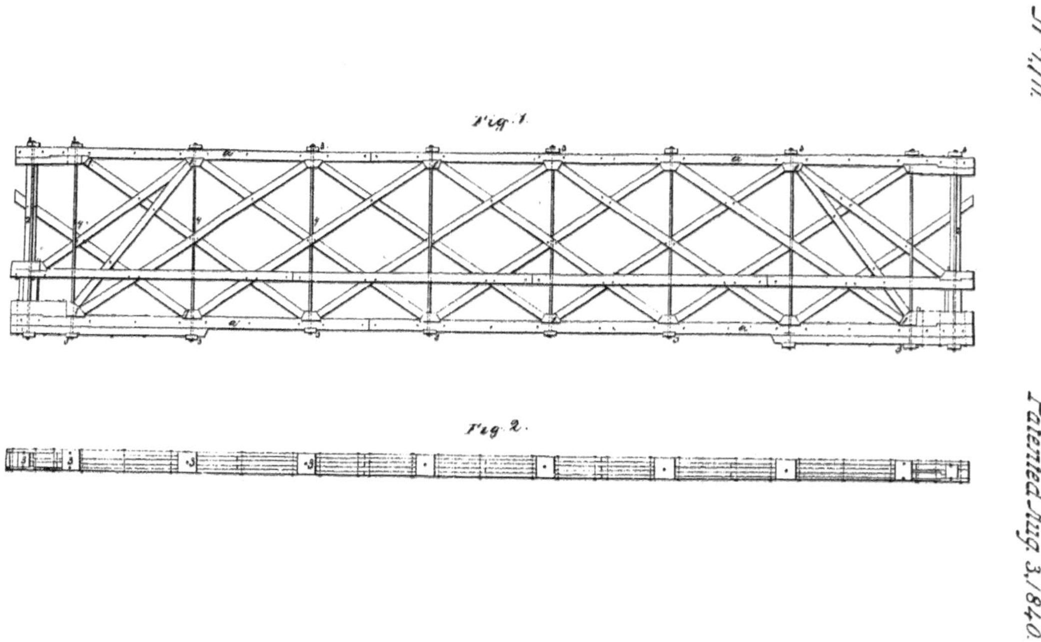

Figure 4.7
William Howe's patent (August 3, 1840; reissued September 3). National Archives and Records Administration.

challenges of the previous bridges while intensifying their constructive design characteristics: relatively small, uniform parts, simpler connections, and a movement toward interchangeability. The use of iron members and connections enhanced the use of tension in the structure. The Howe bridges omitted difficult-to-maintain elements: mortise-and-tenon joints, treenails, and wedges. This removed the need for one type of expert knowledge, though it required another.[32] Of the four Howe-lattice bridges reported built in the United States, the Western Railroad bridge over the Connecticut River at Springfield, Massachusetts (1840–1841) was widely reproduced in European publications; an illustration of it appeared in 1842 in the *Journal of the Franklin Institute* (figure 4.8), accompanied by an article that concluded with a reference to the benchmark: "The writer will add his conviction, that . . . the peculiar truss above described, will be found superior in strength, stiffness, and durability, to . . . Town's double lattice plan." It apparently replaced the Long Bridge as the preferred type on the Western Railroad, a line of great significance for Boston.[33]

Lattice bridges appeared in Europe in the late 1830s, during the initial period of railway construction; by the mid-1840s builders had erected Howe-lattice bridges in countries

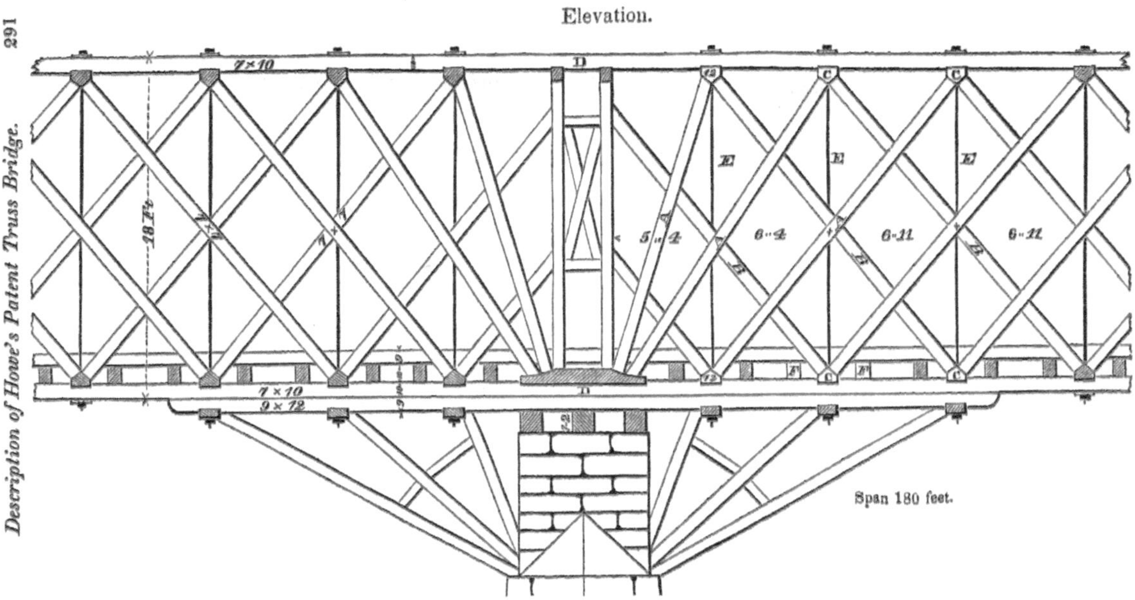

Figure 4.8
Howe-lattice bridge. Lewis M. Prevost Jr., "Description of Howe's Patent Truss Bridge, Carrying the Western Railroad over the Connecticut River at Springfield, Massachusetts," *Journal of the Franklin Institute* 3rd ser. 3 (1842): 290–291. Harvard University.

including Austria, Württemberg, and Russia, although the cost and difficulty of obtaining iron parts meant that it was not the preferred choice everywhere.[34] Howe-lattice bridges were the objects of well-publicized tests of wrought iron, which would become the material of choice for long-span bridges.[35] Though specially crafted and fitted connections were no longer used to join its wooden members, close study of one has revealed that each part was numbered—which indicates that each was fabricated for a specific location—for reasons that likely include the cambering of the bridge.[36] American engineer George W. Whistler, who previously had worked with Stephen Long on the Baltimore & Ohio Railroad, was recruited to work on the St. Petersburg–Moscow (Nikolaev) Railway (built 1842–1851); it included sixty-four Howe-lattice bridges. (Theorist Dmitry Ivanovich Jouravsky [1821–1891], based on study of the Howe-lattice, worked from 1844 for more than eight years on constructional and analytical techniques.)[37] What came to be recognized as the "Howe truss" in the United States, however, was similar to Howe's third patent (1846), which included one set of cross braces rather than a lattice between the iron rods. The patent also contained an arch, which builders did not necessarily include (figures 4.9, 4.10).

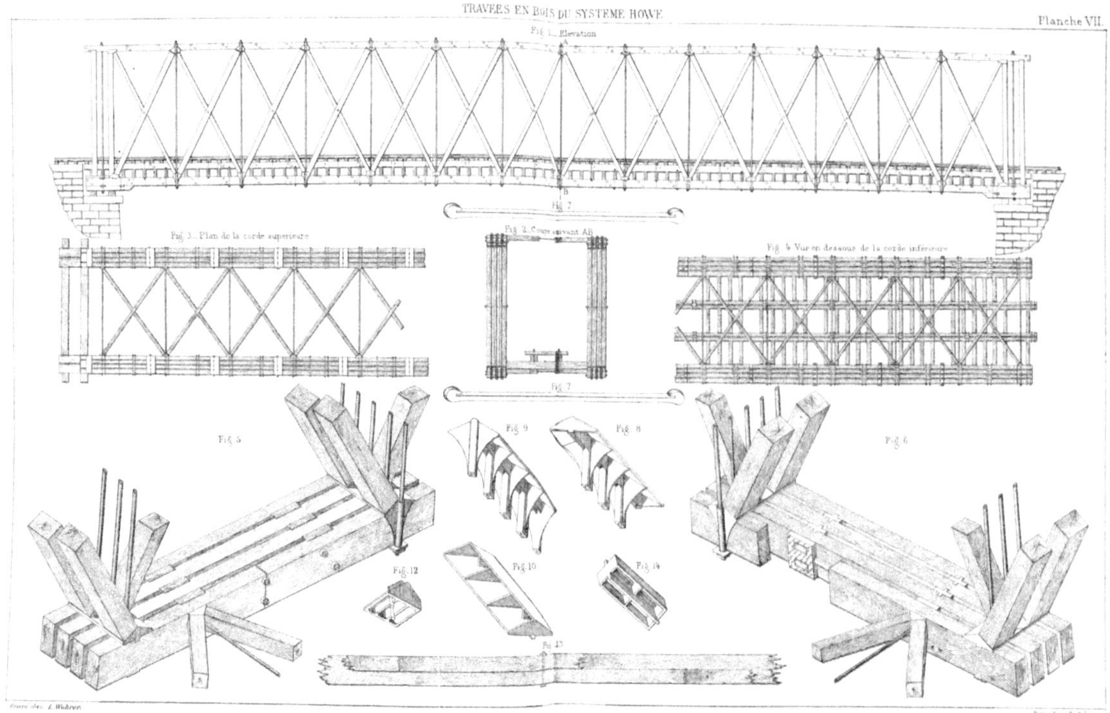

Figure 4.9
Howe bridge. This illustration makes visible the tie rods and cross-bracing usually simplified in one-dimensional depictions. Louis-Antoine Comolli, *Les ponts de l'Amérique du Nord: Étude, calcul, description de ces ponts* (Ambrose Lefèvre, 1879), pl. 7. Courtesy of Smithsonian Libraries and Archives.

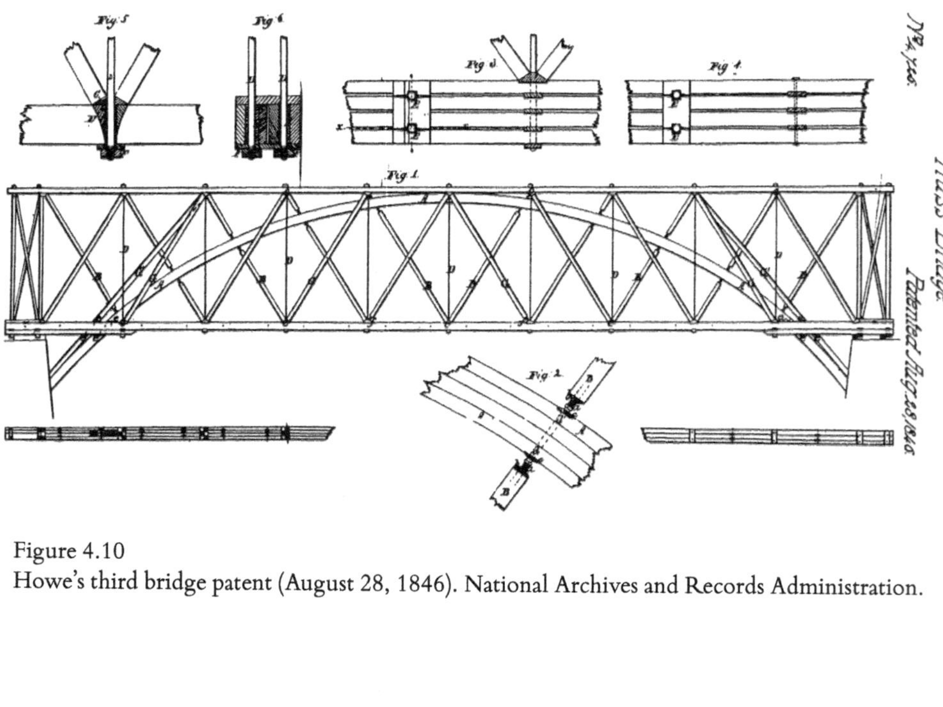

Figure 4.10
Howe's third bridge patent (August 28, 1846). National Archives and Records Administration.

Figure 4.11
US military railroad bridge (Herman Haupt), Bull Run, Virginia, Orange and Alexandria Railroad, ca. 1863. The army rebuilt this tied arch bridge more than seven times during the Civil War. Library of Congress.

Perhaps Howe was not confident that his bridge would avoid excessive deflection and creep, so he resorted to overlay à la Burr (see chapter 2).[38] The combination of arch and truss was an important solution for ensuring structural integrity. During the US Civil War and subsequently, Haupt would employ a structure that embedded an arch within a Howe-like truss, so that the entirety functioned as a beam that rested on the abutments (figure 4.11).[39]

In 1842, the Austrian government sent Karl von Ghega (1802–1860) and architect Moritz Löhr (1810–1874) on a mission to the United States. Ghega, who received his doctorate in 1819 in philosophy and mathematics at Padua, served as engineer in the construction of one of Austria's earliest railroads (Kaiser-Ferdinands-Nordbahn) and became one of Austria's foremost railway engineers. In 1836 and 1837, Ghega had traveled in Belgium and England. In the United States on a state mission from February to August 1842, he saw thirty-nine railroads and traveled 1,500 English miles.

The resulting book examined the Baltimore & Ohio Railroad, whose construction and management he considered the most "instructive" in the world.[40] Ghega noted that, to the common observer, US railways did not appear "beautiful," but they were precisely that to the engineer.[41] In 1845, Ghega's *On North American Bridgebuilding and the Calculation of Strength of Howe Bridges* appeared.[42] This theoretical work was based on Navier's beam analogy.[43] Ghega was mostly concerned with the Howe-lattice bridge at Springfield, Massachusetts; he provided a detailed analysis and discussion of strength of materials, as well as elaborate illustrations. He considered the Howe bridge an outgrowth of the earlier bridges and felt that where stone bridges could not be built, the Howe bridge was the best choice. He included "fleeting mention" and illustrations of Burr, Town, and Long bridges and recommended the books of Poussin and Gerstner for those interested in the lattice. Ghega believed, as others did, that "Howe's construction . . . [was] an improved 'Town's lattice work' and 'Long's truss frame system.'"[44] Jouravsky and Ghega, followed by the theorist Culmann (who we're about to meet), developed methods to calculate the forces in each member of a truss.[45]

Löhr, Ghega's travel companion, became a major figure in the construction of Austria's railways. In 1843, he published a description of a "completely characteristic and original American invention," bridges exemplified by Town, Long, and Howe.[46] He asserted that the lattice bridge "comprises, so to speak, the fundamental principle of all subsequently designed systems," while noting three basic challenges of the lattice bridge: loosening of the connections, absence of a means to prevent or repair sagging, and weakening of the planks because of the many holes bored for treenails. Löhr accurately remarked that these "disadvantages probably moved the inventive Americans to improve the construction, while retaining the advantages of its fundamental principle."[47] As we've noted, in the Long and Howe structures, improvements included putting the vertical members into tension in order to compress the diagonals. The intention was to fully use the capabilities of each material and keep connections simple, tight, and maintainable so that the bridges could reliably function as rigid beams.

"A TABLE LEG OUT OF CABLE"

Admiration for the lattice was not universal. A dissenting voice came from an engineer working in Bavaria, Karl Culmann (1821–1881). The lattice disgusted him. Culmann, trained mostly in Karlsruhe, began his professional life on Bavaria's state railway. In 1855, he became head of the Bauschule at the new Eidgenössische Technische Hochschule Zürich (Federal Institute of Technology), where he developed graphic statics, a method that enabled engineers to analyze structures by creating geometrical diagrams that visually portrayed the forces. Culmann became known through two reports on bridge building that were the outcome of a state-sponsored mission to the United States and England in 1849 and 1850. His reports appeared in the *Allgemeine Bauzeitung*, the widely read journal of building technology and architecture published in Vienna. Culmann's first report encompassed a "chronological sketch" describing the function of bridges and an explanation of a theory of frame bridges.[48]

From at least the 1700s, analysts had focused on gaining a better understanding of structures that were statically determinate, that is, for which mathematical equations could indicate the type and intensity of the forces in each member of a structure. (In a statically determinate bridge, which is a relatively minimal structure, if one member fails, the bridge fails.) The numerous intersecting members of the Town, Long, and Howe bridges meant that they were indeterminate, however. In order to analyze these structures, designers either ignored elements not considered critical or theoretically broke the indeterminate structure into several determinate structures and analyzed each.[49]

Around 1820, engineers analytically viewed what we now call "truss bridges" (formerly sometimes called "philosophical," meaning "scientific," bridges) as beams (figure 2.9) with internal tensile (pulling apart), compressive (pushing together), and shear forces, rather than as structures consisting of members within "panels," as engineers view them today.[50] Trusses are efficient because, in theory, pushing and pulling forces only act along the axis of members, which do not bend (this can cause deformation). The dimensioning of members to take specific forces, and the relationship between the position of each member and the forces passing through it, were concepts beginning to appear in books around the time Town was developing the lattice.[51]

Culmann was one of several engineers, including Squire Whipple and Herman Haupt, who developed methods of understanding structures as composed of individual members connected by pins (that is, free theoretically to rotate relative to each other), each member theoretically in only tension or compression. This involved simplifying if not excluding some bridge members and their stresses from calculations, which made it possible to analyze the loads on key members. It removed ambiguity and made Navier's analogy unnecessary.[52]

Unlike Ghega and Löhr, and likely without knowledge of the historical record, Culmann interpreted the lattice as an anomaly, unrelated to previous and subsequent designs:

> Completely independent from the construction systems described above [Burr, Long, et al.] are the lattice bridges, generally known in Europe, which were invented, patented, and introduced by Town some years ago. . . . (I will omit the complete drawing and description under the assumption that everyone knows it.) This construction, however, has not proven

> itself. The compression and the tension to which the lattice members are subjected, if one regards them solely as braces and cross braces between the upper and lower chords, vanishes in comparison to the magnitude of compression and tension that the pegs exert perpendicularly to the direction of the lattice members. From this arrangement, neither tensile nor compressive, but rather bending strength of the planks is claimed—a circumstance that alone dooms the whole system. For a long time no additional large bridges have been executed according to this system in America.[53]

The members of the indeterminate lattice structure, according to Culmann, worked in bending, not solely in tension and compression, as he felt a *Fachwerkbrücke* (a term Culmann coined, the German equivalent of "truss bridge") is supposed to do.[54] With its diagonals connected at each intersection; the lattice frame was "more or less a homogeneous whole," that is, more or less a beam, which works in bending (figure 2.9). For Culmann, this was not acceptable.[55] Culmann's antipathy was not based on new insight; he admitted that he was echoing existing criticism of the bridge.[56] He cited few sources but appeared to be familiar with publications by Poussin, Chevalier, Gerstner, and Ghega.[57]

While examining the double-lattice bridge over the James River in Richmond, Virginia (figure 4.12), Culmann claimed to have "met an engineer who was astonished that I found something worth drawing, and said '*they are now out of time* [i.e., date], *bridge building has been much improved since.*'"[58] Thirty years after the introduction of the lattice and about a dozen years after the construction of this bridge, the advantages and challenges of the lattice were known; individual bridges had not been designed to support the high impact, rolling loads of increasingly heavier, more powerful cars and locomotives. Plus, inventors such as Howe were developing alternative designs. So it's unsurprising that an engineer deemed the Richmond bridge design outmoded.[59]

Culmann showed the Richmond bridge with auxiliary queen post and vertical posts (in dotted lines), to make clear that the bridge was deficient and had to be reinforced, but still was not strong enough to carry two tracks. This is presented as a failure of the lattice bridge rather than the deficiency of one example. Culmann did not include the lattice in a table summarizing the load capacities of various bridges.[60] From his perspective, the lattice was a flop rather than a constructional, structural, and commercial success and design inspiration. In the year Culmann's first report appeared, Herman Haupt noted the impact of the antilattice attitude:

> Of late years . . . the frequent failures of these [lattice] bridges in consequence of heavy transportation, have produced a revolution of sentiment hostile to the plan, and instead of examining into the causes of failure and providing a remedy for the defects which occasioned it, other modes of construction have been adopted at an expense sometimes double that of an efficient lattice structure.[61]

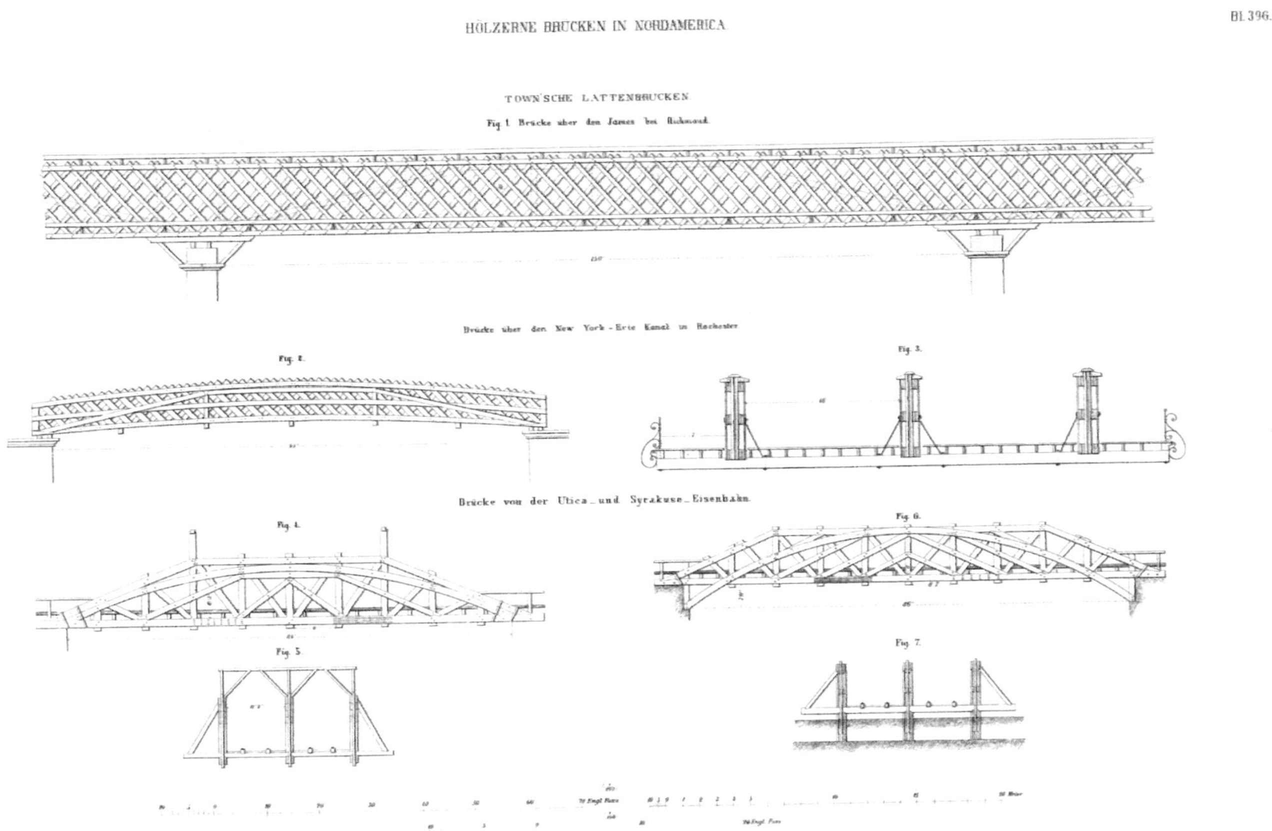

Figure 4.12
Culmann's depiction of the Richmond Bridge (top, see figure 3.8) along with bridges composed of overlaid structures. Karl Culmann, "Der Bau der hölzernen Brücken in den Vereinigten Staaten von Nordamerika . . . in den Jahren 1849 und 1850," *Allgemeine Bauzeitung* 16 (1851): pl. 396. Collection of the author.

Economist Michel Chevalier's report, which Culmann likely knew (see chapter 3), mentioned that the Richmond bridge was "found to be too weak" and then goes on to illustrate in detail the improvements made to a subsequent bridge.[62] Their stories had diverging conclusions: Chevalier regarded the Richmond bridge as part of ongoing invention; Culmann saw it as the end of the line. Town died in 1844, as the "revolution of sentiment" was inspiring innovators to create new modes and, despite Culmann's complaint, translate the lattice from wood to iron.

THEORY, INDUSTRY, AMBIGUITY

Chevalier cherished the world-altering potential of the lattice. Culmann wanted to cancel it due to its statical violations and industrial crassness; he disdained engineers who built bridges that could be "traded by the yard, like a piece of cloth."[63] Culmann's simile was apt; Town's patent fee increased by the running foot, and builders were on the verge of literally rolling long-span lattice bridges into place, as we'll see in the next chapter. For Culmann, the lattice was to engineering what the Big Mac would be to traditional haute cuisine: a mass-producible, inexpensive product whose popularity seemed to defy conventional logic, at least to someone accustomed to a traditional notion of fine dining. But railroads were engineering bridges in the nineteenth century as fast-food managers would engineer burgers in the twentieth, with throughput foremost in mind. Culmann barely hides his contempt:

> New bridges based on [Town's] system are being built only in the far west [of North America]. . . . There they are still the cheapest because the boards and planks cost almost nothing, and the construction itself requires no skill at all; almost any peasant can nail such a framework together.[64]

Architect-engineer Charles F. Zimpel, who spent eight years working in the United States, regarded lattice construction more favorably: "They can be easily assembled by inexperienced craftsmen." Culmann, like America observer Alexis de Tocqueville, saw in quantity production a decline in quality, but regarding bridges, this was not necessarily so. What Culmann also saw in America were enslaved Black people at work.[65] His pinpointing "peasants" hints at the participants in the development of industrialized building, which we'll probe in the next chapter.

Culmann desired theoretical clarity. On the illustration plate that depicted a lattice bridge (figure 4.12), Culmann included the designs of the Syracuse and Utica Railroad's short-span bridges, which featured overlaid systems, and remarked:

> It seems not to have been clear to the engineer which of the existing systems was the best, so he decided to use all of them in his bridge. . . . The whole thing looks like different bridge structures of the same span cut out of pasteboard and stuck together. The builder probably thought: if one doesn't hold, then the other will, or else they'll peacefully share the loads. This, however, is not exactly so . . . one [system] will always carry everything, or almost everything, and the others will by turns come to carry the load, when the preceding system is crushed.[66]

Haupt, a builder and inventor as well as a theoretician, was comfortable breaking apart ambiguous systems and analyzing and revising structures such as the lattice.[67] As an overlaid structure could be theoretically understood only through its individual, determinate, subsystems, they are seen by Culmann as working individually; Haupt was more willing to grapple with them as a unit. During subsequent decades, many engineers in the United States also favored statically determinate trusses with (analytically assumed or actual) pin-connected members (see chapter 7), in line with Culmann's perspective.[68] But this was not the only path designers would take, in the North American or German states.

Culmann pledged in his first report not to refer to Town again, but in his second (1852), on iron bridges, he went back on his word and vented his spleen:

> I have not yet cited a system . . . and would gladly pass over it in silence, if it weren't for the incomprehensible fact that it has so many followers here in Germany . . . I mean the imitation of the Town lattice bridge in iron.[69]

This was more puzzling than the wooden lattice bridge, for "the thin flat iron bars are even less suitable than the somewhat stiff wooden lattices to carry compressive forces from the top to the bottom chords." Culmann cited experimental evidence that demonstrated the flaws of the lattice bridge. In his opinion, it had to be condemned and abandoned. He railed against its use:

> Then why is this design, which is unsuitable in every respect, still so frequently used? Because it's simpler?! If two kinds of forces [i.e., tension and compression] are anticipated, then mustn't one resort to two kinds of measures? Truly, those who want to employ only thin strips show that they have no clear idea of how the forces function within the center portion of beam bridges! Using these iron strips against compressive forces is like wanting to make a table leg out of cable.[70]

Ironically, Town shared Culmann's perspective: he stated that each plank should bear "either a tension or pulling strain, or a thrust or pushing strain."[71]

At the time of Culmann's reports, there were several methods for analyzing the lattice bridge, including design rules, theoretical abstractions, and mathematical theories.

In 1852, an engineer-intern noted that in iron, the "Town system has proven itself to be more advantageous and suitable" than the Howe, yet a method of calculation had been developed only for the latter. He stated that designers could analyze lattice bridges in one of three ways: (1) dimension with rules of thumb, including proportionally (i.e., span to height, as Town recommended); (2) consider it as a beam; or (3) design according to intuition and with simple calculations for dimensioning the parts.[72] The method would be selected according to training, experience, confidence, and knowledge.

Culmann may have nurtured a romantic notion of the refined engineer, but his methods provided designers with analytical tools to build by the yard, whether or not they shared the intuition he called "our technical feeling." Although Culmann dismissed the Town bridge—"Only as historical oddity do these bridges still deserve engineers' attention"—engineers began to construct in Europe a diversity of long-span iron lattice bridges.[73] Among them were the most renowned structures of the era. Still, as we've noted, the analytical tools Culmann and others developed, plus his teachings regarding graphic statics, would shape the design process, becoming the "incarnation" of bridge building.[74] Most engineers were not experts in advanced mathematics; analytical tools enabled them to do their work, that is, design or analyze complex structures and specify the dimensions of each part.[75] Theoretical tools are important, but a designer selects and designs a bridge type in order to optimize for factors that vary with each project; they include local geography, material availability, quantity of materials, conformity to theory, construction cost, construction speed, appearance, maintenance requirements, and relationships with manufacturers, officials, or colleagues.

Culmann and Chevalier represent contrasting understandings of design and building. Culmann saw bridges as subjects for the development of engineering theory, whose application would shape design practice. For Chevalier, bridges represented what nations could achieve through innovation and industrialization. Chevalier was a perceptive observer of industry, politics, and society; Culmann, an insightful structural analyst who developed methods that permitted engineers to understand and compare structures. Neither Chevalier nor Culmann became bridge builders, which might have been a good thing: one loved the lattice too much, the other too little. For each, travels through the American pontiverse became a springboard to a renowned career.

INDUSTRIAL TRANSFORMATIONS

Charles Couche (1815–1879), professor of railway construction and operations at the École Nationale des Mines in Paris for thirty years, confirmed the appeal of the lattice in an 1854 analysis of American bridges (he never visited the United States) in an article entitled "Engineering Works of the Railroads of Germany." It was said to have "caused a great sensation among engineers."[76] After describing its deficiencies, Couche observed:

> It is quite remarkable that lattice bridges, abandoned in the country where they took form, and where their use was so strongly preferred, today find numerous supporters in some areas of Germany—with one condition, it's true: the substitution of iron for wood.[77]

This was part of a quintuple transformation. The first involved nation: a system perceived as quintessentially American was becoming European, although it was always both and more. This transmigration was accompanied by a material change: wood to iron. Engineers based early iron lattice bridges on Town's concept, to which they consistently refer.[78] The third alteration was structural: the variety of connection types and addition, in some cases, of vertical members, meant that the bridge performed differently; sometimes designers prestressed specific elements. Fourth, new methods of analysis inspired new modes of design, evaluation, and dialogue. Finally, all of the above revolved around the rise of the professional engineer, whose roles included design lead, researcher, analyst, manager, critic, and entrepreneur.

Couche's research, sponsored by the Ministry of Public Works, appeared on the cusp of the transition from wood to iron. He noted the strong "influence" of US railways in Germany and criticized German engineers for being so taken with American bridges that they didn't notice that the systems were "often inapplicable in Europe." This remark was primarily aimed at the use of wood in railway bridges rather than at structural configuration.[79] Although engineers working in France were aware of lattice and Howe-lattice bridges, American bridges were not extensively used there until they could be built in iron. This was not merely a matter of maintenance; the burning of bridges during the political upheavals of 1848 had made engineers wary of wood.[80] In any case, Couche asserts that virtually all German engineers had rejected wood as a material for permanent railway bridges, although it still had "great importance" in the construction of high viaducts far removed from waterways. If wooden bridges were to continue to be built, Couche believed that they would take the form of the "remarkable" Howe-lattice.[81]

Couche listed five types of "North American" long-span bridges: arch, Burr, Town, Long, and Howe.[82] More than thirty years after Town's patent, he affirmed the achievement:

> This well-known system, which is generally regarded in France as the classic example [*type par excellence*] of American bridges, is characterized by the extreme simplicity of the elements used for the construction of beams: planks and pegs. [After a description of the bridge, he concludes:] One can, through the repetition of identical elements and identical operations, without joinery, without iron, with no members other than small planks, cross enormous openings.[83]

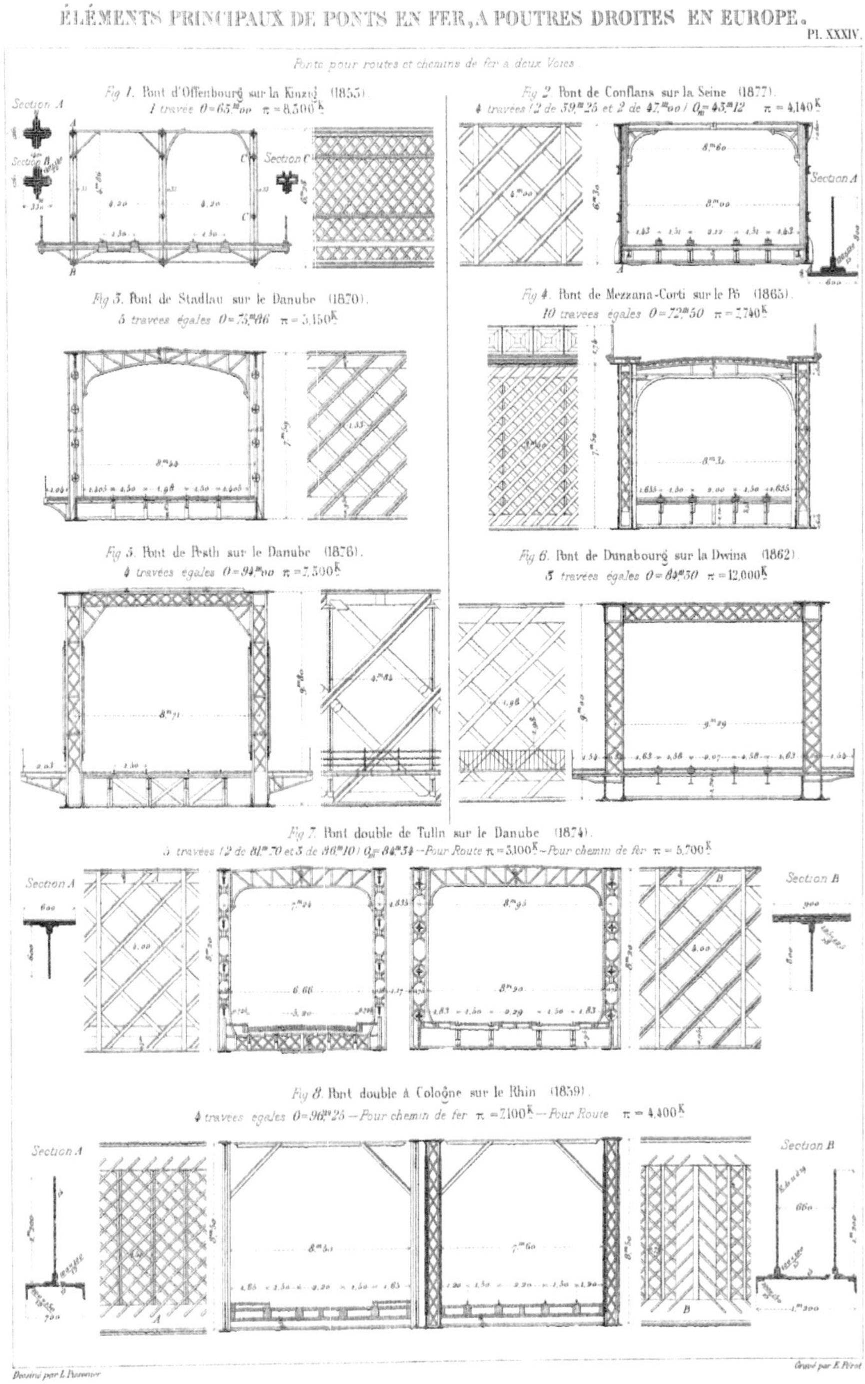

Figure 4.13
Iron lattice bridges (mostly 1870s) built in various countries in Europe. Philippe Croizette-Desnoyers, *Cours de construction des ponts* (Dunod, 1885): pls. 34–35. Courtesy of Smithsonian Libraries and Archives.

Figure 4.14
Lattice beams in the Paris Opéra (Charles Garnier, 1861–1875) roof during construction, after 1865; beams later replaced as the result of a fire). Photographer: Louis-Émile Durandelle.
© Beaux-Arts de Paris, Dist. RMN-Grand Palais / Art Resource, NY.

Couche noted that underdesign, rather than inherent deficiency, had been interpreted as a system defect, so the lattice had been "completely discredited." He compared the bridge to the system of Philibert de l'Orme (figure 2.13): in both, the advantage of simplicity was outweighed by the large number of parts.[84] But what was deemed too large depended on context. For example, Johann Mohnié, an engineer in Bavaria, carefully examined lattice bridges before receiving a patent in 1857 for a type that recalled Haupt's; it seems to have been widely employed.[85] Almost ten years after Couche's article appeared, a paper presented to the British Institute of Civil Engineers reported that "in the case of wrought iron bridges the arrangement most usually adopted [in Germany], when the spans are wide, is that of a lattice construction, in some of its various forms." In the subsequent discussion, engineer and contractor Robert Mallet (1810–1881) remarked that, "In the end, the lattice bridge, in iron, in one form or another, would supersede all others for railway uses."[86]

The desirability of using simple, standard parts and connections to create long, rigid spans kept varieties of the lattice in the mix for innovators and decision-makers.[87] Iron lattice designers could experiment with a variety of configurations, adding vertical members, altering the angle or spacing of diagonals, or increasing their dimensions near the supports to resist shear (figure 4.13). Individual members were themselves often built of smaller pieces (sometimes, like lattice beams themselves); this incorporated kit-of-parts thinking into each member of the bridge. Workers assembled structures using plates and angles with large numbers of rivets and connectors whose behavior in service was not deeply understood for a long time.[88] The replacement of wood with iron, though, did not end challenges posed by flammability, durability, deformation, and inadequate connections.

Designers tested the lattice in new contexts. James Herron developed a railroad track with latticed ties, which he promoted in several engineering journals.[89] Vertical lattice frames were inserted into steam ships.[90] Railroad cars built with lattice frames existed during the late 1830s.[91] In Dresden, suspended wooden lattice beams composed the long-span roof of a temporary auditorium built in 1856.[92] That year, the "American system" of bridges was selected to link the railroad cars that architect Viollet-le-Duc decorated for the emperor of France.[93] Nobody knows how many lattice beams are hidden behind the façades of buildings such as the Opéra (Charles Garnier, built 1861–1875) (figure 4.14), a landmark of traditional architecture erected as part of Haussmann's remolding of Paris. Once designers and entrepreneurs saw the lattice structure, it could not be unseen. Once they understood how it was built, it became a working model for the reinvention of construction.

5

REINVENTING BUILDING

Popular storytellers portray the construction site as a place of urban progress, acrobatic prowess, and physical transformation. Timelapse depictions show structures growing like plants, but biological metaphor tells us nothing about the unnatural processes of building. Who developed construction methodologies that have shaped our environments? How and why? Why don't we know about their work? Art historians and academically minded engineers have typically spotlighted structures as objects for aesthetic and theoretical interpretation. The rise of science as a professional status marker, the working-class connotations of physical labor, and laborers' ethnic and racial identities likely depressed the value that historians placed on construction—a collaborative venture moved forward by figures who in photographs may barely be visible. Working in the mud was not a step forward within a so-called technological evolution driven by civilizing scientification, a momentum that generations of elite professionals wanted to project. Storytellers could not render the construction site placeless or clean, like a laboratory or factory.[1]

Designer-managers of bridge building projects were mobile and agile, working in diverse and sometimes remote locations. Many operated within and across nations and firms, implementing strategies adaptable to a variety of environmental, political, and economic conditions—though not all innovators enjoyed such mobility. While construction was foundational to industrialization, building's strategists sometimes shunned metal and steam (figure 5.1)—industrialization's climactic commodities. Contributors to construction and maintenance, most of whose names and creative contributions are difficult if not impossible to trace, were often transnational—though for enslaved, migrant, and immigrant labor, this term has different overtones than for men welcomed into the ranks of professionalizing engineers. Regarding construction, it's important to acknowledge what we don't know—a lot—and ask why we don't. Why was the development of mass construction a tale never written?[2]

Does it matter that we do not know who developed a technological process that is foundational to all our lives and communities? Judging from how eagerly, for two centuries and counting, individuals, professionals, and media have affixed national labels to

Figure 5.1
Swann Bridge (Zelma T. Tidwell, 1933), Locust Fork of Black Warrior River, Swann Bridge Road, Cleveland, Blount County, Alabama; Jet Lowe, photographer, 2003. Tidwell, with fourteen to twenty workmen who lived at the site, built this structure in 1933 without heavy machinery. Lola Bennett, "Swann Bridge," 4. Library of Congress, Prints & Photographs Division, HAER AL-201-10.

technologies, it seems to matter a great deal. These tags always imply a specific "who" behind technological or industrial change. Evolutionist-nationalist stories, by rousing emotions, concealing context, and discouraging research, create a knowledge vacuum. This reinforces a power imbalance; not knowing who reinvented building determines who will not be on the receiving end of cultural and economic acknowledgements and rewards.[3] The grand narratives also sidestep accountability for the past and future life-altering social, environmental, and political consequences of construction.

Building lies at the heart of *American Bridge*'s work of conceptual-historical reconstruction.[4] The intent is to find a path around evolutionist, machine-focused, metal-worshipping, nation-infatuated perspectives that have shaped the design of history and the history of design, and to reconsider the purported triumph of scientists over carpenters, engineers over labor, and "us" over "them".[5] Business studies typically show designers perfecting a product that then becomes a manufacturing challenge: how to replicate and scale it for the market.[6] Here, we'll focus on the "manufacture" of a replicable process by identifying the early innovation teams, reflecting on patent strategy, and examining the reworking of lattice construction methods for larger scale structures. Our rethink will enable us to discover people and events that standard accounts have eclipsed, while confirming process as the essential product.

INNOVATION TEAM

The Industrial Revolution story can distract us from the changes in international commerce, finance, and manufacture that preceded and enabled the "revolution." During the eighteenth and nineteenth centuries these changes enabled colonization and nation building through the explosive growth of road, canal, and railway construction.[7] By 1824 and, famously, in an 1835 speech, statesman Henry Clay was advocating for an "American System," a phrase that would long resound. He was referring to governmental support intended to ensure the financial independence of the United States, based upon protective tariffs, a national bank, and the building of roads and canals.[8] From this perspective, intertwined federal oversight, financing strategies, and processes of construction would make the nation.

Settlers and immigrants in North America required buildings for shelter and work, and arteries for the passage of people and goods. The number of bridges in England scarcely changed between 1500 and 1750, but between 1800 and 1820, almost 200 miles of public railway were built; within ten years, another 100 were added; in 1847, 6,500 miles were in use or under construction. In France, there were 38 km of track in 1830, 3,870 km in 1852 and 17,000 in 1869.[9] While British industrial production capabilities supported regional railway construction, in the United States and continental Europe, the building of rail lines preceded and helped drive large-scale industrialization, which took place from the mid-nineteenth century.[10]

Speed and economy—whose meanings vary according to changing, overlapping local and international contexts—were paramount considerations in construction design for US canals and railroads, which had to start generating cash as quickly as possible. Already in 1786, an editor of the *Belfast Evening Post* (Ireland) noted, concerning a new bridge over the Charles River in Boston, that it "was completed in a space of time equal to that [the people of Derry, considering a new bridge] lost in thinking of it."[11] In 1818, South Carolina's state engineer in charge of infrastructure noted that projects that took an excessive amount of time were "rendered ruinous" because of "the interest on unproductive capital."[12] This was encapsulated in the catchphrase "time is money," which appeared in English on the cover and title page of the second edition of Guillaume-Tell Poussin's 1836 *Chemins de fer américains* (American Railroads).[13]

In the mid-eighteenth century, it was impressive that the Schaffhausen Bridge "was finished in less than three years."[14] One hundred years later, during the construction of the Portage Bridge (figures 1.3 and 5.2), the *New York Farmer and Mechanic* exclaimed that "this triumph of human invention" was "already attracting a multitude of visitors, gazing with wondering eyes at those lofty cones [i.e., trestles] lifting their proud heads higher than the top of Bunker Hill Monument."[15] This was a poignant comparison, as the Revolutionary War monument, a single 221-foot-tall granite obelisk, took eighteen years (1825–1843) to build. The slightly taller Portage, including fifteen towering trestles, took thirteen and a half months.[16] "Not a building, not a structure of any description, is ever commenced without one constant idea occupying the minds of all those concerned in it, directly or indirectly, which may be expressed by saying, 'get it done as quickly as possible,'" rued an editor of the London journal *Engineer* in 1876.[17] Extended construction periods remained admissible for exceptionally long or high bridges or challenging foundation work, often the most demanding part of a project. But, overall, governments and bridge toll and railway companies wanted structures that could be quickly put into place. Ease of maintenance was important too, because closing a rail line for a day or an hour could be costly or risk military disadvantage.

Innovativeness, efficiency, and ease of assembly are typically recognized as positive industrial values, but the context in which we explore their meanings alters the story we might tell. This becomes apparent with a glance at construction labor, a subject too big for *American Bridge* to uncover, yet too central to silently pass over. Chapter 2 featured a few structures and parts that circulated through early nineteenth-century inventive landscapes. But these didn't tell us how Town developed a construction process that worked. As a dedicated learner and experienced builder, he would have employed scale models to experiment with the novel structure and its assembly.[18] But only real-world iterative implementation would have revealed the challenges and confirmed the potential of a structure whose value lay in its making. With the location of Town's earliest lattice bridges as our guide, we'll reconstruct the circumstances surrounding a critical phase, when a new idea proved itself viable and valuable.

Wealthy North Carolina state official Lewis Beard owned two plantations, worked by thirty enslaved people, on the Yadkin River. As a result of Beard's disagreement with the ferry operator using his land, the legislature endowed him with the right to build a bridge. He chose Ithiel Town to do so—a connection probably made by Beard's son, John, who attended Yale (class of 1817) in New Haven, where Town lived and had recently built two churches on the central village green.[19] Beginning with this opportunity, Town oversaw the construction of lattice bridges over the:

- Yadkin River near Salisbury, NC (known as Beard's Bridge, 1818)
- Cape Fear at Fayetteville, NC (known as Clarendon Bridge, 1819–1820)
- Congaree Creek/Huckabuck Swamp, SC and Goose Creek, SC, both bridges on the state road from Charleston to Columbia (1820)
- Connecticut River at Springfield, MA (1820)
- Mill River at Whitney's Armory, New Haven, CT, on New Haven and Hartford Turnpike (1823)
- Pee-Dee River at Cheraw, SC (original, 1824, and its replacement, 1828)
- South Yadkin River, near Salisbury, NC (1824–1825).[20]

In the twenty-first century, we often hear stories of collaborative design and production efforts: R&D labs, tech incubators, and start-ups intended to produce entrepreneurial innovations. Telling ourselves that such efforts are a new form of practice helps hide the historical roles played by both marginalized and ruling figures in technological spheres. Here, instead, we'll approach the lattice incubation period by observing that Town developed his earliest bridge commissions and achieved proof of concept with the participation of enslaved Black people. In North Carolina at that time, a typical building team might include a combination of enslaved and free Blacks, along with whites. A skilled enslaved worker might have one master during his or her life but participate in a diversity of projects for a variety of clients.[21]

Town likely came to the Carolinas with the structural idea worked out; perhaps he had previously tested it, with models and at full scale. But the only way to develop and verify a process is by actually doing it. Repeatedly. This ensures that it will work in conditions shaped by materials, labor, geography, and politics—often expressed in terms of time and money. So the early bridges enabled Town to hone and integrate structural and constructional strategies in order to demonstrate that, with the application of (Town's) engineering expertise, a professional manager with a small group of trained workers, rather than very highly skilled craftspeople, could assemble long, rigid, durable structures out of relatively small, uniform parts. Town was in Fayetteville (where, he wrote, "there are a number of New Haven people"), guiding construction of the Clarendon Bridge when his patent was issued in January 1820.[22]

As the *Report of the Commissioner for the Department of Public Buildings, to the Board of Public Works for the Year 1820* in South Carolina stated, a key expectation of the lattice was that it "may be understood and executed by any ordinary mechanic."[23] This was acutely significant in the South, where a scarcity of white specialists led local industrialists to pay experts such as Town to come and train less skilled workers, a category that included enslaved men.[24] The understanding of ordinary mechanics was a key requirement for mass construction: simplification of tasks, narrowing of workers' skill requirements, and the elevation of the status of the engineer as a representative of "science" would become the archetypal story of building.[25]

Town was surrounded by people and institutions deeply beholden to the wealth generated by slavery. During the years 1819–1820, the Missouri Crisis and Missouri Compromise, the latter negotiated by Henry Clay, underscored the centrality of slavery and debates about it in the expanding United States.[26] At that time, the employment of the enslaved in craft and industry was widespread; by 1776, in South Carolina they worked in at least twenty-eight trades, including the cutting and transport of trees, the milling of lumber and wood products—so profoundly important for the construction of houses, bridges, and ships.[27] According to historian Catherine Bishir, slavery was "critical to every aspect of building practice in North Carolina"; some enslaved individuals learned informally and others were deliberately trained. Enslavers hired out these expert builders for construction projects; in places such as Fayetteville, some lived independently, hired themselves out, and paid a percentage to their enslavers.[28] The main goal of many bridges in the South, whose tolls could produce wealth for investors, was the transportation of cotton—its production in the United States, which fed Britain's textile industry into the twentieth century, went from 1.5 million pounds in 1790 to 331 million in 1830. Bridges were embedded in and enabled local and international slave-based economies.[29] The participation of enslaved labor in industry was substantial; by the time of the Civil War, three-quarters of the South's railroads employed the enslaved for construction and operation. They often worked in lethal circumstances, characterized by any combination of dislocation, violence, disease, hazardous and toxic working conditions, and inadequate clothing, food, and housing. After Britain and the northern United States, the South in 1860 had more rail mileage than any other region.[30]

The construction of the early lattice bridges provided an opportunity for testing and observation. This meant that, beyond their essential role as labor, enslaved and free workers became part of the innovation teams developing the building process. While patent and myth reinforce the idea of the lone inventor, during the fledgling phase anyone participating in building might have helped invent, that is, devise, test, analyze, and revise, the lattice and its construction methodology. Researchers have not found documentation of the 1818 bridge at Beard's plantation, but if enslaved workers would somehow *not* have participated in the building of Beard's bridge, or in any of the early bridges discussed here, is for future researchers to determine. (Salisbury, site of the earliest lattice and seat of Rowan County, was a slave trade center in a state in which one-third of the population was enslaved.)[31] A belief that for this initial bridge we have to provide a list to prove that enslaved people participated—in a place where the enslaved routinely participated in logging, milling, road building, and bridge construction—rather than prove they did not, reinforces narratives that deny or "actively forget" slavery.[32] The initial projects in the Carolinas and Connecticut shaped development of the lattice bridge. In 1825, the *Western Carolinian* (Salisbury) noted that Town "had greatly improved his plan" since building Beard's bridge (which was "much admired by travellers") and had applied his learnings in Cheraw. Town himself in 1831 confirmed substantial improvements.[33]

Town was present for the 1824 construction of the Cheraw Bridge, during which enslaved workers participated in the fabrication of the wooden parts and in construction. Enslaved worker Horace King (1807–1885), who grew up in the Cheraw area, likely witnessed the building of the bridge, which would have been a significant local event. It was destroyed by flood in 1826; although documentation of its replacement in 1828 is lacking, King and John Godwin, his enslaver from about 1830 and subsequently his partner in a building business, probably participated or at least were able to watch and learn from the work. They knew the man who invested in the construction of the 1828 bridge. In 1832–1833, King's first big project with Godwin was a lattice bridge over the Chattahoochee River at Columbus, Georgia; much of the work force consisted of skilled and unskilled enslaved labor, including King, an enslaved man who was especially valuable because, beyond his skills as a carpenter, he supervised other enslaved people. The bucolic scene depicted in a lithograph (figure 5.2) represents the fundamentals of the industrial economy: Black labor, cotton, river navigation, and transportation infrastructure. The bridge was built to bring cotton to the market. Godwin and King would build five bridges across the Chattahoochee; they erected many other structures in this frontier region. King, who eventually was freed, managed the building of many lattice bridges in the South during the following four decades. Except for his presence or participation at the sites of Town's initial structures, there is no plausible explanation for his acquiring expert knowledge of lattice bridge building.[34] (As photography was new in the late 1830s, there are no photographs of early construction.) It would have been in Town's interest to encourage King since the bridges he built would have generated patent fees. Town's involvement in the slave economy also includes the construction of the North Carolina state capitol (1833–1840), for which many enslaved people provided labor; he and his partner Davis were the main architects, though not onsite managers.[35]

As "superintendent" of the Clarendon (an alternative name for Cape Fear) River bridge at Fayetteville, North Carolina, where in 1820 half the population consisted of enslaved people, Town would typically direct the building contractor; for example,

Figure 5.2
Dillingham Street Bridge (Horace King, 1832–33, demolished 1910) over the Chattahoochie River, between Columbus, Georgia and Girard (now Phenix City), Alabama. Over the years, King repeatedly repaired the bridge, with a rebuilding c. 1870. François de Laporte, Columbus Bridge, ca. 1838. Courtesy of the Columbus Museum, Georgia; Museum purchase G.1983.75. Photograph, ca. 1900, Alabama Department of Archives and History.

Captain Grafton at Clarendon and Congaree Creek and Samuel Lemley at South Yadkin. He continued to monitor and maintain the performance and reputation of the Clarendon Bridge; during May 1823, Town, in a repeatedly published announcement in a local newspaper, confirmed the ongoing "stability and safety" of the structure.[36]

These early bridges were not merely local news. On March 4, 1825, when the sixty-eight-year-old Marquis de Lafayette stopped in Fayetteville, North Carolina, during a celebratory tour of the United States, he was greeted at the Clarendon Bridge.[37] Karl Bernhard, duke of Saxon-Weimar-Eisenach, mentioned it in his book on his travels through the United States during the years 1825–1826. Engineer Guillaume-Tell Poussin, who between 1817 and 1832 traveled through and worked in the United States, including North Carolina, was the author of a well-known 1834 publication that contained a section devoted to the lattice (see chapter 3).

Town also embedded the lattice in his home region, where it would have been visible to the people who populated thriving local high-tech and academic networks. He built bridges in Springfield (1820) and at Eli Whitney's armory (1823). The cotton gin, with which Whitney has long been associated, had enabled cotton production to massively expand—and with it, slavery. This transformative tool was also almost certainly a product of many designers, including enslaved people.[38]

Previously, bridge innovators typically worked as builders; the money was in construction rather than design. Town, however, began to operate as what we today would call a consulting engineer, in this case, an inventor who sold patent rights to and licensed contractors who could build without his direct intervention.[39] The change in practice was in line with the precepts of Taylorism, which had its roots in plantation management. Expertise, control, and profit resided with the engineer/manager, while the workers' role was to implement.[40] As desirable as interchangeable parts would become, as significant was the ability to exchange highly skilled workers with workers of lesser, even if still specialized, experience. Deskilling, routinized work, fewer workers, lower labor costs, and rapid timeframes were fundamental to the reinvention of building. Town became wealthy by supplying and approving bridge designs, providing construction oversight, earning patent fees, and, in the case of the Clarendon Bridge, partly owning the bridge toll company. He helped establish professional ways of working that would become standard for architects and civil engineers.[41]

In the past, many historians presented slavery as a regional phenomenon whose impact ended with the Civil War. The organization of enslaved labor on cotton plantations in the South, however, provided industrialists and engineers in the North and Europe with a model of worker control (that is, brutal domination) that generated immense wealth. (Karl Marx provided incisive analysis of the barbaric economic and industrial significance of slavery and its legacy). Slavery also helped shape the disposition of workers involved in the development of the structures and strategies of mass construction. Enslaved labor provided a template for the treatment of construction workers, who have long worked in unsafe conditions and contributed significant expertise without earning an equally significant share of the profits. The fact that enslaved Black labor participated in construction teams at key sites during a founding moment of reinvention has been "actively forgotten" in evolutionist stories of innovation, liberty, and national achievement. As industrial capitalism is founded on slavery, it is no surprise that the ingenuity, labor, and unspeakable suffering of enslaved people are enmeshed in building strategies essential to all industrializing nations.[42]

Engineers and entrepreneurs celebrated the efficiency of "American" industry—even if, in some cases, workers were enslaved, dehumanized, and excluded from American citizenship. Concerning the building of the Long bridge, for example, in 1830 an editor of *Dingler's Polytechnisches Journal*, a Stuttgart periodical that published favorable accounts of "American" technology, remarked, "One sees . . . that the American carpenters are somewhat nimbler than the European, about whom, in almost all European languages, the expression goes: 'pretty slow, like a carpenter.'"[43] While German-American carpenters would in the United States continue their slow building traditions and adopt faster new ones, observers were fascinated by novel industrial building and management practices.[44] This was top of mind for one of the world's most experienced railroad engineers, Franz Anton von Gerstner (see chapter 3), when he stated in an 1839 report that he had solved the "mystery" of how US railways were so cheaply built, although US construction was twice as costly as in the German states and Russia. He listed seven points:

1. *Planning*. Railways are carefully planned, usually by several engineers, during a preparatory period, which lasts two to three years, that is, at least as long as the construction time itself. In Europe, neither money nor patience is available for such planning.
2. *Consulting engineers*. A construction engineer and a consulting engineer, who works on several lines and therefore has broad and deep experience and expertise, are in charge of construction. In Germany, engineers are selected on the basis of low salary and birthplace.
3. *Design*. Earthwork, high bridges, tunnels, and inclined planes are avoided. Instead: "undulating" railways, steeper climbs, sharper curves.
4. *Mechanization*. Because the cost of labor is very high, machines are extensively used, for example, steam-driven pile drivers and excavators.
5. *Bridges*. More bridges are required in the United States than in Europe. Included are bridges of "completely new design," great length, and low cost, with many made of

wood. Gerstner provided measurements and prices for the lattice bridge over the James River at Richmond.[45]

6. *Flexibility*. Design conforms to the available capital. For example, if the money is lacking, cheaper, lighter rails are used, to be replaced later when traffic increases.
7. *Efficiency*. Construction of buildings is kept to a minimum and implemented at minimum expense. Fewer locomotives and cars are employed than in Europe on lines with equivalent traffic; there are no reserves. "The locomotives, like the entire [American] people, are always busy."[46]

Engineers working in the United States did not necessarily agree with this assessment, especially the first point. But Gerstner's evaluation emphasizes elements—planning, expertise, mechanization, simplicity in design, and financial techniques—that were basic to industrial capitalism and attracting the interest of industrialists in Europe.[47] Bridges were an essential part of the immense geographic and capital-intensive scale that railroad builders inhabited, so it is not surprising that Gerstner spotlighted the lattice as a model of creativity as he noted the importance of flexibility, a concept also central to recent digital innovation and growth. Gerstner summed it all up by attributing low-cost construction to the "practical sense" of the Americans.[48] Although he referred to enslaved people in his report, and likely witnessed slave labor, perhaps during the construction of the Raleigh & Gaston Railroad, which included five lattice bridges (figure 5.3), he did not cite slavery as he unraveled the mysteries.[49] This omission helped many to forget that, like America, American bridges were the product of liberty and slavery.

PROCESS AND PATENT / PUBLIC AND PRIVATE

During the early nineteenth century, the US government promoted patents as a way of confronting the labor shortage and stimulating the economy. Technologists patented numerous bridge designs; the construction of transportation infrastructure held the promise of wealth to those skilled and fortunate enough to be able to profit from it.[50] For many inventors, the value of a patent lay in a product, that is, an object that could be sold, but for some, it was about a process. In the latter case, in order to succeed, it was necessary to navigate complex interactions among patent, product, process, and profit, and do so within changing legislative, enforcement, and market conditions. Here we'll consider why Ithiel Town, a pioneer in his professional use of patent licensing, remained silent about a key element of his intellectual property. We'll also ponder why storytellers maintained engineering omertà.

The manner in which inventors exploited their creations depended on whether they viewed it as a process or a product. A process can be used to create a range of products;

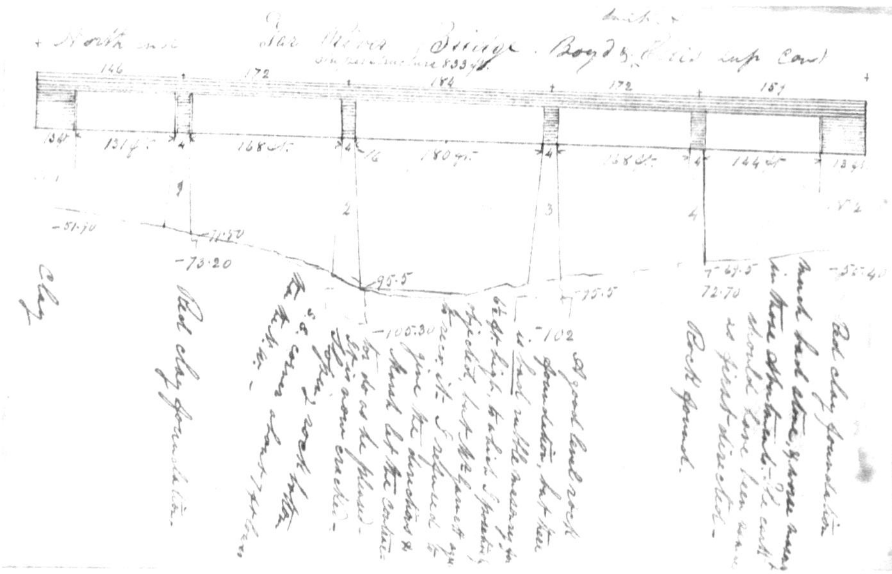

Figure 5.3
Tar River Bridge (James Herron, 1836–1840; possibly specified by Moncure Robinson), North Carolina. Engineer James Herron kept detailed notes regarding products used in the construction of double-lattice bridges on the Raleigh & Gaston Railroad, one of the first lines in North Carolina. Lattice bridges were also built over the Roanoke and Neuse rivers and Cedar and Crab Tree creeks. James Herron, "Location Book: Raleigh & Gaston Railroad" (1838–1840). David M. Rubenstein Rare Book and Manuscript Library, Duke University.

for example, iron casting can produce a multitude of patentable objects, such as storefront elements or a pot for cooking. A product might be manufactured in a variety of ways; for example, a gun whose parts have been made by hand might be comparable, structurally, to one that is machine-made, although the process and cost would be different. An inventor might patent a machine that cuts and heads a common nail in one operation, even if the product, the nail, cannot be patented. As historian Christine MacLeod explains, an inventor might decide to keep a machine for exclusive use in their own factory or opt to sell it to others. Whether or not an inventor seeks a patent at all is a significant decision because patenting reveals information whose diffusion and exploitation the individual is not be able to control.[51]

If the lattice bridge had been a one-off or highly complex product, it would not have been prized by builders. The patent's value lay in its method of assembly, that is, its constructability. The lattice reduced construction challenges and costs by using uniform parts, lowering skilled labor needs, enabling preassembly, and requiring less costly supporting structures during building (we're about to explore how it was built).

Town could patent the lattice because its structural design could be deemed novel, even if its components, planks and pegs, could not. The methods of building the lattice, in and of themselves, were not novel, either! In fact, a considerable part of its value lay in the use of ordinary parts, tools, machines, and procedures.

In terms of process, Thomas Blanchard's lathe (a machine that shaped wood products) provides a striking analogy to the lattice bridge. Blanchard's patent (first issued on September 6, 1819, and in its revised, final form registered on January 20, 1820, eight days before Town's patent) omitted what distinguished his lathe from existing lathes: the fact that the cutter, rather than the piece being shaped, moved. As historian Carolyn Cooper explains, the "lathe's *kinematics* were original, but its *components* were not, and components may have been more salient to observers at the time, as they were later to many of the witnesses testifying in his patent suits."[52]

How was the original lattice bridge built? Assembly was simplified and systematized, but there are few contemporary descriptions and none with adequate detail. Likely, it occurred as follows (figures 5.4 and 5.5). A small crew of workers laid out the parts of each lattice frame flat on the ground. They clamped the planks together, drilled the holes, and hammered the treenails into the holes. Each lattice frame was hauled, flat, onto a temporary scaffolding previously built in the waterway and raised with basic machinery such as block and tackle, winch, or windlass, capstan, derrick, crane, gin pole, or raising pole. After each lattice frame was moved individually into place, builders constructed the floor and roof structures that joined them at bottom and top; then they installed cladding and roofing.[53] A possible alternative would have been to lay out and number the pieces, drill the holes, and then move all the pieces to the site and assemble them in a vertical orientation on lightweight scaffolding built at the location. This was how David Stevenson described it in his rare 1838 account:

> The timbers of which Town's and Long's bridges are composed, are fitted together on the ground previous to their erection on the piers. They are again taken asunder, and each beam is put up separately in the place which it is to occupy, by means of a scaffolding or centering of timber.[54]

The Long bridge (see chapter 4), which relied on mortise and tenon connections as well as wedges, was likely constructed in this manner. Stevenson had read Long's *Description*, in which Long indicated that, supported by temporary scaffolding, the "frames . . . may always be put together in an erect position, by adding piece after piece, till the frame is complete."[55] Concerning the Town bridge, he may have extrapolated. Stevenson was in the United States for only three months; Town observed that Stevenson had not seen good examples of the bridge. He likely had not seen one being built or was unable to recognize the significance of what he witnessed.[56]

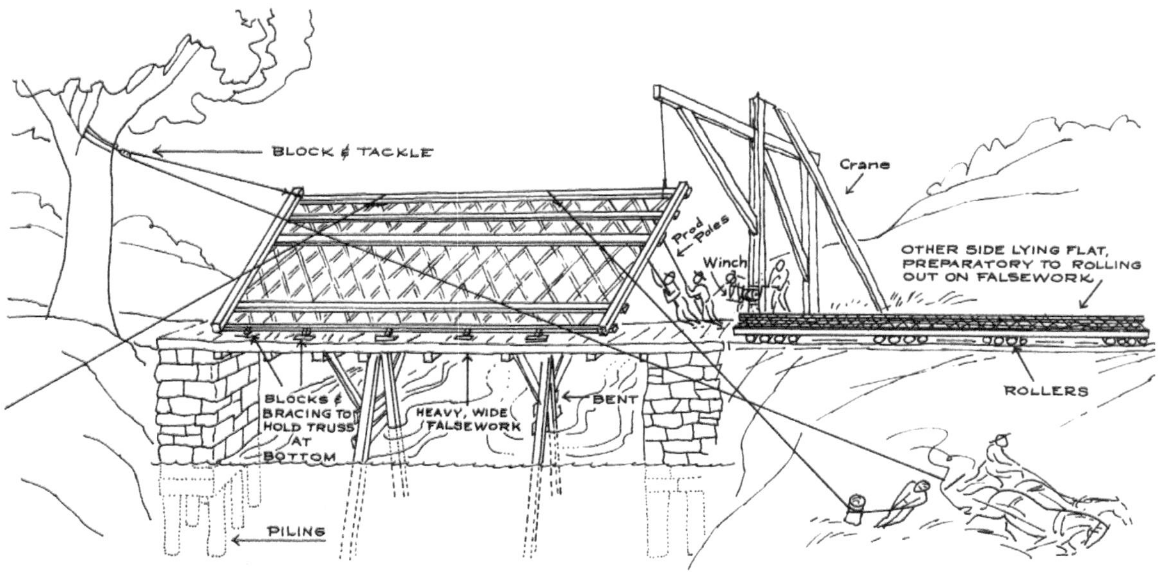

Figure 5.4
Raising the lattice frame (presumed method). The lattice's significant out-of-plane strength (which prevents bending) enabled ease and speed of construction. Illustration by George Daly; from *Covered Bridges of the Northeast*, by Richard Sanders Allen, copyright © 1957, 1974 by Stephen Green Press. Copyright renewed 1985 by Richard Sanders Allen. Used by permission of Viking Books, an imprint of Penguin Publishing Group, a division of Penguin Random House LLC. All rights reserved.

Unlike an arch bridge, the lattice was not integrated structurally with its piers and abutments; it was placed on top of them. This meant that, during construction, temporary supports and scaffolding could be less massive, complex, and expensive than required to support the building of a stone arch, for example. Because each lattice frame would carry itself immediately, temporary scaffolding did not have to be strong enough to support more than the initial installation or assembly of the frames.

Colonization and attendant patterns of mass migration meant that structures were shipped around the globe for settlers, soldiers, and prospectors, who required immediate shelter but lacked the skills, time, or materials to build. During the early eighteenth century, wooden houses were transported in parts from North America to the West Indies; these included beams fabricated by carpenters.[57] They created the mortise-and-tenon connections, fit parts together, numbered, and then disassembled them for shipping. Apparently, at the end of the eighteenth century, manufacturers in Britain shipped cast-iron arch bridges to distant locations, including the Americas; they shipped structural members as a kit of parts.[58] During the 1820s, troops of the Royal Engineers in

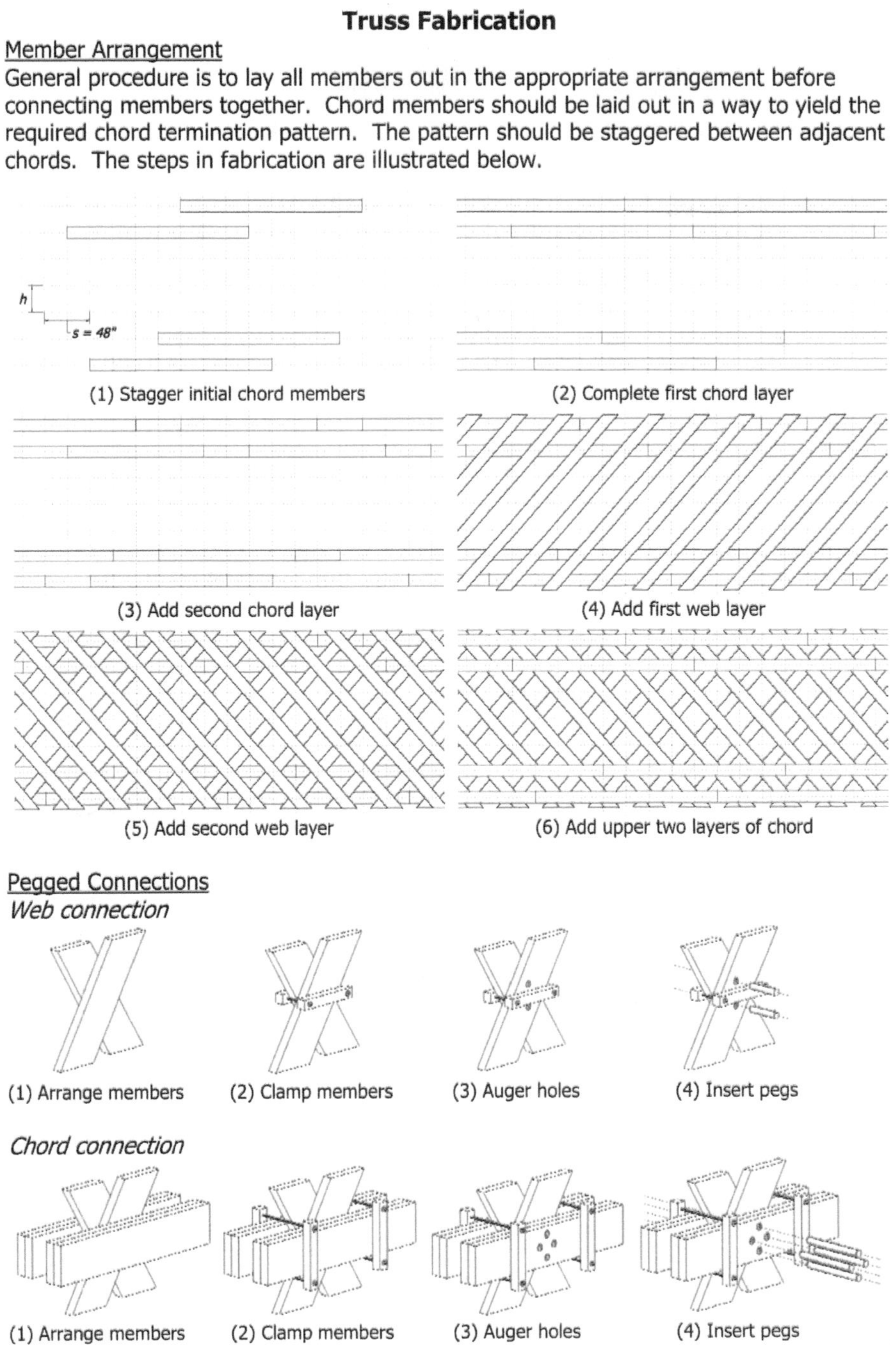

Figure 5.5
Reconstruction of lattice assembly process. Todd Craig Radford, "Town Lattice Truss" (PhD diss., MIT, 2010), 295. Courtesy of Todd Radford.

the West Indies lived in structures of iron made in Birmingham. Settlers from Britain brought prefabricated structures to South Africa and Australia in the 1830s and 1840s.[59] Enormous quantities of lumber manufactured at the Cape Fear steam sawmill, in the region where Town built his earliest lattices, were shipped to islands in the Caribbean.[60] California Gold Rush fortune hunters lived in houses of lumber or iron transported from "the United States, England, France, Germany, Belgium, even China, New Zealand and Tasmania."[61] Portability and the ease and speed of building were as important to colonists as they would be for railroad companies.

Outside of colonial constructions, how common was it for builders to preassemble structural components? We don't know. Historian John Fitchen asserts, "There is complete silence in all written records of early timber construction."[62] For centuries, carpenters cut joints, ensured that walls or roofs or trusses fit together, then disassembled them for reassembly in place. For half-timbered buildings, builders assembled portions of the frame on the ground and then raised or "reared" them with ropes and poles; it appears, though, that most rectilinear frame structures were built piece by piece.[63] In eighteenth-century North America, quick construction was crucial to barn raising; by the end of the century, barns were made of preassembled frames and builders were raising house frames in New York City.[64] Carpenters might design roof structures, for example, so that two or three strong people could raise them.[65] Preparing the members might take an extended period, but labor scarcity was a prime reason to design for speed of construction, minimize the number of workers, and reduce required skill level. Fitchen explains:

> Both the ordering of the work, and much more significantly, the very structural system itself, were pervasively affected . . . by the necessity to concentrate all group activity with respect to the barn's erection into the shortest possible period of time: two days at most, and perhaps only one long day. . . . In this respect, the American situation was quite different from that of contemporary Europe, where communities were not so widely separated and there was usually an adequate number of nearby neighbors who could be called upon for assistance as needed.[66]

What was Town's experience with pre-assembly? Burr bridges (figure 2.17), which Town reportedly built early in his career, could be preassembled, but this apparently was exceptional rather than routine.[67] Town's early work includes the construction of church roof trusses that were likely lifted and placed on relatively light timber frame walls (figure 5.6).[68] This indicates that he understood the value of independently built superstructure and supports, with the latter designed to take compressive forces.[69] At Ithiel Town's first major building project, Center Church in New Haven (constructed 1812–1814), workmen used a windlass, pulley blocks, and rope to lift the spire built within the tower's base, thus using the building itself as scaffolding.[70] In 1826, after

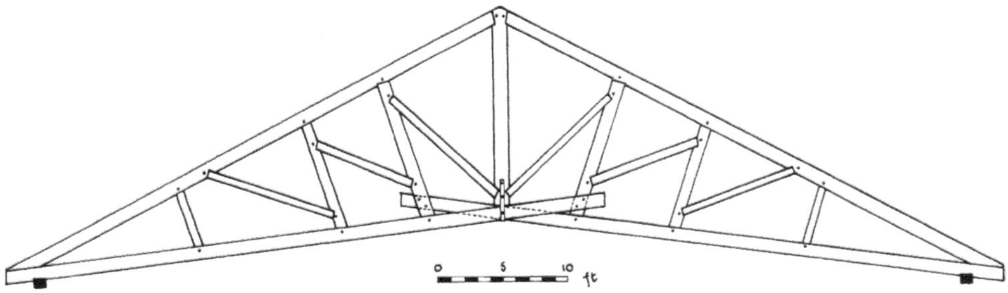

Figure 5.6
Center Church roof truss (Ithiel Town, 1814). David T. Yeomans, *The Trussed Roof: Its History and Development* (Ashgate, ©1992), 133.

completing his earliest lattice bridges, Town worked as architect of the New-York (a.k.a. Bowery) Theater in New York City (figure 5.7). When this renowned entertainment venue opened in October, a roof structure consisting of lattice trusses sitting atop two-foot-thick brick side walls bridged the 75-foot-wide interior of the playhouse. Although construction drawings have not survived, the lattices (likely triangular) were probably assembled on the ground and lifted into place.[71] This means that Town understood the lattice as a long-span beam useful in many contexts. For him, the Bowery Theater's architecture represented good taste, modularity, functionality, and moral/stylistic economy, all bundled in a "Grecian" mode that, appropriately for the United States, referenced a tenuous reconciliation of democracy and slavery.

So how did workers actually build the lattice bridge? A crucial shred of evidence concerning construction of one of the earliest comes from Town. He states in a letter written in Fayetteville, North Carolina, dated January 16, 1824, to Roger S. Baldwin, who was probably his lawyer:

> I have been at Cheraw considerable part of the time [?] my return and the Bridge is already framed and the workmen are now raising it—it bids fair to be a very Excellent Bridge and will be ready for Public use in about 60 days from this time.[72]

The lattice beams (or frames) had apparently been assembled ("framed") and were being moved into place ("raised").[73]

In a letter sent from New York, dated June 5, 1837, addressed to "Jonathan Livermore, Bridge-builder," Town wrote:

> In raising your Bridge, you need to be very careful to put temporary braces to keep the winds [or?] sudden *gusts*, from blowing it over side-ways—in raising the Bridge here at Harlem Hayward had one of the spans blown over in this manner and gave him a great deal of trouble & hindrance.[74]

Figure 5.7
Bowery Theater (Ithiel Town, 1826), New York. The roof was supported by what may have been Town's earliest use of lattice trusses in a building. The image, which does not show neighboring structures, was probably created by Alexander J. Davis, who began to work for Town in 1826 and would become his business partner. The Miriam and Ira D. Wallach Division of Art, Prints and Photographs: Print Collection, The New York Public Library.

This may confirm that individual lattice frames were assembled beforehand, moved into place, and then lifted (that is, tilted) into a vertical position before being connected to each other with deck and roof structures.

Systematization and simplification did not mean that inexperienced labor could undertake construction. The men who successfully designed and built bridges were experienced professionals.[75] Even with increasing uniformity, construction demanded considerable know-how and skill, encompassing preparation of the site and design of the construction sequence. Builders had to understand the characteristics and behavior of specific species of wood such as pine and oak. Expertise included, for example, how to size and assemble parts so that the deck arched slightly upward in the center—which would flatten under the bridge's own weight and when a load passed over. This technique prevented the structure, over time, from sagging in the middle. Achieving this upward (convex) shape, called "camber," meant that parts in the upper chord may have had to be fractions of an inch longer than the parts in the bottom chord; the placement of bore holes was also important.[76]

Although the construction process formed the soul of the lattice bridge concept, shaping every element and its overall configuration, Town never publicly revealed the construction method. He did refer to it obliquely in his advertising brochures:

> All persons are particularly requested not to make any use of the improvement without first obtaining my particular directions and specifications; for, in the event of any such violation, the persons violating will be certain to sacrifice at least twenty times as much as I should lose, even if I were not afterwards to receive any thing for the right: such is the importance of having the directions, &c. of one who has had several years experience in the practical execution of this kind of Bridges. And further, it will cost no more for using the right of the improvement *with* all the suitable directions, &c. than to use it *without* such directions.[77]

Keeping mum on what made the patent valuable was a process-inventor's strategy.[78] Town had to register a patent because he would not have been able to keep the lattice design hidden, unlike an inventor who might try to keep a machine from view behind factory walls. Plus, he needed to make the lattice public in order to market it to potential clients. Demonstrating the success of the lattice in service, especially in the early years, was essential for selling it. So Town navigated the patent paradox by selectively sharing and dissembling.

Town's clients paid for access to a trade secret: construction reinvented. If he had tried to retain for himself the exclusive right to build, it would have likely required substantial investment to develop administrative and building capabilities, limited the number of bridges that could have been built, and reduced the possibility of other builders improving the lattice and becoming invested in building more.[79] It might also have encouraged unauthorized building. Licensing was the solution: Town earned one dollar per running foot of bridge, with a minimum fee of $100.[80] Unfortunately, his business records are gone. We do not know how he managed his affairs, though his brochures indicate that he actively marketed.

Town claimed that his bridge was from one-half to one-third cheaper than comparable bridge types.[81] His warning that a rogue builder would "sacrifice twenty times" the cost of obtaining the rights indicates the patent's hidden value: how to build. And a century later, engineers Robert Fletcher and J. Parker Snow confirmed that, though the bridge was "often criticized as wasteful of material, such waste is much more than balanced by the resulting simplicity of framing and erection."[82] "Waste" is a debatable term, as material that might be labeled redundant provides rigidity required during construction and enhances safety and durability. Town was also aware that it was important to prevent unauthorized construction, as a poorly built bridge would harm his reputation and damage the patent's value; defects might be attributed to design rather than workmanship. He seemed dedicated to providing a structure that would endure.[83]

While Town's reticence regarding process appears rational, why the silence on the part of expert observers and communicators such as Poussin, Ghega, and Culmann (see chapter 4)? Many technological travelers probably had not witnessed construction in progress. For academically trained observers, some of whom had limited experience in construction, the process was of less moment than scientific or theoretical concerns. This is reflected in professional engineering journals, where in-depth discussion of construction as a separate subject was rare; when described it was usually in relation to specific projects.[84] (Numerous construction accounts are cited in this chapter, but this should not give the impression that such accounts were commonplace. They constitute a small fraction of the vast quantity of contemporary published material on bridges.) It may be that some observers understood the value of remaining silent; they did not want to reveal a trade secret. Historians' later silence might be ascribed to the bridge's wooden birthmark and the fact that it did not require specially designed or complex machinery; this would not have fluffed evolutionist progress narratives. Still, as we've documented, contemporary observers did note the significance of the lattice bridge structure, if not its construction.[85] Considering Herman Haupt's expertise (see chapter 4), his 1851 endorsement carries a lot of weight:

> No plan of bridge construction has met with more general favor amongst engineers and builders than the lattice. Its great simplicity, the ease with which it can be framed, and chiefly its economy, have secured its introduction for viaducts of almost every class. . . .
>
> The most ordinary carpenter who is able to bore a hole with an auger is capable of constructing it, and the timbers being all of the same size are delivered from the mill in the state in which they are put together; hence no preparatory labor is required, no carefully fitted joints, no bolts, straps, or ties of iron, and consequently it is fair to presume that the lattice principle is the cheapest upon which a truss-bridge can be constructed.[86]

Decades after first encountering the lattice bridge—indeed, on his deathbed—Michel Chevalier still enthusiastically recounted that a man could easily carry on a shoulder one of the "absolutely similar" boards that made up the bridge.[87]

Engineers with a theoretical bias told a different story, though. West Point Professor D. H. Mahan confirmed in 1846 that the lattice was "of very easy mechanical execution [i.e., easy to build], the beams being of a uniform cross section and length"; he noted that the bridge eventually sagged and concluded that "it is more recommendable for its simplicity than scientific combination."[88] In a similar vein, civil engineering professors Mansfield Merriman and Henry S. Jacoby explained in their late-nineteenth-century standard *Text-book on Roofs and Bridges*: "The lattice truss introduced by Town about 1820 consisted of planks pinned together, and was important only on account of ease of construction."[89]

Unconcerned with the industrial, economic, and cultural significance of the lattice, the professors acknowledged its value, even if qualified with "only." Such stupendous understatement might seem improbable, but academic engineers used the lattice as a foil for bridges that they believed embodied the "scientific" approach on which they based their status. To this day, stories that segregate and rank knowledge and skill types shape the politics behind what is labeled "technology" or "science."

A MACHINE THAT BUILT ITSELF

The lattice's pivotal contribution, its construction process, encompassed the making and mobility of parts that workers assembled into long, rigid components intended to be maneuvered into place. The lattice frames (that is, large beams), or in some cases the entire bridge superstructure, could be preassembled and moved onto its final supports; this procedure minimized or replaced temporary scaffolding. Designers sometimes went further: they used the bridge as a crane for the construction of its own supports, the piers upon which the bridge would rest.

New, complex machinery's absence from the lattice program, as we've noted, likely constitutes one reason for its exclusion from the evolutionist canon of technological history. But let's also observe that the purpose of a machine is to enhance a specific ability, or it "may simply be to modify and transmit forces and motions."[90] This definition might inspire us to wonder why "bridge" is not categorized under "machine" (figure 4.2). So classified, it might have altered the narratives that shape our understandings of bridges and their designers and builders. But such musing is beyond the scope of this chapter. Here, we'll recover evidence of a constructive, mechanical, and structural reinvention present in long lattices and towering piers. We'll consider bridges as structures and as machines whose power was supplied by the people who built them.

Builders inspired by American bridges developed a variety of lattice bridge construction techniques; the earliest may have been modeled directly on what we've surmised of Town's own. Engineer William Moorsom as early as 1837 installed preconstructed lattice beams during the building of at least twelve bridges on the Birmingham and Gloucester Railway (see chapter 3).[91] When engineer Joseph Hyacinthe Garella supervised the construction of his first lattice bridge over the Rhône in 1838 (figure 3.10), builders lifted lattice beams, previously assembled and lying flat on temporary supports on boats tied up in the river, and put them into place.[92] Some engineers employed traditional building methods to produce the novel structure. For example, George W. Hemans, engineer in charge of the work for the Royal Canal bridge (John Macneill, 1844–1845) on the Dublin and Drogheda Railway reported:

> The erection of the whole structure was carried on like a common building operation, every bar, plate, and rivet being brought separately to the spot, and built together on a timber scaffolding, piece by piece. . . . The mode of proceeding was . . . somewhat similar to that of building an iron ship.[93]

Workmen erected temporary scaffolding, on which they built the bridge; parts were delivered by boat. After assembling the bottom chords, they installed the diagonals with temporary bolts. Each man wielding a rivet hammer completed at least one hundred rivets per day. While the scarcity of wood in England may appear to have been the impetus behind iron construction, Hemans observed that "a timber bridge, answering the same purpose, would have been considerably cheaper, but it will be acknowledged, not so durable."[94] Yet, already in 1857, the iron bridge and railroad were in "alarming" condition, the former apparently because of a lack of lateral bracing.[95] Factors such as material quality control, structural details, or maintenance practices, rather than intrinsic characteristics of wood or metal, determined durability.

Engineers developed and implemented strategies that enabled mass construction. Richard Boyse Osborne (1815–1899), an English engineer who worked from 1838 until 1845 on US railroads, including the Philadelphia & Reading, noted, "At Pottstown the superstructure of different kinds of bridges were kept ready for immediate use, and bills of timber for the longer spans were stored in sheds for any emergency."[96]

In 1848, smaller scale 15 meter lattice frames were shipped by boat from the factory of a major manufacturer, Borsig, in Berlin to a site over the Saale at Grizehna on the Magdeburg-Leipzig railroad. Each frame was lifted onto a car which ran on rails laid on a temporary wooden bridge built on the permanent piers; the car was then wheeled to its destination, where the mobile crane lifted the frames into place.[97]

The rapid erection strategies desired for railways had long been important for armies; as we've noted, military engineers played important roles in early railways.

Figure 5.8
Yard at Alexandria, Construction Corps Finishing Portable Bridge Trusses, ca. 1862. Andrew J. Russell, photographer, United States Military Railway Department. *Photographs Illustrative of Operations of Construction and Transportation. As Used to Facilitate the Movements of the Armies of the Rappahannock, of Virginia, and of the Potomac* (Wright & Potter, 1863), photo 42. Library of Congress Prints and Photographs Division.

During the Civil War, short-span bridge trusses, assembled ahead of time and stored, were a readymade product (figure 5.8). Herman Haupt reported in 1863:

> A large number of bridge trusses were prepared [by the Construction Corps] in spans of 60 feet to be transported on flat cars, hauled by oxen to the sites of the bridges, and hoisted bodily into position by suitable portable machinery. These trusses were called "shadbellies" [i.e., fish bellies] by the workmen from their peculiar shape. A plan was also designed for a military truss bridge, the parts of which were interchangeable, and which could be put together without previous fitting, and with so much rapidity that, as my foreman, E. C. Smeed, expressed it, he could put the bridge together about as fast as a dog could trot.[98]

Figure 5.9
Pont Saint Michel, temporary bridge (Lagalisserie, ingénieur en chef and Vaudrey, ingénieur ordinaire, 1857), over the Seine, Paris, Auguste-Hippolyte Collard, photographer. École Nationale des Ponts et Chaussées.

Builders employed preassembly in a variety of bridges. Railroad engineer James Millholland (1812–1875), who designed one of the earliest iron plate girder bridges (1847), typically used for shorter spans, had it built in a train depot; subsequently, it "was put in its place."[99] The renowned plate bridge at Asnières (Eugène Flachat, 1852), which carried heavily traveled rail lines that led to Paris, was built in the shop in 50-foot sections, then moved to the site. The provisional wooden bridge built there in 1848, inspired by Town and Long bridges, was used as scaffolding for the installation of the plate bridge (figure 1.8). This minimized interruption to railway traffic.[100] "American bridges" offered advantages for temporary and permanent structures: light weight, rigidity, portability, quick installation, low cost. French state engineers would employ temporary wooden lattices during the 1857–1858 reconstruction of the Pont Saint Michel and Pont au Change in central Paris (figure 5.9). A photograph of this American bridge (as it was called) was displayed at the Paris Universal Exposition of 1867. Images that documented the reconstruction of Parisian bridges helped build an evolutionary story of urbanization that supported Haussmann's restructuring of the French capital.[101]

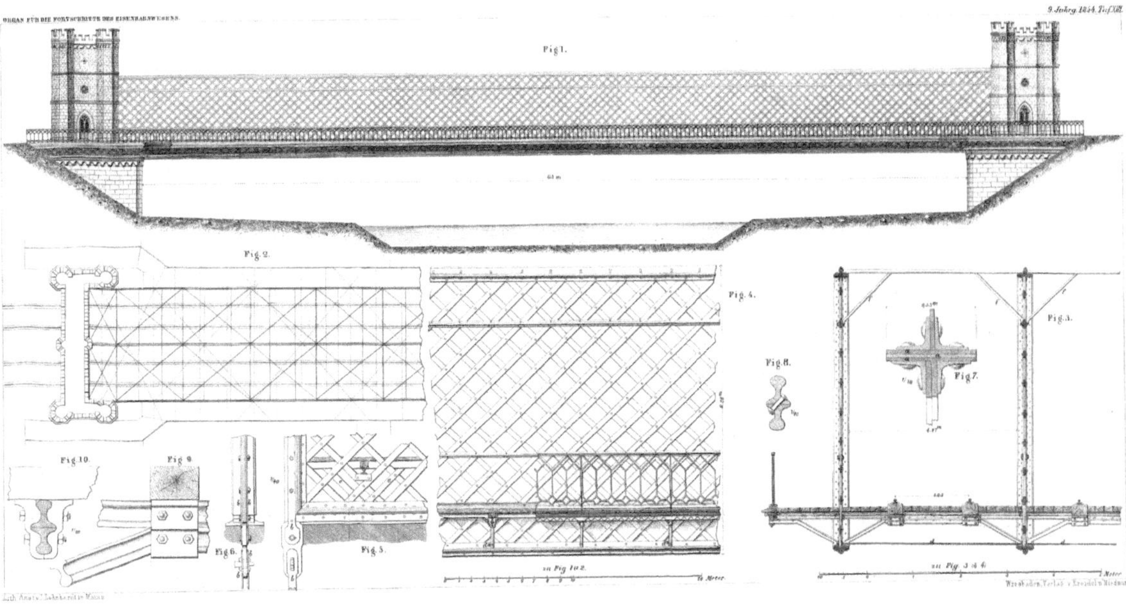

Figure 5.10
Bridge over the Kinzig (Carl Rupert, 1852–3, demolished 1883) at Offenburg, Baden. Edmund Heusinger von Waldegg, "Notitzen auf einer Reise durch Süddeutschland und die Schweiz über die Tannus und Hessische-Ludwigs-Eisenbahn, die Badische-Bayerische, und Würtemburgische Staats-Eisenbahnen," *Organ für die Fortschritte des Eisenbahnwesens in Technischer Beziehung* 9 (1854): pl. 13. Deutsches Museum Library.

Engineer Carl Ruppert (1813–1881), responsible for one of the earliest long-span lattice bridges in Europe, over the Kinzig River at Offenburg, near Strasbourg (figure 5.10), appears to have based his work directly on Town's methodology (and if so, this would confirm what's been conjectured). Ruppert was a graduate of the Polytechnical School in Karlsruhe and a prominent railroad engineer in Baden and Austria-Hungary. Planning apparently took place during late 1851; riveting of the first lattice began in May 1852; the bridge was open a year later.[102] This was surprisingly quick when compared to the Dirschau and Vistula bridges, two of the biggest engineering projects at mid-century and highly visible lattices—whose construction lasted from 1850 to 1857 (see chapter 6).[103] Engineer Émile With reported in 1854 that "travelers come daily from all countries to examine [Ruppert's Kinzig bridge], and all French tourists who go to Germany stop in the charming land of Baden to see the railroad trains pass over this new transparent bridge."[104]

Ruppert realized his project as Culmann was condemning the lattice (see chapter 4). The widely read journal *Allgemeine Bauzeitung* featured the opposing views: Max Becker

(1817–1884), engineer and professor at the Polytechnical School in Karlsruhe, announced that the Kinzig span would be "a wrought-iron lattice bridge based on Town's system" not long before the publication of Culmann's second report (1852) on bridges in the United States and England; undoubtedly this stimulated Culmann's antilattice bile.[105] When a detailed description of the Kinzig bridge appeared the following year, Ruppert stated that he "couldn't let himself become confused by another technical man of letters' recent disparaging judgement."[106]

The bridge over the Kinzig replaced a five-arch cast-iron bridge that had been carried away because its piers obstructed flood-driven wooden debris from log-rafts that had also destroyed upstream bridges. In order to avoid a repetition of the catastrophe, a bridge with fewer piers (that is, with longer spans) was desired. The railroad administration considered neither stone nor cast-iron arch systems, and they rejected wood as insufficiently durable. They considered only two structural systems: an iron tubular or "tunnel" bridge and "lattice bridges in accordance with the system of the American, Town."[107] The tubular bridge was rejected—part of a pattern we'll explore in the next chapter. As an economy measure, old "Vignoles" railroad rails were used for the cross-bracing; these might be considered standard members, employed at a time when standard wrought-iron structural beams were just beginning to be manufactured.[108] The lattice design at Offenburg featured large openings and no vertical members. Ruppert was careful to design a rigid structure. His specifying strong diagonals and rivets enabled him to increase the distance between the diagonals; fewer members made up for the increased weight of each. He stated that, analytically, he considered the bridge as a beam.[109]

As we've said, the construction of the Kinzig bridge resembled the presumed method employed in the original Town bridge (figures 5.4 and 5.5). *Illustrirte* [*sic*] *Zeitung*, the popular journal read throughout the German states, stated that "the home of this new bridge system is America, where the engineer Town first brought it into use in wood."[110] The 71.1-meter long and 6.3-meter high frames were assembled in two workshops of 90 × 11 meters (with 150 workers in each), erected especially for this purpose.[111] Workers laid the iron lattice bars on a wooden frame support, clamped the bars together, bored holes, and then riveted the bars together. As with treenails, the chief concern was that each rivet would completely and tightly fill the bore hole; to this end "hundreds of tests" were conducted. As with a wooden lattice, they built camber in. The first lattice took about fifteen weeks, the second about thirteen, and the third about seven weeks to construct. For the erection of the bridge, a temporary wooden scaffolding bridge on piles was built in the river, next to the temporary railway bridge (figure 5.11). The first lattice frame was put on a train of twelve flatbed railcars and transported on rails that ran from the workshops onto the provisional bridge.

Figure 5.11
Raising of the bridge over the Kinzig at Offenburg. *Illustrirte Zeitung*, March 25, 1854, 196–197. Deutsches Museum Library.

Workmen then raised it from a horizontal position in one hour with twelve winches and six mobile cranes; they installed the middle frame the same way. Because of a lack of space, they moved the third one, supported by temporary braces in an almost upright position, next to the middle frame, and then moved it into position. During erection, the crew was "organized in units in a military manner," and the incremental raising of each lattice frame was synchronized with trumpet blasts. When the frames were in place, the cross-bracing and floor beams were installed.[112]

Single lattice frames were unstable and had to be carefully moved; lifting them from a horizontal to a vertical position was a delicate job. On the Saint Gallen Railway, which crossed the Sitter, Glatt, and Thur rivers, the French firm Schneider installed lattice spans (1853–1856) over three deep valleys in Switzerland with a different method. In charge was preeminent railroad engineer, architect, administrator, and editor Karl Etzel (1812–1865), who would work on the railways of Württemberg, Austria-Hungary, and Switzerland; his background (like Ithiel Town's) included carpentry.[113] Etzel's decision-making process for the bridge over the Sitter (the first to be built, figure I.6) gives us an idea of the options from which engineers could choose. He rejected a suspension bridge as too difficult to maintain and a stone arch as too expensive and time-consuming. A Howe-lattice, preferred at first, was eventually rejected because wood maintenance, along with periodic tightening of the tension rods, high over a valley where winters could be severe, was seen as too challenging. A lattice in iron was settled on.[114] During the construction of the Sitter, Glatt, and Thur bridges, lattice frames were launched onto a temporary bridge supported on the permanent cast iron piers, built beforehand. The lattice frames were assembled at the site, hung from the side of rail cars (one on each side), and wheeled to their final locations. This was done one span at a time. During the construction of the Thur bridge (the last of the three), the first span was temporarily bridged; on installation of the lattice frames for the first span, the temporary bridge was dismantled and moved to the next span; and so on, four times. Modular thinking reduced the cost of temporary supporting structures, as well as the bridge itself.[115]

In 1856, an anonymous "thinking engineer" who claimed that he had no connection to the construction of the Sitter, Glatt, and Thur bridges, which he had visited "like a tourist," described them in the *Allgemeine Bauzeitung*. The level of detail provided, however, indicates that the author was more than an errant sightseer. He appended a "Suggestion for Launching Lattice Frames without Scaffolding," something that apparently had not yet been done for a long-span bridge; the idea appears to have landed with the construction designers of the bridge over the Aar at Bern (Karl Etzel, 1856–1858, figure 5.12).[116] Scaffolding consisted of one temporary pier of pine, which supported the bridge during launching; the bridge served as a crane for the erection of

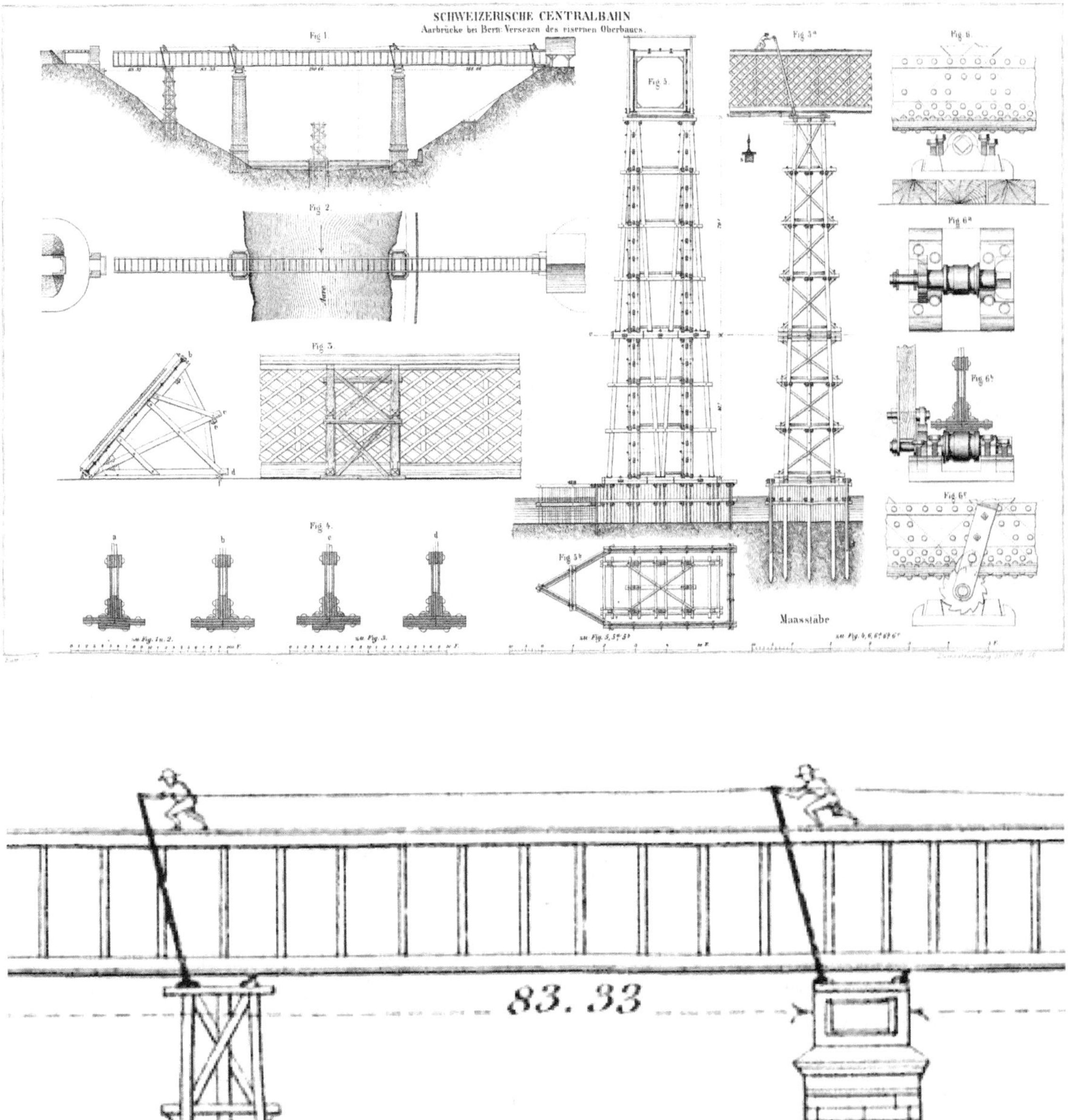

Figure 5.12
Launch of Aar Bridge (Karl Etzel, 1856–1858). Workmen (see detail, below) advancing the structure over permanent and temporary, movable piers. (The lattice is not shown in the bridge elevation drawing.) R. Merian, "Gitterbrücke über die Aare bei Bern," *Eisenbahnzeitung* 16 (1859): pl. 6. Deutsches Museum Library.

Figure 5.13
Construction of Grandfey Viaduct (Wilhelm Nordling and construction engineer C. F. Mathieu, 1857–1862) over the Sarine, Freiburg, Switzerland by the firm Schneider (Le Creusot), see figure. 8.14. *L'Illustration*, June 21, 1862, 392. University of Michigan.

the permanent pier. As the launch advanced, the temporary pier was moved two times, from the middle of the first span to the middle of the second, and then to the middle of the third, with the help of a movable crane running on tracks on the bridge. Above each pier workers operated rollers with a ratchet (comparable to jacking up a car, but horizontally instead of vertically); for the final span, twenty men pulled with "moderate effort" on each of five devices to move the 1,800,000-kilogram superstructure, which they advanced 15 to 18 meters per day.[117] Launching meant that the structural assembly work could be completed on land, rather than high in the air. Workers connected the lattice frames with floor beams (at the bottom) and wind bracing (at the top) and launched the entire tube-like bridge.

For the construction site of the Grandfey Viaduct (Wilhelm Nordling and C. F. Mathieu, among others, 1857–1862) over the Sarine (Saane) near Fribourg (Freiburg)

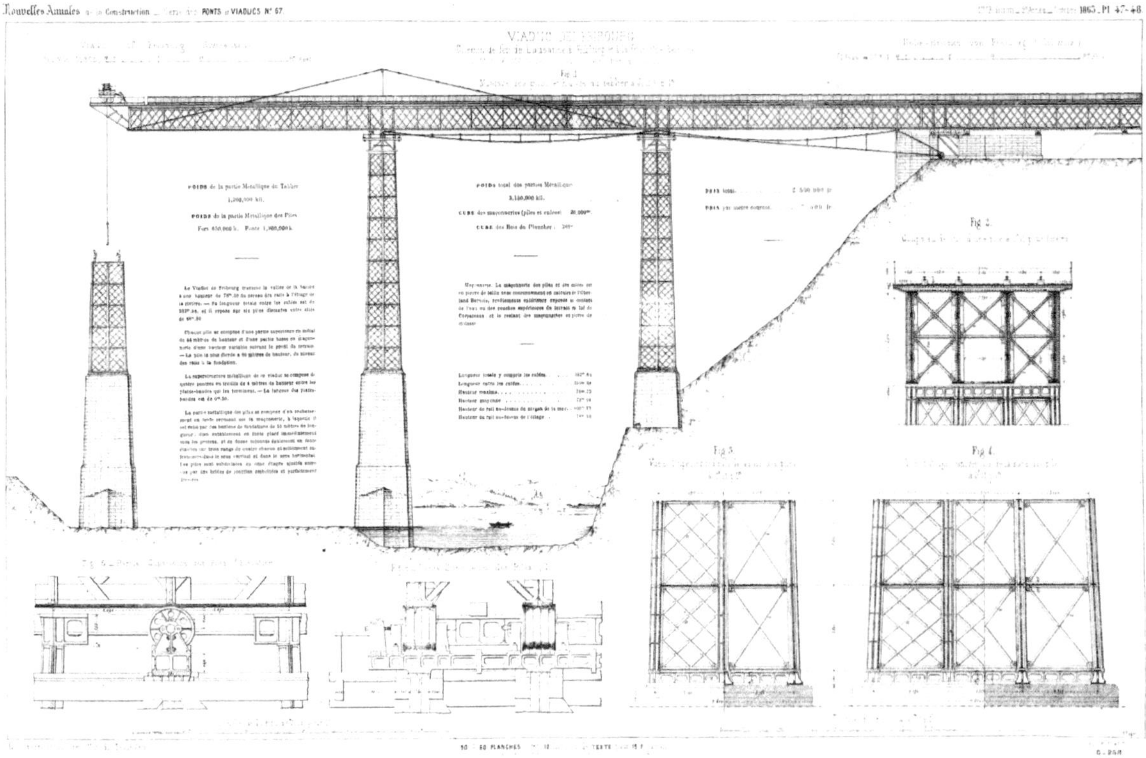

Figure 5.14
Launch and construction of third pier of Grandfey Viaduct. *Nouvelles Annales de la Construction* 9 (1863): pls. 47–48. University of Illinois at Urbana-Champaign.

in Switzerland on the Lausanne to Fribourg railway line, engineers made the bridge a machine that built itself (figures 5.13 and 5.14). They dispensed with scaffolding and used the bridge as a crane. Because of its great height, they rejected wooden scaffolding as too time-consuming, costly, and dangerous. So workers built the length of the first three spans on land (and continued to build it even after the launching phase had begun); the superstructure was supported by temporary cables and chains to prevent excessive downward bending of the forward portion of the bridge as workers launched it. Using a winch composed of three small gearwheels and three large cogwheels, sixteen men took six hours to move each structure 48.80 meters, that is, the length of each span (eighteen men, eight to ten hours was also reported). The deck was designed to be able to cantilever the full distance between supports; this enabled the bridge span to be used as a crane for the construction of each pier; a temporary beak added to the front gave it extra length. Each time the superstructure reached a pier location, workers sent a rail car loaded with pier components to the front end, where they lowered them into place with a winch and built the pier up to the height required to support the bridge deck. Temporary rollers were installed at the top of each completed pier so that workers could advance the deck to the next pier location. As described at the time: "The parts come from the factories by railroad right up to the assembly site and are then carried to their final location without inaccurate moves and almost without the use of motor force."[118]

Ferdinand Mathieu, with the firm Schneider (Le Creusot), was the engineer in charge of construction; he apparently designed the building process.[119] In 1862, Nordling reportedly invited Gustave Eiffel to the construction site in order to introduce him to launching. This was years before he hired Eiffel to construct the two viaducts that launched Eiffel's business career (see chapter 8).[120]

With the launch strategy, designers and builders synthesized structural, mechanical, and logistical knowledge that they had accumulated and applied it to the construction of long-span bridges in challenging environments. Assembly on solid ground was easier and safer than working in the air over waterways and valleys; costly scaffolding was reduced or unnecessary; there was less impact on the environment beneath the bridge. With the superstructure constructed as one continuous piece, designers avoided the labor and maintenance issues that accompanied the making of connections among bridge segments. A continuous structure, rather than a series of single spans supported at each end, enabled the design of lighter bridge decks, as a continuous span does not have to be as deep (that is, tall). It can be longer between supports, because bridge spans that are continuous over piers deflect downward (sag) less between those piers. To this day, these remain advantages of launched bridges.[121] Using the deck as a crane for pier construction reduced the need for additional power, by taking advantage of the force of gravity: parts were lowered into place rather than lifted from below.[122] Firms could re-use launching devices in multiple projects, unlike scaffolding, which was usually custom-built at and for each site.[123]

Launching was carried out in at least three variations; winch on the abutments (Grandfey Viaduct, 1857–1862); winch on the bridge deck itself (Creuse, 1863–1865; Cère, 1863–1866; and Bouble, 1867–1869, figures 5.15 and 5.16a-c); and lever-operated rollers on each point of support (Thur, 1854–1855; Aare at Bern, 1856–1858; Neuvial, 1868). The launch of the viaduct over the Cère proceeded at 8–12 meters an hour. Engineers calculated precisely how many men it would take per crank to apply the force that would move the structure. Installation of one span and construction of one column took about twenty-four days.[124] There was a failed attempt to launch with steam power during the construction of the bridge over the Rhine at Kehl (1858–1861; figure 5.17); in the

Figure 5.15
Launch and pier construction of Cère Viaduct (Wilhelm Nordling, 1863–1866). *L'. Illustration*, August 19, 1865, 121. Collection of the author.

end, two teams of forty men alternately cranked the bridge into place.[125] Hydraulic rollers were used by Gouin to launch lattice bridges over the Scorff at Lorient (1861–1862); over the Loire at Briollay (1862–1863); and over the Vistula at Warsaw (1864).[126] It is unknown how common launching became, but it was reported in the 1870s in France that "a very large number of bridges" had been installed in this manner.[127]

This account does not imply that launching was appropriate for most bridges, or that it was universally accepted. Some designers objected to bridges designed as a single, continuous girder. Although the lattice was a rigid structure, which helped minimize sagging during installation, launching put additional stresses on the lower chord of the bridge as well as its connections; stresses on individual members varied during launching, as points of support changed, and these would be different than the stresses the bridge was designed to undergo for the rest of its service life.[128] Because deformations during installation could compromise structural integrity, engineers had to design launchable lattice bridges to

Figure 5.16a
Three views of construction and launch of Bouble viaduct (Wilhelm Nordling, 1867–1869), Hélios, photographer; pier construction, Berthaud, photographer. École Nationale des Ponts et Chaussées.

sustain expected everyday loads as well as one-time conditions during the launch. Factors such as unexpected winds could cause collapse.[129] The well-known Crumlin Viaduct was not launched; its engineers estimated that they would have had to design a 50 percent heavier bridge to stand up to installation stresses (a conclusion that other engineers did not necessarily agree with).[130] So workers bolted together its tall piers first (figure 5.18) and subsequently lifted truss spans with a steam winch, a method that turned out to be challenging. One worker was killed and two seriously injured during the process.[131]

Drawings and photographs rarely convey the essential role of construction labor. Building is a hazardous job, with threats intensified by pressure to perform quickly, lack of preparation, carelessness, inexperience, (in design, planning, and implementation), and material failure, all amplified by distance above or below ground and by local

Figure 5.16b
Construction and launch of Bouble viaduct (Wilhelm Nordling, 1867–1869), Hélios, photographer. Éccole Nationale des Ponts et Chausssées.

geography, weather, waterways, and disease. Building is still dangerous work. Despite safety codes, in 2019 the United States reported deaths during construction of more than 1,000 people—20 percent of all worker fatalities.[132] Through the nineteenth century, numbers were not widely reported or recorded. Still, when describing bridge construction methods in 1880, engineer Théophile Seyrig, who designed several of the Eiffel firm's most prominent bridges, said, "In erecting girders [i.e., bridge superstructure] the motto ought to be: first, safety; next, cheapness; third, rapidity."[133] How compatible are these three? "Safety," like "American," is a word perpetually under construction, with ever-mounting body counts behind malleable meanings, as we'll explore in chapter 7.[134] The narrative of awe inspired by workers' physical and technical prowess, noted at the start of this chapter, masks conditions for which uncounted workers have paid with their lives.

Figure 5.16c
Pier construction, Bouble viaduct (Wilhelm Nordling, 1867–1869), Berthaud, photographer. Éccole Nationale des Ponts et Chausssées.

Just as mechanized watch production was the result of technologists' creating "virtually a new product," designers reimagined bridges for mass construction.[135] For example, the engineering and construction firm Cail, "thanks to the improvement in its machine tools and precision of its methods," manufactured standard parts, including rivet holes, for the lattice viaduct at Busseau d'Ahun over the Creuse (1863–1865) (figure 8.3), one of several launched under Wilhelm Nordling. The structure required no preassembly in the factory or reworking of parts on site. Each of the tall piers was built atop its stone base within five working days.[136] Uniformity of parts, and eventually precision manufacture, became integral to the merging of structure and machine, not to mention the emergence of high-rise construction. We call them "bridges" before and after, but it makes sense to ask if we should. Engineers such as Ferdinand Mathieu

Figure 5.17
Workers launching the Kehl bridge (Fleur Saint-Denis and Vuigner, 1858–61) over the Rhine. Courtesy of Deutsche Bundesbahndirektion Karlsruhe.

(1819–1895), who worked for Schneider (Le Creusot); Félix Moreaux (1828–1890), who may have played a major role as Nordling's collaborator and who worked for Cail and Fives-Lille (companies that partnered and reportedly built more than 800 bridges in many countries between 1860 and 1870); and Émile Nouguier (1840–1897), who worked for Gouin and Eiffel, were involved in pathbreaking construction projects. The companies built thousands of structures under engineers including Karl Etzel, Wilhelm Pressel, and Wilhelm Nordling, who knew each other, although we know next to nothing about their relationships.[137]

In 1858, the French firm Gouin signed a contract in St. Petersburg to construct lattice bridges for Russian railways. The iron was manufactured and bridges assembled in Paris and shipped 3,000 kilometers to Russia, where work was completed by 1862.[138] When describing this project, engineer Ernest Cézanne (1830–1876) stated:

Figure 5.18
The first pier of the Crumlin Viaduct (Charles Liddell and Lewis Gordon, 1853–1857) (see figure 6.11). Photograph, J. Paul Getty Museum, Los Angeles, 84.XA.871.6.38.

Figure 5.19
Launch and pier construction, Sioule viaduct (Wilhelm Nordling, 1869). Hélios, photographer. École Nationale des Ponts et Chaussées.

> The towing [i.e., launching] of decks offers little chance of an accident; it permits assembly of the truss on solid ground, with a rigor not possible on scaffolding; it requires only simple devices that are easy to transfer from one construction site to the next; as for scaffolding, it allows great economy, by doing away with the need for intermediate supports. It seems to be coming into general use at all construction sites where conditions permit.
>
> A viaduct whose deck rolls over tall piers to which it had offered a point of support . . . is witness to the progress accomplished since the beginnings, and under the influence, of the railroad. The art of the engineer has been transformed.[139]

It's exciting to connect with Cézanne's exhilaration. Imagine standing on one of these structures, surveying the landscape, as workers inch it forward (figure 5.19). Is this where we end the story, at the apex of a progressive arc that has taken us from skilled enslaved Black workers embedded in construction teams to professional engineers managing multinational industrial firms?

PARTS, PROCESS, POWER

The reinvention of building incorporated and propelled an ongoing development that would eventually become unexceptional: the manufacture of standard parts. Standardization is usually regarded as a global imperative, but it may be more helpful to consider how in each case a standard was a choice of participants in the development of industrial practices intended to achieve goals related to, for example, schedule, maintainability, or profit. In the swelling wave of bridge and nation building, the possibilities, flexibilities, and constraints of uniformity and standardization came to configure the work of designers and builders.

The challenges and narratives of standardization were on conspicuous display in London in the 1851 Great Exhibition of the Works of Industry of all Nations, whose major focus, despite its name, were products made in Britain. The exposition, housed in a Crystal Palace that appeared to announce radical change in building, provided rich opportunities for citizens of industrializing regions to see what the competition was up to, while reinforcing stories about progress that merged technology with national identity and pride. (In 2025, Wikipedia still tells us that the design of the Palace "embodied the spirit of British innovation.")[140] Engineering works such as bridges, dams, ports, and lighthouses were underrepresented at the fair.[141] Cyrus McCormick's reaper, Samuel Colt's revolving pistols, and a "full-sized model" of a bridge designed by Nathaniel Rider, made of standard iron parts, provided evidence of US firms' labor-saving innovation and interchangeability. Mechanized manufacture and standardization of parts in US industries attracted intense interest; soon made-in-America machines would be operating in England.[142]

The Crystal Palace (Joseph Paxton, William Barlow, and the firm Fox Henderson, 1851), the Great Exhibition's renowned container, became the historian's barometer of standardization and systematization and an invitation to glorify metal.[143] It was erected quickly using components, many machine-made, that could be added to or stacked, including some that appeared to be standardized but were not; it served to some extent as its own scaffolding during construction. The framing of the glass exterior and the ribs of the vaulted roof of the acclaimed cast-iron-and-glass structure were in fact made of wood, as were flooring, truss components, and critical details such as wedges (made of oak, figure 5.20). (Paint helped hide the large percentage of the structure that was wooden.) The Crystal Palace suffered from dysfunction due to expansion and contraction related to changes in temperature (a condition made more complex by combining wood and metal) and the resulting stress in its connections. At the conclusion of the exposition, when it was dismantled and moved to a new location, the many fixes meant that the re-erected structure might be considered a new Palace.[144]

Flaws that pushed the limits of safety in a temporary world's fair shed would have been particularly risky in a railway bridge. So let's consider two rapidly built wood-and-iron structures, each aspiring to be composed of standard components: the still widely

Figure 5.20
Crystal Palace (Joseph Paxton, William Barlow, and the firm Fox Henderson, 1851). During construction, workers prepared lumber within and for an iconic "iron" structure that contained a significant quantity of wood. Metal girders were lifted into place, a practice employed with lattice beams. *Illustrated London News*, January 11, 1851, 25. UC Berkeley Library, University of California.

celebrated "iron" (and wood masquerading as iron) Crystal Palace, which opened May 1, 1851, and remained standing for about half a year, and the rarely remembered "wooden" (with 49 tons of wrought iron connectors) Portage Bridge, which opened August 14, 1852, and was destroyed twenty-three years later (figure 5.21; see chapter 1). The latter appears to have rigidly served its purpose while the former performed much more briefly and shakily. Although promoters and historians (and architects, particularly in the twentieth century) considered the Crystal Palace an object for (national) commemoration, it is less clear that it was a meaningful model for structural and constructional designers. The Portage Bridge seemed to have been an exemplar for builders, but storytellers relegated it to obscurity. Workers built each trestle (pier) by lowering wood members from a crane on the deck; then installing a new portion of deck truss (possibly built in advance), deck

Figure 5.21
Images and reports of the railroad bridge (Silas Seymour, 1851–1852) at Horseshoe Falls, Portage, New York, broadcast constructional and structural possibilities to an international audience. Library of Congress.

floor, and rails; then advancing the crane on the rails to continue.[145] To this day, the Crystal Palace remains a touchstone—perhaps more accurately, a touchiron—an artifact of delight and study for architectural historians focusing on industrialization.[146] Regarding the Portage Bridge, however, the inapplicability of the concepts "architecture" and "style," along with its apparently wooden nature, meant that its historical cancellation would be almost complete despite its visibly high degree of modularity; reportedly, any part could be removed and replaced without affecting the whole.[147] The narrative palace guard would not direct our attention to developments in uniformity with wood and in the woods. Material (iron, not wood) and location (urban, not rural), an association with technological heroes and national progress, and proximity to art historical values: these factors determine why one structure becomes a star while another fades away. Storytellers have the alchemical power to turn wood into metal and vice versa. They determine symbolic performance and historical longevity.

Historians once claimed that the application of the term "American system" to manufacturing, characterized by interchangeable parts, emerged at the 1851 fair, but as historian David Hounshell has observed, the etymology of the term is hazy and its meaning unclear.[148] The "system" seems to have been as much American as British, French, or German, to mention only three nationalized contributors.[149] Long before the erection of the Crystal Palace, de l'Orme's dome, Town's lattice, and Timothy Palmer's arches (figure 5.22) were designed with an overriding objective: ease of construction. As Mahan wrote during the 1830s, "Simplicity should be regarded as an essential condition, so that any part may be easily taken out, and be replaced, without deranging the rest." For builders focusing on the rapid construction of a large quantity of structures, uniformity and portability were key.[150]

Michel Chevalier observed:

> Lattice bridges are made up of members all of the exact same model and of the simplest form, so that, through completely elementary mechanical processes, nothing is easier than to cut all the diagonals with perfect precision, and, with no less precision, to bore the holes which will receive the pegs.[151]

Thousands of holes needed to be bored, but that was rote work compared to the skilled carpentry required to make traditional joints.[152] (When carpenters crafted the latter, they had to fit the pieces together ahead of time, disassemble them, and then put them back together onsite.) Chevalier recognized the elements that defined the reinvention of building: simplicity, a deskilled work force, precision, and uniformity if not interchangeability of parts and materials. These principles remain the focus of designers whose goal is to create

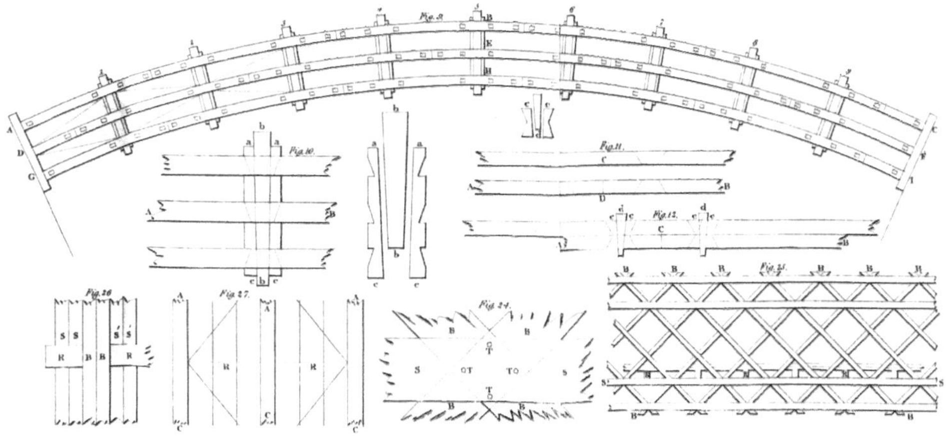

Figure 5.22
Portsmouth Arch Bridge (also called Piscataqua Bridge, Timothy Palmer, 1794). Depicted here with the lattice bridge is a structure seven miles from Portsmouth, New Hampshire known for its long span and inclusion of uniform, wedgeable parts; Palmer also built the Permanent Bridge (figure 3.2). Howard Douglas, *Essay on the Principles and Construction of Military Bridges*, 2nd ed. (Boone, 1832), pl. 11. Courtesy of Sutro Library, California State Library.

large, lightweight, high strength, rigid structures with a small number of small part types (figure 5.23).[153]

Standardization was (and is) seen as both liberating and inhibiting—it opened new possibilities as it closed others. The creative potential of standard parts is balanced by the limitations it sets regarding what might be built, how, and where the profits will flow. Standardization made things less (apparently) complex and put boundaries around professional practice in design and construction. It offered a menu of choices to experts grappling with the unique conditions of each project along with general conditions shaping many. It altered the balance by placing economic and engineering dominance in fewer hands. It enabled corporate managers to pursue ambitions on a global scale.[154] The "industrial liberty" that Saint-Simon fought for in the Revolutionary War was financially remunerative for industrialists, but it did not necessarily enhance the freedom of workers or most inhabitants of industrializing nations.

Was launching as transformational as Cézanne believed? It might be viewed as a performance of the lattice's promise: the reinvention of beam, truss, bridge, and tower construction. Launching's mobilization of strategies for precise planning and management, including structural design and analysis, manufacture, transport, preassembly, and efficient use of power sources, illuminates why the lattice can be said to rival the Model T as a transformative industrial phenomenon. Rather than a single solution, however, the lattice

Figure 5.23
Prototyping of high-performance structures for deep space. Scaling Omnidirectional Lattice Explorer (SOLL-E) robots perform simple, precise, repeatable tasks in order to build at a large scale. NASA / Dominic Hart.

represented flexibility. Builders might assemble a structure at a distance, on-site, or some combination. When, for example, the Oderberg and Kaschau Railroad in Austria-Hungary replaced twenty-eight bridges of wood with iron lattice bridges, launching was employed so that each replacement would take only a few hours (figure 5.24). Some bridges were assembled on a platform adjacent to the existing bridge and then moved sideways into place.[155]

Business historian Alfred D. Chandler Jr. concluded that the building and construction industries in the nineteenth and early twentieth centuries remained "traditional": "There was little need to build a complex organization to coordinate the flow of goods from one process to another. These industries long remained labor-intensive."[156]

Figure 5.24
Tursocz Bridge over the Waag River, 1871. Alexander Schild, photographer. Albertina Museum, Vienna.

Historians and economists excluded bridge building and mass construction from the industrialization narrative, which "became synonymous with mass production."[157] They awarded a fixed facility, the factory, a central place in industrial history. Bridge building, however, was temporary and contingent (although a long-term project sometimes included the establishment of a temporary town for workers and families at the construction site). The manufacture of bridge parts was factory-dependent; specialists who designed and constructed bridges worked in close coordination with iron manufacturers such as Le Creusot.[158] Building did conform somewhat to Chandler's criteria for mass-production industries: "capital-intensive, energy-intensive, and manager-intensive."[159] But essential was the ability of mass constructors to adapt to changeable conditions impacted by economy, topography, materials, and labor. Flexibility, not "industrial" in terms of evolutionist narratives describing the nineteenth century, might be considered a prime industrial value in the twenty-first.[160] We can appreciate this by thinking of the over four million "lattice" telecommunication towers built in recent years, with millions more projected globally. While ongoing construction at this scale attracts little attention, technologists continue tower research and design, as they did with the lattice bridge.[161]

Regarding the application of new power sources to production, lattice bridge construction collides with standard evolutionist conceptions of industrialization.[162] Historians and corporations pushed us away from a celebration of the manual—it was too close to craft. It didn't help sell more machines and the fuel that makes them move. Just as the lower rungs of technology's evolutionist ladder are made of wood, muscle power was seen as less sophisticated, if not less civilized, and more working-class and immigrant, if not redolent of slavery, than machinery driven by steam or electricity. Today, we might consider the application of human power to launching as a tour de force of environmental sustainability. For some challenges, people work better than steam.[163] The bridge launch, an ingenious synthesis of diverse knowledge and skill types, enabled the efficient deployment of machines and people. Workers kept the enterprise in motion, with nationality, race, and gender overriding factors in the roles they were allowed to play.[164] Engineer John Gregory, who left Ireland to settle in Wisconsin, noted a study by scientist Adolphe Quetelet on the relative strength of workers (figure 5.25). Gregory believed that "uneducated brute force can effect little, as compared with the same amount of power under the guidance of scientific skill."[165]

"There are many kinds of mass production," confirms historian Ken Alder. Designers, managers, and workers blended the traditional and novel, and muscles and machines, in order to achieve construction goals.[166] Their reinventions, told and untold, remade the world, bridge by bridge.

OF WISCONSIN. 251

main at home. Every working man will improve his condition by leaving Ireland and coming here. The Irish laborer is well adapted to the laborious work of this country. When he is well fed on good American fare, he proves himself to possess a greater share of animal power than perhaps any other foreigner to be met with here, which is an additional proof of Professor Forbes' observations respecting the relative strength of natives of different countries in Europe.

Professor Quetelet tried experiments in the University of Brussels, to ascertain the relative strength of the students, as indicated by pulling out the stem of a spring of a dynamometer :

	Average height, inches.	Aver. weight, lbs.	Av. strength. lbs.
English,	68½	151	403
Scotch,	69	152½	423
Irish,	70	155	432
Belgians,	68	150	339

In corroberation of the above, an eminent engineer in London had occasion to ascertain the relative animal powers of English and Irish laborers in raising weights by means of a crane, and he found that the utmost effort of a man lifting at the rate of one foot per minute, ranged as follows :

Englishmen,	from	11,505 lbs.	to	24,255 lbs.
Irishmen,	"	17,325 "	"	27,562 "
Welshmen,		11,112 "	as utmost effort.	

These results prove that where only mere animal power is required, no one can be found to surpass, or indeed equal, an Irishman.

As connected with the subject of labor, I might mention that uneducated brute force can effect but little, as compared with the same amount of power under the guidance of scien-

Figure 5.25
Worker strength. Products, machines, and power sources, including human, were nationally comparable. John Gregory, *Industrial Resources of Wisconsin* (Langdon & Rounds, 1853), 251. University of Wisconsin.

6

ENGINEERING NATIONS

A name makes one technology appear to be viable, if not a triumph, while another fades away. A name, which defines a category or makes a reputation, confers professional distinction, generates wealth, and produces international reverberations.[1] That's why being seen as the inventor is a story to struggle for. Technological categories rest upon a narrative infrastructure that blends tangible objects with intangible cultural claims. "Technology transfer," for example, describes invention originating within individuals bounded by a (national) culture and then spreading like the rays of the sun into other places.[2] Within this universe, national tags such as "British" and "American," material identifications such as "wood" and "iron," and structural types such as "lattice" and "tube," enable storytellers to answer fundamental questions about people and technologies before asking them. Categories enable technology's raconteurs to dispense with credible evidence as they sow creation myths and reap economic, political, academic, and cultural power.

Here, we'll forsake the traditional tale in order to probe invention as a meaning- and identity-building, rather than -confirming, activity. We'll view bridges as products of "discourse and physics, myth and materiality," in the words of historian Amy Slaton.[3] By highlighting the ambiguity of categories and recognizing their cultural roles, we'll see technologists producing and interweaving engineering knowledge, national identity, wealth, and prestige. Treating taxonomy as fluid rather than fixed helps us begin to dismantle nationalist and evolutionist accounts of technological change.

Inventing and promoting a technological type means creating perceptions. A new story paves the way for acceptance. While bridge classifications may appear to be as solid and incontrovertible as bridges themselves, classifications are shapeshifters. The equivocality of bridge types rests upon interpretations that identify or neglect material, structural function, appearance, analytical theory, designer, manufacturer, or place. Inventors' and entrepreneurs' perspectives, founded upon their own interests, may be intensified by colleagues, historians, and popular media; these perspectives then shape ours. Since classifications are so potent, we might endow our approach with the label Critical Engineering Theory.[4]

This chapter examines debates among the technologists who reinvented building through the erection of thousands of lattice bridges in Europe, the United States, and elsewhere, and the developers of the tubular bridge, who built five, in Wales, England, and Canada. We'll locate a "British" icon, the tube, within the context of its creation: a multinational landscape shaped by an "American" invention, the lattice. We'll see creators, clients, and commentators creating categories as they define their own and others' professional, technological, and national boundaries.[5] Decision-makers considered an apparently local question—"Which type of bridge to build?"—as technologists transnationally devised solutions that altered building possibilities. Much of this revolved around a quintessential constructive component, the beam.

INVENTING THE TUBE

Today, most chroniclers present the lattice as a superseded bridge type—at best a step in an evolutionary scheme of technological development. The Britannia and Conway tubular bridges, on the other hand, appear in numerous publications as a remarkable historical engineering achievement. Like the American bridges, the rectangular wrought-iron-plate tubes enabled trains to pass through their interiors. Robert Stephenson (1803–1859), one of the most lionized of Britain's nineteenth-century engineers, developed the tubular bridge type with William Fairbairn (1789–1874), an equally revered engineer with extensive experience in shipbuilding and the use of wrought iron.[6] The two bridges, each of which consisted of a pair of parallel tubes, lay about fifteen miles apart on a line that enabled travel between Dublin and London. The construction of the bridge over the Conway River lasted from May 1846 to December 1848; it was a test case for the Britannia, which opened in March 1850, after four years of work (figure 6.1). There was no precedent in a bridge-building project for the use of wrought iron on this scale or for such long spans in any material.[7]

What was on Stephenson's mind as he gave shape to the bridge project? The goal was to span the treacherous currents of the Menai Strait without obstructing shipping with temporary scaffolding or elements of the completed bridge. After the Admiralty rejected his plan for a cast-iron arch bridge, Stephenson, according to his own account, considered how a suspension bridge roadway might be made rigid enough for railway traffic; he considered several bridges in which lattices were used as lightweight railings.[8] Stephenson studied, too, a canal aqueduct built by engineer John A. Roebling; he objected to its lattice structure (in combination with a suspension system) because he believed that "a much stronger and more ponderous system" was required, and because "the direction and amount of the complicated strains [i.e., stresses] throughout the trussing become incalculable." (The stresses would be just as incalculable in a tube with solid sides.) Finally, he judged wood "inadmissable" because of concerns about durability and flammability.[9]

Figure 6.1
Britannia Bridge (Robert Stephenson, William Fairbairn, et al., 1846–1850) with Menai Suspension Bridge (Thomas Telford, 1819–1826) in the distance. *Illustrated London News*, March 23, 1850, 192. UC Berkeley Library, University of California.

During the period that Stephenson settled on the tube, American bridges were the most widely built bridge types in tubular, rectangular form. The renowned late-eighteenth-century box-like bridges built in Switzerland by the Grubenmann brothers (see chapter 2) would also have been familiar to Stephenson.[10] While consideration of structures such as the lattice were instrumental to Stephenson's inventive process, as they were for many bridge-system designers of the time, he did not devise a structure that would be named a "truss." According to his own account, he thought of a proposed wrought-iron plate bridge over the River Lea at Ware, and then mentally went back to a bridge designed by Roebling, "substituting for the vertical wooden trellis trussing, and the top and bottom cross braces, wrought-iron plates riveted together with angle-iron. . . . The first arrangement . . . of the tubular structure was exactly similar

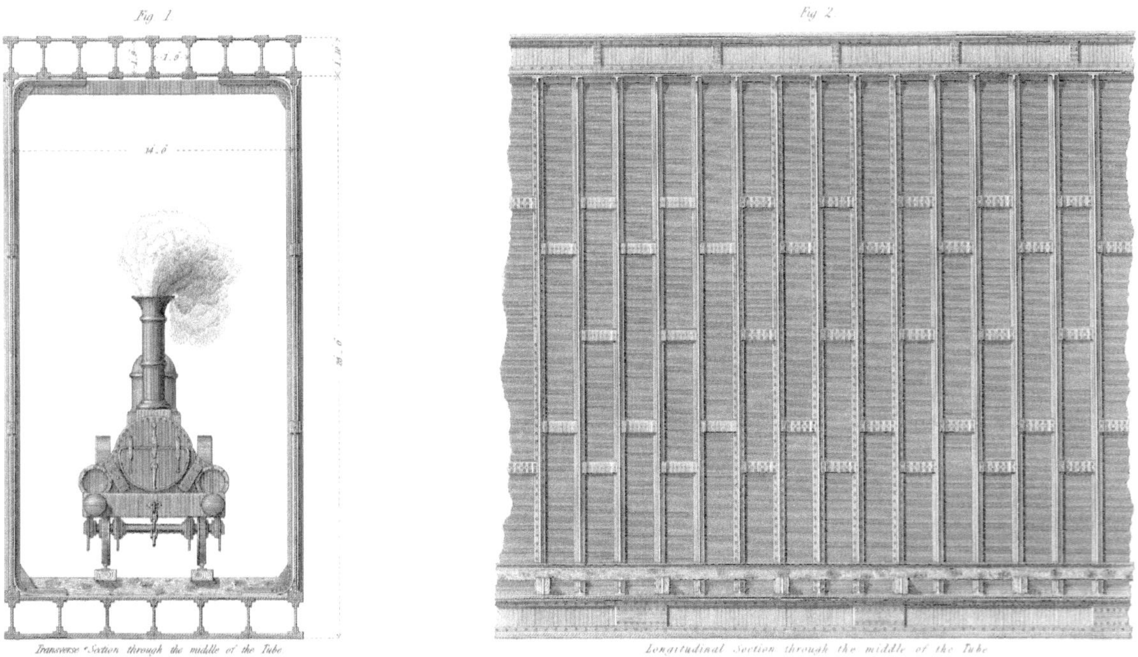

Figure 6.2
Britannia Bridge, section, and elevation. William Fairbairn, *An Account of the Construction of the Britannia and Conway Tubular Bridges* (John Weale, 1849), pls. 4 and 7. University of Chicago Library.

in form to the trellis trussed wooden design . . . but it was evident that the action of the top and bottom of the tube, composed of thick wrought-iron plates, would be infinitely more efficient than the top and bottom braces, whose duty was chiefly to keep the side trusses in their vertical position."[11] Stephenson reimagined the "trellis" (that is, lattice) as a solid plate that would form the walls of a "tubular structure" (figure 6.2).

TUBE VERSUS LATTICE: DEVELOPMENT AND DISCOURSE

Members of the tubular analysis, testing, design, and construction team included mechanics of materials expert Eaton Hodgkinson (1789–1861), along with Fairbairn, Stephenson, and supervising engineer Edwin Clark (1814–1894), who told the tale of the design and construction of the Britannia Bridge through publications and lectures.[12] Questions of technological authorship—which man deserved the credit?—led to detailed public accountings, as well as Stephenson's ejection of Fairbairn from the project. It may be that Stephenson had conceived of the concept of a wrought iron tube, but he was far less than honest in acknowledging the role of others, especially Fairbairn and Hodgkinson, in turning the vision into a viable project.[13]

From the beginning, lattice and truss bridges loomed behind the tube. At a public lecture in London on the last day of 1850, soon after the Britannia Bridge's opening and five months before the inauguration of the Crystal Palace (see chapter 5), engineering professor Edward Shickle Cowper of King's College London gave a public lecture in the still-under-construction exposition building, during which he showed the capability of a tube (with reference to the tubular cast-iron columns of the exposition building) by piling weight on a board until the (tubular) quills supporting it failed; the demonstration "drew forth large applause." He rolled a sheet of tin to demonstrate how "weak" a "trellis-girder" would be in comparison to a tube.[14] As we'll see, this appraisal would be at odds with the response from most of the professional engineering community. An examination of the tensions behind the development and reception of the tubular bridge reveals the channels of discourse that shaped it: technological and national, professional and personal.

The chief inspecting officer of railways General Charles William Pasley (1780–1861), who approved the tubular project when Stephenson was examined before the House of Commons in May 1845, found Stephenson's initial scheme for a cast-iron arch bridge to be "crude and ill-digested . . . and either impracticable, or nearly so"; he proposed instead that Stephenson build a suspension bridge "trussed to advantage by wrought-iron lattice work, such as Sir John Macneill had adopted for passing the Dublin and Drogheda Railway over the Royal Canal near Dublin" (figure 6.3). He misgivingly approved Stephenson's tubular plan: "Whether it would be better than a latticed or a trussed bridge, I cannot pretend to say."[15] Stephenson and his allies subsequently ignored or discredited lattice and truss structures, which would have included types such as the Howe-lattice. Engineer Richard Boyse Osborne recounted a June 1845 meeting during which he showed Stephenson a ten-foot model of an iron Howe bridge. Osborne was flabbergasted: "He pronounced my model a Lattice!!" "Mr. Stephenson had evidently '*Tube on the brain*,' he would investigate *nothing else*."[16]

The tubular engineers attempted to eliminate the lattice as a category of long-span bridge construction. In 1847, five people had been killed in the collapse of the River Dee railroad bridge at Chester near the English-Welsh border. The bridge, which contained cast-iron beams reinforced by wrought-iron bars, was designed under Robert Stephenson, who had pushed boundaries with the longest beam of this type. Stephenson presented false and misleading evidence in order to avoid liability for a structure designed with insufficient capacity and made of parts whose quality was questionable.[17]

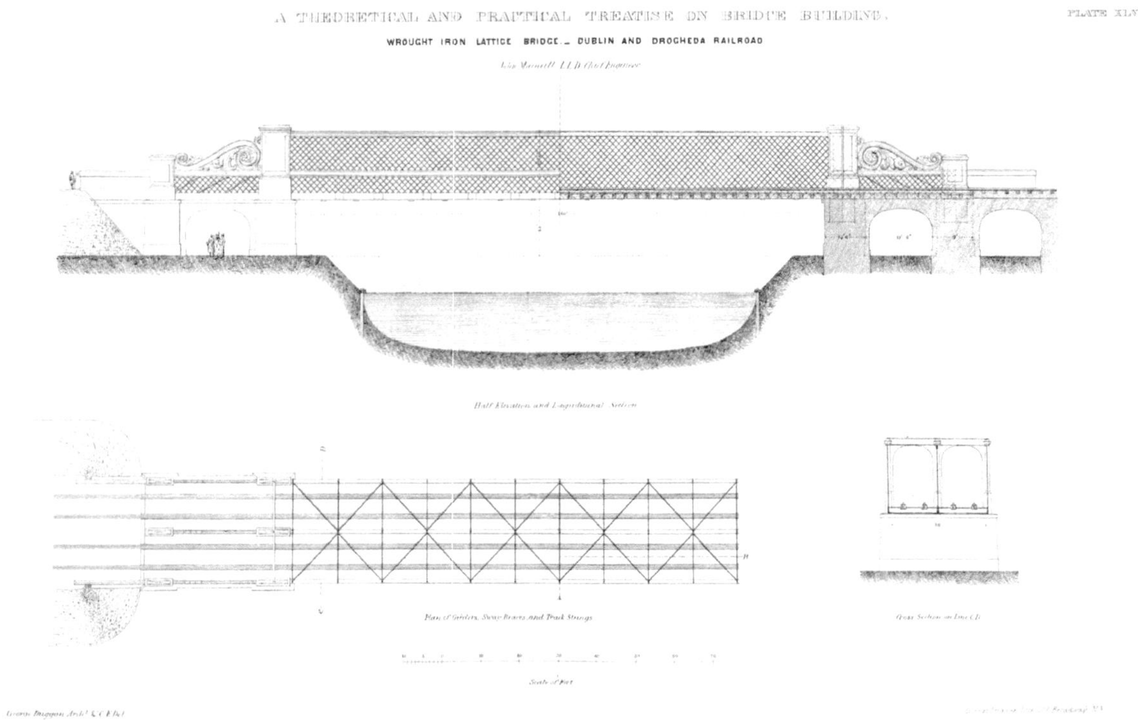

Figure 6.3
Bridge over the Royal Canal near Dublin (John Macneill, 1844–1845). George Duggan, *Specimens of the Stone, Iron and Wood Bridges, Viaducts, Tunnels, Culverts, &c. of the United States Rail Roads* (Appleton, 1850), pl. 45. UC Berkeley Library, University of California. Also in Julius Pollack, "Metallbrücke zur Uebersetzung des Royalkanals bei Dublin für die Dublin-Drogheda-Eisenbahn nach Town's nordamerikanischem Konstrukzionssystem gebaut im Jahre 1845," *Allgemeine Bauzeitung* 13 (1848): pl. 148.

The inquest led to a broader investigation, published as *Report of the Commissioners Appointed to Inquire into the Application of Iron to Railway Structures* (1849).[18] Stephenson, his reputation in jeopardy, answered about 140 questions on March 16, 1848. At the very end of the interrogation, the subject turned to the lattice.

> *Have you had any experience as to the stability of lattice bridges?*
>
> —No, I have never used them, because my experience in railway bridges has always led me to something more substantial than wood. [Stephenson then quoted a Mr. Stevens of New York on the problems of wooden lattice bridges and stated that he preferred to use wood in "whole logs" rather than "slices."]
>
> *Does that objection equally apply if you make lattice work of wrought iron, on something like the principle of Sir John Macneill's?*

—No, it does not apply certainly; but a new difficulty there starts up, that it is utterly impossible to convey compression through that bridge when made of thin bars of iron; the base upon which the strains have to pass is so narrow, that it becomes, to use a north country expression, "wabbly"; the lattice bridge is the same as the sides of the tubes in the Conway [which was under construction]. Suppose you went on with a lattice work, until you closed it all up; it then becomes the sides of a tube.

There is no provision for the compression of the top of the lattice girder?

—Yes, that is just the objection. At Dublin, Sir John Macneill has remedied that by putting a cast-iron top to it.

[The Witness withdrew.][19]

Stephenson acknowledged the structural kinship of American bridges and his tube. But his strategy was to construct a tubular category distinct from the lattice, which he portrayed as irreparably defective, while casting aspersions on early iron lattice designs. He associated the lattice with wood, which branded it flammable and obsolete. When confronted with the existence of an iron lattice, he noted the use of cast iron. The failure of a cast-iron beam, however, was the cause of the Dee Bridge collapse (figure 6.4), which led to the royal investigation; a jury had subsequently declared cast-iron bridges "unsafe." So Stephenson's portrayal of a lattice bridge with cast-iron members may have been dubiously regarded. Most subsequent analyses indicate that Stephenson was most likely responsible for the collapse, although there were at the time a variety of understandings about why the beam failed. Our tendency today may be to provide excuses for engineers during an era of structural innovation driven by a railway construction boom, but if we look at the context, what was known, what engineers cared about, and what governments did and did not require, it may make us quite a bit more cautious about forgiveness.[20] Many of the figures involved were both engineers and entrepreneurs, with conflicting responsibilities to public, investors, and themselves.

The extensive tubular bridge design-phase testing program that Fairbairn conducted in 1845–1847 with Hodgkinson became an essential part of tubular legend and a noted event in the history of strength of materials and understanding of structural behavior.[21] Testing focused on the form and proportions of the tubes and the cells that would best provide for rigidity, lightness, and strength; the need for suspension chains; and the strength of iron. For about a year following May 1845, the designers tested cylindrical, elliptical, and rectangular tubes, as well as built-up I-beams.[22] But an astonishing fact remains. The designers of the tube did not test their system against its inspiration and rival, the lattice. Apparently, they did not wonder, as General Pasley had, how the tube compared. Others were investigating the lattice, however: between 1844 and 1853, Macneil was having lattice experiments conducted and built at least five.[23] In April 1844, the month before Ithiel

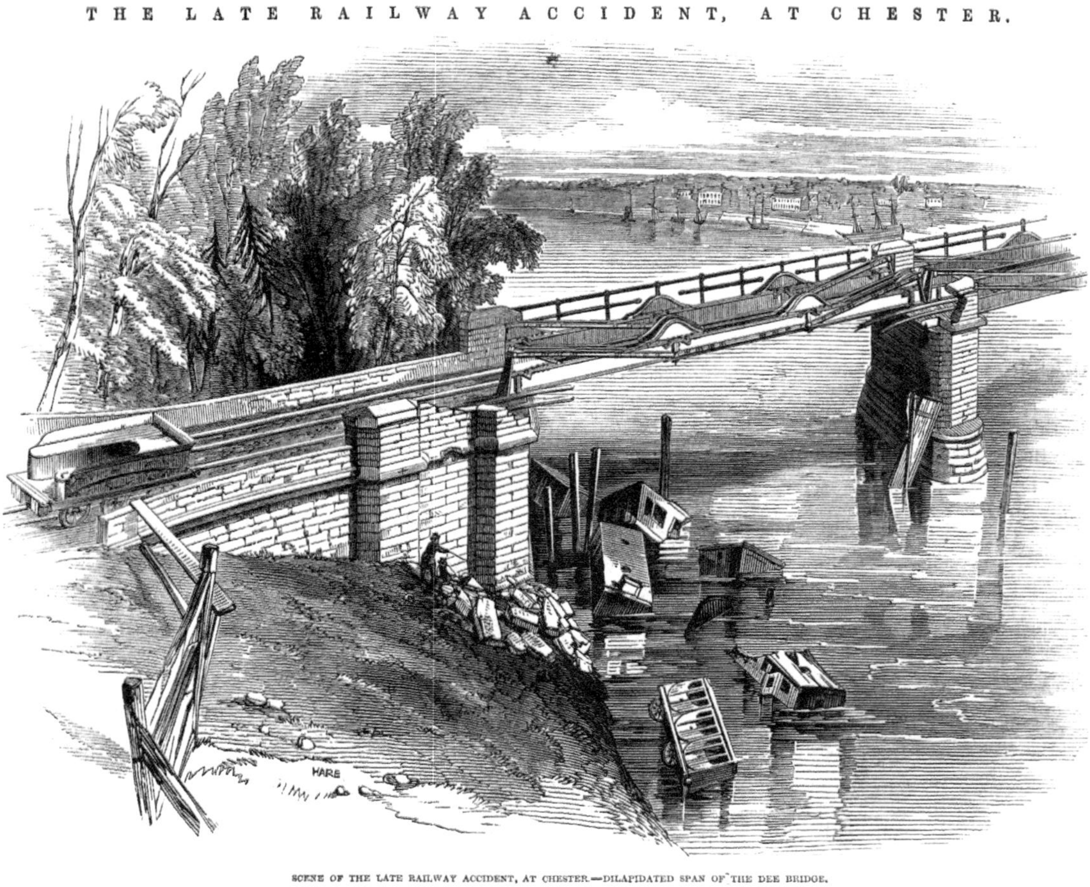

Figure 6.4
Dee Bridge collapse. Hare, *Illustrated London News*, June 12, 1847, 380. UC Berkeley Library, University of California.

Town's death, Macneill's development of the lattice bridge in iron was reported in the Philadelphia-based *Journal of the Franklin Institute*, reprinted from the Edinburgh-based *Proceedings of the Royal Scottish Society of Arts*. This summary of engineer James Thomson's presentation, which included "numerous drawings and models," shared information about testing, design, and construction. Tests appear to have begun in 1843, if not before.[24] Subsequently, however, historians adopted the tubular engineers' perspective. They neglected the context and biases of the tube's development; there was no celebration of the lattice tests. They made the tubular engineers' construction of history everyone's history.

But the tubular engineers' view of their own achievements did not correspond with contemporary technologists' perspectives. In an exploration of reactions to the tube in Britain and the United States, historian Stanley Smith wrote, "Within 15 years the tubular solution had started to fall into disfavor, and after 1860 very little more is heard about large span structures of this type." Smith claimed that "change . . . had taken place with regard to economics" during the years after the construction of the Britannia. Money was certainly an issue, but engineers were dubious from the beginning about the enormously costly and controversial project.[25] One month before Stephenson's testimony, quoted above, John Macneill of Ireland, who formerly was one of Thomas Telford's assistants and enjoyed a status comparable to Stephenson's, had reported positive outcomes regarding the testing and the construction of lattice bridges.[26] Between 1846 and 1851, Fairbairn was in the running to build a bridge over the River Boyne at Drogheda in Ireland, a project overseen by Macneill; Fairbairn proposed a tubular bridge and contractors bid on both tube and lattice.[27] Macneill favored and ultimately chose the latter. So, the year after the completion of the Britannia and its attendant attention and jubilation, a prominent engineer rejected the tube. This became the norm.

Although the Conway and Britannia were regarded as unprecedented, they seemed to upend the tenets of industrial engineering. Efficient use of materials and money seemed not to apply. (British engineers since the 1840s had been encouraged to use metal in bridges in order to support local manufacturing; industrialists such as Fairbairn profited handsomely from use of "his favourite material.")[28] Stephenson and his colleagues were attempting to establish standards for engineering practice and British technology that contrasted with the views of his contemporaries, who remained unconvinced that the tube was a viable bridge type. Eugène Flachat, whose importance to railway construction in France rivaled Stephenson's in Britain, wrote admiringly in 1851 of the tubular project and the information its engineers provided regarding the use of iron plate. He did not, however, defend the tube as a type. (His major new railroad bridge, which opened in 1852 over the Seine at Asnières near Paris was a box girder, not a tube, and was considered the turning point for iron bridge building in France.) Karl Culmann compared the tube unfavorably to other bridges. At a July 1853 meeting of the American Association for the Advancement of Science, engineer Herman Haupt criticized the Britannia for not proportioning the thickness of the metal according to the stress it would sustain along its length, which, for Haupt, meant that it contained much "useless material." In 1855, D. I. Jouravsky in St. Petersburg, who only skimmed "the well-known flaws of the system," felt that the tubular builders had made a "grave error in the arrangement of the iron plates in the parts of the tube subject to tension," which would lead to rupture. In his railroad handbook of 1857, Émile With explained the gravity of potential challenges: rust, since the cells couldn't be examined; the tubes presented more of a surface to the wind than lattice bridges, a condition that would loosen the rivets; and the rivets added great weight—30 percent of the total. Alfred Bommart, a French member of the jury at the 1862 international exhibition in London, stated that the iron walls "by themselves offer no resistance to buckling," "they imprison the view while crossing the bridge," and that their "heavy and unsightly appearance [is] completely in conflict with the impression that should result from the nature of the material used." He went on to state that for the last ten years, most metal bridges were constructed of lattice "suivant le mode américain."[29]

Debates among engineers working in Britain had illuminated differing viewpoints that sometimes divided a younger generation of engineers trained in the use of structural analysis from elders who had learned on the job, though some lattice builders, such as William Moorsom, were no longer young. It also seemed to pit Irishmen against Englishmen; the Britannia Bridge enabled trains to proceed to Holyhead, where passengers could board a ship to Dublin, thus intensifying possibilities for Irish-English collaboration and competition.[30] Tension surfaced in a meeting of the Institution of Civil Engineers (ICE) on April 23, 1850 in London, when William Thomas Doyne (1823–1877), who earlier in his career had worked under Stephenson and under Macneill, presented his recently built wrought-iron lattice bridge on the Rugby and Leamington railway. Referring to Stephenson and Isambard Kingdom Brunel (both famous sons of renowned engineers), Doyne noted that the "prejudice amongst engineers against this mode of construction" was based on poorly designed examples. He explained, in detail, the lattice's advantages of "economy and facility of construction," particularly in relation to iron plate (the tubular material). Reacting to obfuscatory arguments that Stephenson put forth, Doyne clarified: "All the parts [of the lattice] perform the same duties as in boiler-plate girders, and are subject to the same rules for calculating their strength." In response, George Parker Bidder (1806–1878), an eminent railway engineer who enjoyed close personal and professional relationships with Stephenson, stated, in short, that the tube was better. He gave his opinion, he said, "with a desire to guard the younger members of the profession from adopting [the lattice] form and disposition of material." Bidder apparently feared that the lattice would subvert, perhaps Americanize, the training of engineers in Britain.[31]

Engineer James Barton, under Macneill, was in charge of the design and construction of the 264-foot-long central portion of a continuous span bridge that would enable a railway to link Dublin and Belfast. This structure over the Boyne River (figure 6.5) near the town of Drogheda, built 1851–1855, considered in Ireland as an iconic achievement and the most prominent iron lattice in Britain at the time, was the subject of a paper presented by Barton that ICE engineers discussed during portions of four evenings in April 1855, just after the opening of the structure. Early in his talk Barton paid homage

Figure 6.5
Boyne Viaduct (John Macneill with James Barton, built 1851–1855). Courtesy of the National Library of Ireland.

"to the genius which designed and executed the Britannia and Conway Bridges." (As noted above, a proposal for a tube had been rejected for this site.) His "anxiety for the advancement of practical science" drove him, however, to claim that the lattice structure was more in line with the action of tensile and compressive forces in a beam, that it required substantially less material than tubular beams, and that it was cheaper than the tubular and even the Warren—a parallel-chord frame bridge whose structure consisted of isosceles triangles; this was the configuration of the Crumlin Viaduct, still under construction, as we'll see below. The Warren required expensive connections for its few, heavy bars in order to prevent buckling, whereas the lattice's numerous interconnections served this purpose. Barton concluded that "the lattice-beam . . . is, for carrying heavy loads over wide spans . . . theoretically and practically superior both to the plate-beam and the Warren-beam" (figure 6.6).[32]

For the Boyne Viaduct, engineers employed "principles of mechanical science" that enabled them to proportion structural members according to the amount of stress each was to bear. They designed tension and compression members differently, and they conducted tests to ensure that their design worked in practice. Rather than design the structure to be uniform along its span, Barton and his team, with Fairbairn's experiments in mind, designed it to respond to the loads it would carry along its length. (Note that the Britannia tubes' height varied along its length to accommodate forces and that it was cambered.) Some consider the Boyne lattice to be one the earliest bridges designed according to stress calculations. This was the view of the London journal *The Engineer* in 1867, when it noted, regarding the "disparaging" and "condemnatory" testimonies about the lattice of Stephenson and colleagues in the 1849 *Report of the Commissioners*, that "men of scientific attainments are not exempt from those foibles and prejudices which some people fondly imagine are confined to the less educated and instructed world at large."[33]

Stephenson took part in the Barton discussion at the ICE, during which rancor occasionally broke through the polite locutions, and the conversation revealed a variety of beliefs about the behavior of forces in a beam. Several of Stephenson's and Bidder's arguments today seem confusing if not deliberately misleading. They seemed misleading to their contemporaries, too. This was a debate about bridges but also about who would be the engineering authority. Startling is Stephenson's seeming misunderstanding of the neutral axis. When a beam is loaded, along its length the top half is in a state of compression (shortening or pushing together) and the lower portion is in tension (pulling or elongation); these increase from the unstressed middle—the "neutral axis"—to the upper and lower portions of the beam (figure 2.9). Stephenson, however, claimed that every "particle" of the web (the middle portion of a beam) might be subject simultaneously to tension and compression. Hodgkinson had published papers on the neutral axis in 1824 and 1830 (it was not then a new concept), and this had reportedly led to its correct use by engineers in England. Stephenson seemed to disavow a theoretical fundamental—one used in the design of his tubular bridges.[34] The cancellation of the lattice by Stephenson, not known as a designer, and his former collaborator, Fairbairn, known to have a poor understanding of mathematical analysis, may have stemmed from the fact that contemporary designers of lattice and truss structures employed methods of analysis that Stephenson and Fairbairn were not comfortable with.[35]

WROUGHT-IRON BEAMS. 449

The geometric considerations, on which the calculations were based, will be understood, on reference to Fig. 6.

Fig. 6.

Diagrams, representing by Areas, the Material required in the Vertical Portion of different Systems of Wrought-Iron Girders.

Relative Areas.

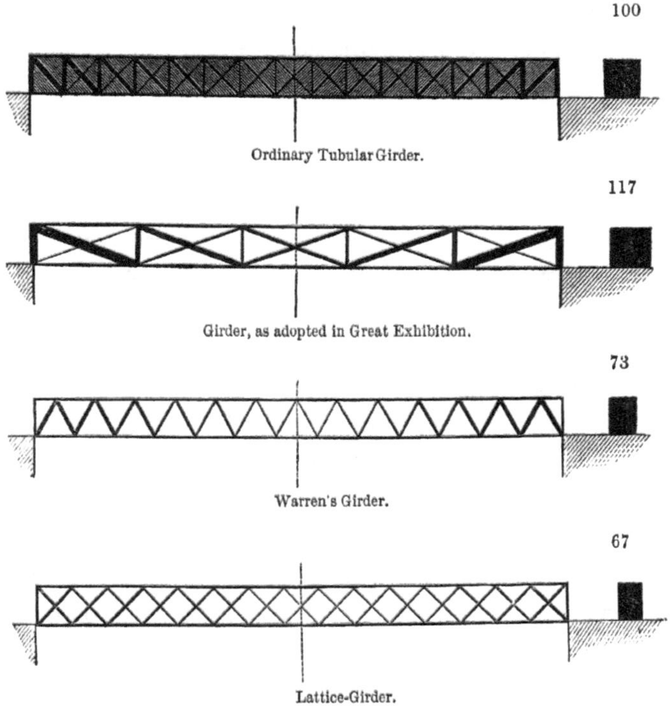

Calling the amount required for a tubular-beam 100, the Warren-beam will only require 73, and the lattice-beam, 67 : or, in other words, there will be a saving of 27 per cent. of the material, by the use of a Warren-girder, and of 33 per cent., by the use of a lattice-girder. These results have been arrived at, by supposing the case most favourable to the tubular-beam, for in the calculation for this beam, the Author has assumed, that the vertical pillars are at such distances, as to allow the diagonal strains to act at an angle of 45°, which is the angle of economy; but the beam, as usually constructed, would give a relative result, still more unfavourable than that which has been stated. The importance, in an econo-

[1854-55. N. S.] 2 G

Figure 6.6

Comparison of the quantity of iron required by each beam type. James Barton, "On the Economic Distribution of Material in the Sides, or Vertical Portion, of the Wrought-Iron Beams," *Transactions of the Institution of Civil* Engineers 14 (1854–1855): 449. UC Berkeley Library, University of California.

As they experimented with novel structures and materials, designers engaged in discussion and debate about how forces (tension, compression, and bending) work in a bridge and the optimal way of accommodating them. Indeed, at that moment, there was a variety of understandings about beams and the forces in the web, which might be built in solid, cell, lattice, or other configurations and in a variety of materials and forms (figures 6.6 and 6.17). And engineers had not yet developed a clear, shared understanding of shear.[36] The outcomes of these debates would help determine cost, safety, and potential market.

Controversies around personal credit swirled around the Britannia and the Boyne; as historian P. J. Geraghty observes, this was about receiving public credit and renown for what was a team design effort. Macneill and Stephenson, respectively, assembled teams whose objective was to design (or invent) a specific type of structure. The two men bore ultimate responsibility for the projects, but testing, analysis, design, manufacture, construction, and promotion were collaborative undertakings. Was the chief engineer really the designer?[37] It can be exciting to imagine Edison establishing in 1876 the "first" industrial research lab in New Jersey, but three decades earlier, entrepreneur-engineers such as Macneill and Stephenson assembled diverse teams of experts to create revenue-generating technologies.

It was through disputes that engineers forged identities. Debates about authorship and practice were about professional status and individual legacies, business opportunities, and national pride. Macneill, who previously was responsible for the construction of at least three smaller iron lattice bridges, inspired by his knowledge of what he called the "American lattice bridge," was unequivocal in his belief in the future of the iron lattice.[38] His repudiation of Fairbairn's tubular bridge proposal for the Boyne in 1851, the year after the Britannia's completion, foreshadowed its imminent demise as a type.[39] Still, opposing Stephenson on his home turf was a daunting battle; Barton had written to Macneill in 1854, asking if the latter might attend his upcoming presentation at the ICE: "Perhaps if you are in London when it comes on, you will go and fight the Stephenson party."[40]

In March 1856, Stephenson defended his final tubular project, the Victoria Bridge (which we'll discuss below) against the lattice, the Warren, and Roebling's Niagara suspension railway bridge. He asserted that the lattices would collapse when disconnected from their upper and lower chords, and thus were adding stress to the top and bottom members.[41] Barton, who had recently completed the Boyne Viaduct, responded to this and several of Stephenson's claims with restrained incredulity. This was not the first time that Stephenson spoke in a way that appears intended to confuse listeners.[42] Roebling had pointed out already in 1855 that the tubular bridge required a vast weight in wrought iron to resist compression, necessary because the material operated much better in tension, which is how suspension bridges employed it: "For extraordinary long spans, the tube cannot compete on any terms."[43] Historians Nathan Rosenberg and Walter Vincenti have acknowledged Stephenson's challenges by emphasizing the unknowns, the pressure, and the novelty of the tubular project; one should cautiously assess the team's decisions.[44] On the other hand, the omission of the lattice in testing and arguments that appear to defy the (engineering) logic of the day, along with the designers' predominance, wealth, and self-promotional efforts, raise questions about historians' assessments of the tubular bridge's impact on engineering. The tube appears to be a peculiar, highly publicized case of bridge building with a substantial materials testing program rather than a turning point in the development of structures or construction. Certainly, Stephenson's team was not the only one taking inspiration (or umbrage) from ships, boilers, and American bridges and producing new forms and knowledge. Fairbairn, Clark, and Doyne all published reports on the testing of rivets that held metal parts together.[45] Baltimore and Susquehanna Railroad engineer and inventor James Millholland built in the winter of 1846–1847 and installed in Baltimore in April 1847 a 50-foot iron-plate box-girder bridge.[46] The fact that technologists had previously developed metal box girder beams might easily appear in the contemporary hype and subsequent nationalist historical haze to be a consequence of Stephenson's tubes! Just as nationalist accounts postdate the lattice relative to other bridges. As we'll see, if the tubular bridge was a turning point, its significance—with its one-off character, bewildering justifications, and monumental physical mass—was as much about the end of an engineering era as the start of a new one.

LATTICE VERSUS TUBE

The story of the lattice and the tube illuminates a decisive moment in the development of industrialized engineering and materials. Is this a traditional story of the success and failure of competing technological types and national engineering cultures? Or does this tale of two bridges depict a multinational collaborative process of invention? To begin to answer, let's first examine how decision-makers judged the bridge categories.

The immense lattice bridge projects carried out (1850–1857) in Prussia over the Vistula at Dirschau and over the Nogat at Marienburg (today part of Poland, in cities now called Tczew and Malbork) (figures 6.7 and 9.2), are comparable in scale to Stephenson's tubes. Karl Lentze (1801–1883), the engineer in charge for the Prussian Ostbahn, who had planned to build two suspension bridges, examined structures in France and England, including the Conway and Britannia bridges, before concluding that a lattice bridge would be lighter, cheaper, and easier to build. Rather

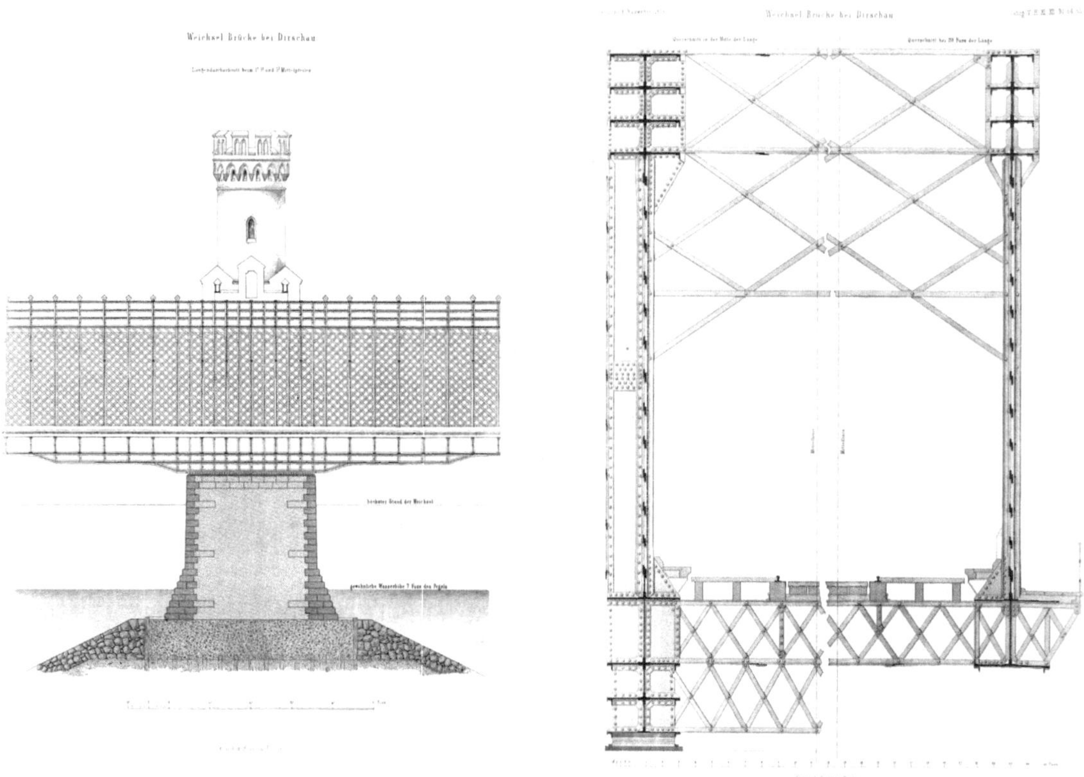

Figure 6.7
Bridge over the Vistula at Dirschau (Tczew) (Karl Lentze, 1850–1857; structural analysis, Rudolph Eduard Schinz), elevation and section. Karl Lentze, "Die im Bau begriffenen Brücken über die Weichsel bei Dirschau und über die Nogat bei Marienburg," *Zeitschrift für Bauwesen* 5 (1855): pls. 63, 65. Courtesy of the Library of the Deutsches Museum.

than replicate the Britannia testing program, the designers in Prussia cancelled a test after analyzing information provided by Edwin Clark in an 1850 talk in London; they employed the theories of Culmann and Schwedler to calculate the design.[47] The Britannia also provided a model of infrastructure as nationalist monument; over the Vistula and Nogat, the elaborate piers, portals, and sculptural reliefs expressed a vision of Prussian dominance. Train travel over the bridge, through neighboring landscapes and past local railway station architecture, constituted an immersive national experience.[48]

Carl Ruppert (1813–1881), a prominent railway engineer who worked in Baden and Austria-Hungary, was responsible for one of the earliest iron long-span lattice

Figure 6.8
Bridge at Cologne (Herman Lohse, 1855–1859). The structure's axis aligned with the cathedral. The lattice's geometric transparency, a pattern comparable to stained glass windows, may have made it more palatable to its ecclesiastical neighbor. One roadway served pedestrians and vehicles; the other, the railroad. Cologne, Rhine, and bridge, ca. 1896. Collection of the author. Bridge and cathedral. Pedestrian roadway, with cathedral visible. Both courtesy of Smithsonian Institution, National Museum of American History

bridges, over the Kinzig at Offenburg (1852–1853, see chapter 5, figures 5.10 and 5.11, which opened two years before the Boyne Viaduct. Ruppert believed that (1) the tubular bridge was ugly; (2) the lattice bridge was lighter and without question cheaper; (3) maintenance was easier, because all parts were accessible for inspection and painting; and (4) it offered less resistance to wind.[49] Moreover, a massive tube was not viable for launching.

Factors such as these meant that the tubular engineers had difficulty convincing decision-makers. When in 1848, after the completion of the Conway Bridge, officials in Cologne were planning a bridge that would join the banks of the Rhine and be a symbol of growing Prussian supremacy, they invited Fairbairn to Berlin. Despite the intervention of renowned scientist Alexander von Humboldt, who got Fairbairn an audience with King Friedrich Wilhelm IV, his plan was rejected: "The Cologne people had declared they would *never* consent to a 15-foot-high bridge that would obstruct the view of Cologne from Deutz!"[50] An international competition that drew sixty-one entries was held, 1850–1851. Moorsom received second prize for "a lattice bridge on the American plan"; a model of it was on display in London's Great Exhibition in 1851. Construction of a lattice bridge at Cologne began in 1855 and ended in 1859 (figure 6.8).[51] The German nationalist architecture of its stone portals complemented the aura of the cathedral, completed in 1880 as a grand national gesture. In 1850 Fairbairn pitched another project, for a tubular bridge over the Neva in St. Petersburg, directly to Russian emperor Nicholas I, but a commission did not result.[52]

Figure 6.9
Bridge over the Rhine between Kehl and Strasbourg (Fleur Saint-Denis and Vuigner, 1858–1861). Firms based in Germany and France collaborated on the construction of the bridge that linked the two nations. Courtesy of Smithsonian Institution, National Museum of American History.

The Franco-Baden Commission was responsible for the decision concerning the railway bridge over the Rhine at Kehl (Fleur Saint-Denis and Vuigner, built 1858–1861), a politically sensitive link between France and Germany. It considered three options: a seventeen-span wooden (arch?) bridge, a thirteen-span plate-beam bridge, and an iron "American" bridge of three long spans. They selected the last (figure 6.9). By this time, the choice was unexceptional. The Kehl bridge's foundation work (directed by the railway company based in France) seemed to have attracted more attention than its superstructure (the work of a Baden-based firm).[53] By the early 1860s, engineers considered four options for the tall viaduct at Busseau d'Ahun in France (figure 8.3): stone arches, cast-iron arches, iron lattice with stone piers, iron lattice with iron frame piers; they selected the last. It is unlikely a tube was considered.[54]

While the Britannia's monumentality aroused awe, its method of construction did not seem to respond to the needs of industrialized building. It was not integral to the conception of the structure, as it was with the lattice. Stephenson himself criticized "the precarious and expensive process of floating and afterwards lifting the tubes"; his final tube, the Victoria Bridge (see below) was built on scaffolding rather than lifted.[55] On both ends of the Britannia, between the land-based abutments and the towers, workmen assembled massive temporary scaffolding (the overall construction required 500,000 cubic feet of wood), floated each of the central spans on massive barges and lifted them assisted by three hydraulic presses powered by steam engines. Multiple capstans employed fifty men each to guide barges into place and lift the tubes (figure 6.10). Commands were given via wooden letters, colored flags, trumpet blasts, and arm movements. During the raising of the tubes, there was one fatality (in addition to six others) due to mechanical failure; "the number of men employed was 685; of these, 386 were sailors, hired for the occasion."[56] Thousands of people watched. "At the conclusion of the unprecedented and unrivalled floatation, a Lady standing on the Engineers' Platform exclaimed, 'What nation can vie with Englishmen?'"[57]

A key reason behind Lentze's rejection of the tubular bridge for Dirschau and Marienburg had been that conditions did not allow floating and lifting; due to a long winter, there wasn't enough time to build on long-term scaffolding, so the bridge was built on movable scaffolding.[58] When after thirteen and a half months of construction the reputedly world's largest wooden Portage Bridge (figure 5.21) opened in New York State in 1852, a Canadian journal favorably compared its cost to "a wrought iron tubular bridge."[59] While this judgment did not take into account vastly different geographical conditions, it was clear that divergent mindsets about cost and construction were in play. Just as engineers were developing strategies such as launching, Stephenson's team designed a difficult-to-lift, unlaunchable structure. Unbuildability made it more unlikely that engineers who were responsible for large projects, such as Etzel, Ruppert, and Nordling, would choose tubular bridges.

Perhaps most mortifying for the tubular team, due to its propinquity and visibility, was the Crumlin Viaduct (1853–1857, Charles Liddell and Lewis Gordon were probably the engineers) across the Ebbw and Kendon Valleys, South Wales (figure 6.11), near the Britannia Bridge. This might be considered ground zero of tube versus lattice and truss. The Crumlin engineers passed on a lattice design in favor of the Warren truss, a configuration that reduced the lattice to a minimal triangular pattern. The builders claimed that the Crumlin was ten times cheaper, although it was of comparable length and height to the Britannia. At the conclusion of a gala dinner marking the opening of the Crumlin, railway official William Fitzmaurice called it "one of the

Figure 6.10
Installing a Britannia tube. Thomas Jackson, *The Tourist's Guide to Britannia Bridge* (Smith, 1851), 20, 28, 33. Ricker Library of Architecture and Art, University of Illinois Urbana-Champaign.

Figure 6.11
Crumlin Viaduct (Charles Liddell and Lewis Gordon, built 1853–1857), Ebbw and Kendon Valleys, South Wales, not far from the Britannia Bridge. The Crumlin became a tourist attraction, like two nearby suspension bridges and the Britannia, Portage, Niagara and Grandfey bridges. William Humber, *A Practical Treatise of Cast and Wrought Iron Bridges and Girders, and to Buildings Generally* (Spon, 1857), frontispiece. The New York Public Library.

greatest works the genius of the country ever produced," and "a most fitting wife" for the "massive grandeur" of the Britannia Bridge. He claimed that the Crumlin "was all elegance and beauty, and true to her nature she had no bonnet upon her head, according to the modern fashion. (Much laughter.)" This was more than a cheap shot aimed at a competitor, the iron-sheathed Britannia; conspicuous aesthetic dissimilarity exposed the profundity of industrial and technopolitical differences within one small region.[60]

The tube failed as a bridge type, but three more were built—directly or indirectly under Stephenson. One was in England: the Brotherton Bridge across the Aire (1851), where he finally erected a bridge with an elliptical top (always his first choice). The government, however, forced him to cut the top open and make it square because it was judged too narrow—a danger for passengers who stuck their heads out the window (figure 6.12).[61] Apparently, a smaller-scale tubular bridge was built in Montreal, Canada, during the construction of the ultimate tube, the Victoria Bridge over the St. Lawrence River. Alexander Ross, who had been engineer on the railway whose trains passed through the Conway and Britannia bridges, developed the project in collaboration with his friend, Stephenson, who was named chief engineer in order to attract investors.[62] It gave Stephenson the opportunity to realize his invention on a grand scale—and to do it in North America.

Four years passed between completion of the Britannia and the start of the exceptionally long Victoria in 1854 (figure 6.13). The initial proposal was to build long approaches with twenty-three wooden-member Burr bridge spans (figure 2.17) and one long iron tubular span across the central, deepest area of the river; after the British industrialists became involved, the railway line tripled in length and tubes replaced the American bridges. The tubes weighed about a ton per linear foot. They were manufactured in England and sent on twenty-five ships to Canada, where additional rivets and bolts were made and workers assembled the individually numbered pieces (4,926 for most tube spans although the center one consisted of 10,309 pieces and required the punching of half a million holes for rivets). Again there were disputes about engineering authorship.[63] A lavish book published in 1860, the year the bridge officially opened after five years of construction, referred to the structure as the "eighth wonder [in] the world's museum" and "another trophy to the power of mind over matter."[64] The encomia were no more convincing this time around; in the tribute publication, the appendix featured a letter from Stephenson, apparently written in mid-1856, well after construction was under way, defending the tube against the lattice, the Warren truss (figure 6.6), and Roebling's Niagara suspension bridge (1855) (figure 6.14).[65] Headquartered, financed, and manufactured in Britain, the costly Grand Trunk Railway, along with its Victoria Bridge, which made a critically important rail link in Canada,

Figure 6.12
Tubular bridge over the Aire at Brotherton, North Yorkshire (Robert Stephenson, 1851). William Weaver Tomlinson, *The North Eastern Railway: Its Rise and Development* (Andrew Reid, 1915), 502. UC Berkeley Library, University of California.

was in the twentieth century judged "an embarrassment almost from its inception." Intended to compete with US railways, the company was an "organized mess" that did not turn a profit. Already in 1861, a book review that began, "Are tubular bridges costly blunders?" called the Britannia and Victoria "ruinous to the companies that constructed them."[66] A slot eventually had to be cut into Victoria's roof so smoke could disperse; rather than reinforce the bridge, in total one of the world's longest, in 1897–1898, engineers scrapped the tubes and replaced them with trusses. Many older bridges required reinforcing, but the Victoria was single track, too confining to accommodate large locomotives, and was difficult to maintain. But the masonry piers and abutments were upgraded to support the new structure.[67]

Figure 6.13
Victoria Bridge (Robert Stephenson and Alexander Ross, 1854–1859) over St. Lawrence, Montreal, ca. 1865. Courtesy of Smithsonian Institution, National Museum of American History.

During the construction of the Victoria Bridge, debate had exploded into the open, revealing ongoing pushback against Stephenson and his tubular idée fixe. An 1856 publication printed in London, addressed to the shareholders of the Grand Trunk Railway, exposed the issues in devastating detail.[68] Robert McCalmont, an investor in the Grand Trunk, and Charles Liddell, a designer of the Crumlin Viaduct (manufactured by Kennard), commented directly on statements by Stephenson, Brunel, Clark, and Ross. (Liddell provided a point-by-point rebuttal of Stephenson's claims; this would later appear in the Victoria Bridge commemorative publication, but sans Liddell's commentary.) Even by today's standards, McCalmont's vehemence is noteworthy. He wrote

Figure 6.14
Niagara Bridge (John Roebling, 1852–1855). The Miriam and Ira D. Wallach Division of Art, Prints and Photographs: Photography Collection, New York Public Library.

that "an overwhelming preponderance of authority" went against the tube and that the railway company was being "led on blindfold."[69] Brunel countered that there really wasn't much difference between a lattice, Warren, or tube—a remarkable response, as there were significant cost, construction, maintenance, railway-traveler-experience, and aesthetic differences. Liddell in his reply noted that "it is well known that [Stephenson's close colleague Brunel] has never applied a tube for any of his bridges of large span." Liddell called out the "looseness, inaccuracy, and fallacious character" of Brunel's "utterly untrustworthy" statements; bold language applied to a prominent engineer whose celebrity was perhaps unrivalled except for Stephenson's.[70] Liddell quoted from statesman George C. Lewis's *On the Influence of Authority in Matters of Opinion* (1849): "Let it, however, be borne in mind that whatever deference is due to great names and competent judges, they are not to be regarded as infallible—as oracles of a scientific religion, or as courts of philosophy without appeal."[71]

Not every day does one find in an engineering discussion about bridges an appeal to a classic treatise on opinion, fact, politics, and the uses and abuses of authority. Certainly, there was a variety of debatable uncertainties hovering around structural analysis and iron. Certainly, Liddell was not a disinterested party; he was advocating for the use of a Warren bridge instead of a tube and it wasn't yet apparent, because the Crumlin was not complete, what the deficiencies of the long-span Warren might be.[72] But his statements align with those of other engineers, as does his observation that "universally other systems have been preferred to that of the tube."[73] Brunel, Clark, and Ross (like Stephenson) employed wordy, befuddling arguments, a strategy that today might be called "gaslighting." Stephenson's friends rallied to support him and protect the substantial business interests involved. According to Liddell's detailed analyses, including comparisons to other construction projects, the Victoria tube cost at least one million pounds more than a lattice or Warren would have.[74]

Stephenson was asked to bid on a railway bridge near Niagara Falls, but Roebling won the project and built a suspension bridge with a deck stiffened by a wood and iron frame (figure 6.14). A railway suspension bridge was quite rare; it sparked debate and comparisons that, as usual, threw the tube into unfavorable light. Roebling's novel structure became an object of immense national pride and a world-renowned symbol of America: another rebuke to Stephenson's tube.[75] Stephenson and Ross each visited the completed suspension bridge and received information from Roebling, which they used to explain why a suspension structure was not appropriate for the Victoria project.[76]

By any measure, the lattice was immeasurably more successful than the tube, but it was also frequently rejected in favor of other types it had inspired. There are always multiple solutions.[77] Even while paying homage to a British nabob, decision-makers considered the tube massively heavy, expensive, constrictive, difficult to build, and ugly. In 1907, an editor of *Engineering News* reflected, "The old Victoria tubular bridge at Montreal stood till it was taken down; but that fact could not save the type it represented from condemnation by every intelligent structural engineer."[78] The tube was once a novelty, competing briefly with alternatives recognized as lower-cost, lightweight, good-looking, American. Contemporary judgment against the tube was swift. Still, as we conclude this story, we wonder: Does a focus on competition between bridge types merely confirm hand-me-down perspectives? With transnational flows in mind, might we tell this tale in a different way?

LATTICE IS TUBE IS LATTICE

An historically based view of lattice-tube opposition has dominated our narrative thus far. These categories help us understand how historical figures—engineers and their chroniclers—thought, or at least how they presented their thoughts. The lattice-tube dichotomy revealed divergent understandings of cost management, structural behavior, construction, and aesthetics. But debates surrounding these issues often obscured a physical and theoretical commonality. This means that boundaries dividing the names "lattice" and "tube," and "beam" and "truss," just like "wood" and "iron," and "British" and "American," must be explained; they do not explain. Let's dive deeper into a story that might be told from a perspective of creative cooperation rather than as a struggle to survive.

TUBE IN NAME

Why design a bridge in tubular form? With material concentrated at its top and bottom, where compression and tension forces are greatest, a tube can provide increased stiffness for less weight than a conventional beam.[79] Lightness can significantly reduce the cost of the superstructure along with its supporting piers and foundations, often the most challenging part of bridge building. The special qualities of a bridge that functioned as a tube were recognized well before Stephenson's mind alighted on the idea. For example, we noted in chapter 4 that engineer Herman Haupt, who admired Town's bridge, analyzed its defects and in 1839 patented an "improved" version, which included the addition of vertical members. He thought of the structure as a tube: "This arrangement increases the stiffness on the same principle that a hollow cylinder is more stiff than a solid one."[80]

Within the discourse about beams and tubes and the behavior of forces within them—Stephenson's belief in the "efficiency" of a tube's top and bottom as sources of its bending strength notwithstanding—the tubular engineers did not emphasize the advantage of the tubular bridge as a tube! *Gentleman's Magazine* described the term as "almost a misnomer," because the structure was not round, but had acquired the name through the testing of tubes of various shapes. Britannia construction engineer Edwin Clark stated that "tubular" signified only "hollow." He and Stephenson did understand that (round) artifacts such as quills, straw, and bird bones showed that "a thin tube of any material is far stronger than the same quantity tightly rolled up into a solid rod of less diameter than the tube"; also that the bridge's bottom and top were composed of smaller tubes, where they believed "its whole strength" lay. Historian Tom F. Peters has pointed out that "they built the bridges as rigid box beams and called them 'tubes,' but everyone still thought of them as plate girders and neglected their three-dimensional behavior."[81] We might consider Stephenson's tubular bridge, in terms of design, as a hollow concept, more visual than structural. It indicated the form's potential.

STRUCTURAL AMBIGUITY

While the Britannia designers created a tube-like structure that they didn't understand three-dimensionally, it is unlikely that lattice bridges behaved as tubes, even if they appeared tubular in cross-section (figure 6.15). So the structural difference between bridges labeled "lattice" and "tube" was not necessarily distinct.[82] For example, the assessment of École Nationale des Mines professor Charles Couche appears to be accurate: "the substitution of the lattice for solid walls is only a detail; the Dirschau bridge [figure 6.7] will be, even if this label is rejected, nothing more than a *tubular bridge*."[83] Fairbairn in 1852 asserted that there wasn't a substantial difference between plate and lattice construction; he believed that former would be more rigid and that he would build with whichever proved superior.[84] An engineer involved with the Boyne Viaduct noted that its "cross section . . . is the same as that of the tubular girder, if we substitute intersecting bars for the continuous plate."[85] As engineer-historian Martin Trautz recently observed, the Cologne bridge "merged tubular and beam behavior." At Kehl, the bridge's lattices functioned more as beams along its length and did not form a tube.[86] Despite discussions about lattice vs. solid webs described above, there was apparently general agreement on how forces in the top and bottom chords should be understood and the members dimensioned.[87]

Lattice-tube affinity is evident in the inclusion of vertical members as a design solution. In the tubular bridges, the sides experienced shear load and would have buckled if too thin or if not held in place; this was resolved by riveting vertical bars every two feet.[88] Earlier, in the Peacock Bridge (1839–1840) over the Schuylkill near Reading, Pennsylvania, Moncure Robinson also had included intermittent verticals as "reinforcement" in the wooden structure.[89] (Already in 1851, engineer Herman Haupt criticized the double lattice [Town's 1835 patent] for employing in its central portion so

Figure 6.15
Contoocook Bridge, Contoocook, New Hampshire (J. Parker Snow, 1889), Boston and Maine Railroad. The bridge, displaced on two occasions by floods, displayed tube-like rigidity. See fig. 7.5. Courtesy of the Contoocook Riverway Association.

much material that did not increase structural strength.)[90] Ultimately, it was the "great waste of material . . . [that] arises from the great multitude of rivets and laps" required to make the structures rigid (whether lattice or tube) that led engineer John Roebling, whose aqueduct was an early inspiration for Stephenson, to conclude in 1869 that "Lattice bridges, like Tubular Girders, have seen their best days."[91]

Beyond similarities of function and design, lattice-tube opposition is difficult to maintain because there was never one type of lattice bridge.[92] In addition to Town's two patents, others substituted bolts for pegs, added vertical members, or combined lattice with arch. The translation into iron included more rigid connections (rivets), allowing builders intentionally or inadvertently to make lattices more monolithic and perhaps a bit more tubelike in performance. The profile of the iron lattice bars might be flat, square, round, L-, U-, or T-shaped, and these might be used in any combination, the spacing between them designed to reflect anticipated loading (figure 4.13).[93] Like most structures, the lattice was a "bundle" of innovations, and engineers might direct their attention to any of its parts.[94] Lattices in wood and metal offered numerous opportunities to customize or test a variety of details; that is, to be innovative. Like "tube," "lattice" was as much a visual as structural term—and symbolic of a way of thinking about building.[95]

ANALYSIS

Emerging methods of understanding structures played a role in lattice-tube ambiguity. Both were indeterminate structures, which required more sophisticated analysis; engineers who lacked that ability considered them as large beams—determinate structures.[96] The widely known beam theory of Claude-Louis Navier did not specify how the top and bottom chords of a beam were connected—they just had to be solidly connected. As we've noted, structural analysts and designers, following Navier's lead, used this theoretical model when evaluating single-span parallel chord trusses in the 1830s and 1840s.[97] Joseph Hyacinthe Garella, for example, an engineer who in 1838, near Lyon (see chapter 3), built two provisional "ponts américains" with the addition of bolts and vertical members, presented calculations in which he considered the lattice as a "hollow solid" consisting of upper and lower chords.[98] During the late 1840s and early 1850s, the formative period of iron tubular and iron lattice development and construction, Whipple, Schwedler, and Culmann were sharing with engineers analytical tools that enabled them to calculate the forces in individual members of a structure. (In 1847, in Utica, New York, just a few months after the Dee collapse, Whipple, in his epochal publication detailing how to analyze stresses in every member of a truss bridge, devoted two pages to the failure. He seemed to accurately assess its cause, which led to his "respectfully offering these humble essays to the attention of the engineering profession"—which included Stephenson, though his awareness of Whipple's analysis is unknown.)[99] In a theoretical analysis of the iron lattice bridge over the Royal Canal, near Dublin, Ireland (John Macneill, built 1844–1845), engineer Julius Pollack in 1848 referred to its lattice frames as "walls, or rather, beams." He noted how similar Stephenson's bridge was to the "American bridge systems," when one imagined a solid where the lattice was.[100] Some engineers used the lattice to understand metal plate, while others used plate to understand the lattice. Stephenson himself considered each side of the tubular bridges as the theoretical equivalent of a member "in which the trellis bars may be considered infinite

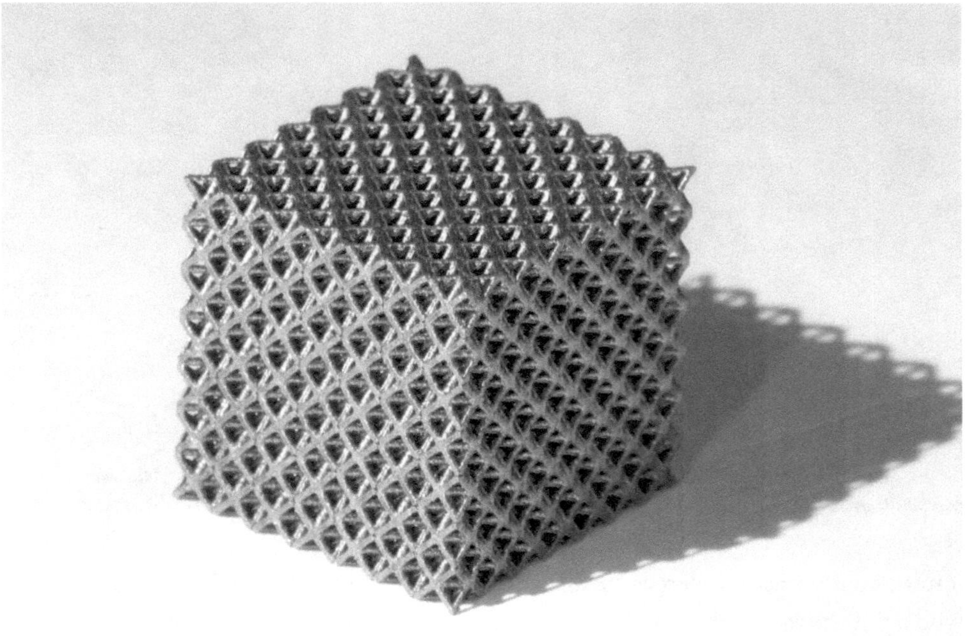

Figure 6.16
Octet truss lattice (2015), originally proposed by Buckminster Fuller, made of stainless steel using laser powder bed fusion. Courtesy of Dr. Thomas Tancogne-Dejean.

in number."[101] Far in the future, designers with a firmer theoretical grasp would find it useful to blend structural perspectives. In Chicago, the John Hancock Center's engineers (Bruce Graham and Fazlur Khan, 1969) merged tube and truss, and more recently, researchers have developed extremely light and rigid designs, stiff and strong in three dimensions, by combining lattice geometries with plate (figure 6.16).[102]

As engineers treated the Britannia, theoretically, as a beam rather than a tube, the debate about tubular bridges merged with debates about beams. A question for I-section beams became: Should the web be lattice or plate (figure 6.17)? (This was discussed along with other fundamental questions of the period: Should bridges be composed of one continuous span or separate spans? Should two tracks be placed on a wider single span or should each track have its own span, as in the Britannia Bridge?) Charles Couche diplomatically dissolved the lattice-or-plate dilemma: "It seems certain, when all is said and done, that the two systems are about equal"; the conditions in each case would determine the decision. Couche noted, "It is true that most of the renowned engineers of Great Britain are against the lattice, but with them, the

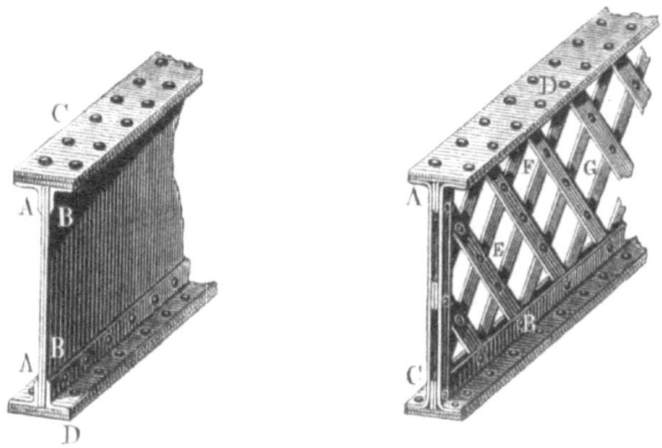

Figure 6.17
Beams with plate web and lattice web. Julius Weisbach, *Lehrbuch der Ingenieur- und Maschinen-Mechanik*, 4th ed. (Vieweg, 1865), 2:159 (also in 3rd ed. of 1857). University of California.

result of a discussion is determined at the start. Consequently, the Prussian engineers, who attribute the double advantage of economy and a more elegant appearance to the lattice, ascribe the marked condemnation of the lattice in England to the unfriendly prejudice that welcomes in that country all ideas of American origin."[103] Analysis of these public-facing structures comprised structural design, nationality, and looks. Regarding the latter, even a staunch supporter of the Victoria Bridge admitted, "It cannot be said that the appearance . . . is picturesque. . . . Possibly it mars the landscape" (figure 6.18).[104] This applied to wooden covered bridges, too! They could be perceived as unattractive, as charming as they may today appear.[105] For Stephenson, the lattice may have been a more effective choice, in terms of construction cost as well as his business interests, but it did not align with his cultural and technological prejudgments. Identifying the stories and ambiguities behind engineering decisions, though, can help designers and all of us better understand intuitions and the possibilities they deny and offer.

MATERIAL FORM

Stephenson and Brunel called attention to the woody origins of the lattice in order to flag the structure's supposed backwardness. When Clark stated that the "novelty in the tube consists of its being a constructed beam," he dismissed the prior novelty of innumerable constructed wooden beam bridges, with the lattice a renowned paradigm.[106] The tubular engineers' cancellation of wood was a platform for the promotion of their own iron-founded interests; Fairbairn's business, for example, was based on the sale of iron plate. And they were communicating what they wanted the local engineering

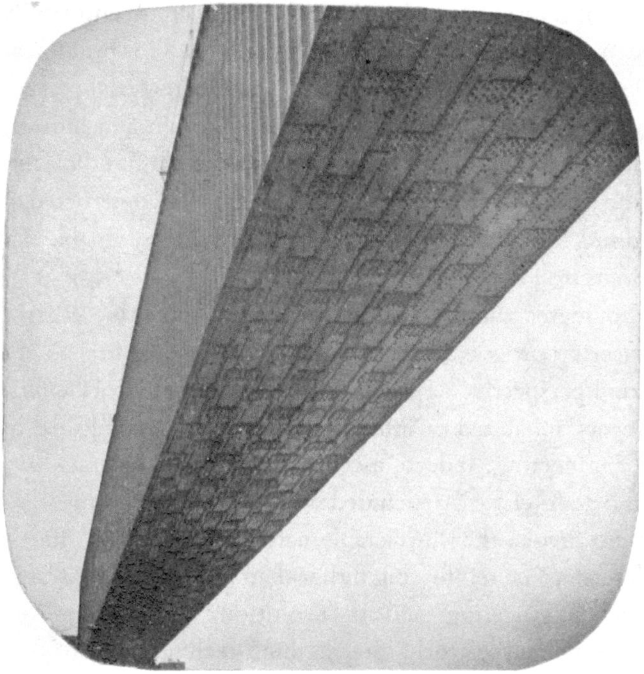

Figure 6.18
Victoria Bridge, Montreal, Quebec, from below, ca. 1876. William Notman, photographer. J. Paul Getty Museum, Los Angeles, 84.XC.873.8592.

community to concentrate on. Karl Culmann pointed out that the focus on iron was based on a prosperous iron manufacturing industry. He observed:

> What is befitting of the wealthy Englishman, who goes everywhere fully conscious of the idea, "I am in possession of the iron and do not need to worry about the statics," is less fitting for the poor devils on the continent; they have to fiddle and experiment, stake out and estimate many solutions for every railway to be planned in order to discover the cheapest, and draw various force diagrams for every bridge to be built so that no material is wasted and only the essential is used. . . . From the viewpoint of a national economy, it is the American who treads the right path: he uses no more than is absolutely necessary and preferably a little less.[107]

A project of outsized proportions, named "Britannia," made of a cutting-edge, seemingly impervious material and assembled in a tubular form, embodied the idea of an iron-willed empire with an iron constitution.[108] Still, engineers under Stephenson did build a few lattice bridges, even on the same line as the Britannia Bridge, and they built stone bridges, too—not to mention the hundreds of wooden bridges built in England during the nineteenth century.[109] Like many US states, England built with imported wood. This was not unprecedented; during the eighteenth century, Britain imported iron from Sweden and Russia.[110] Designers adopted metal for bridges, as they would in ship, clock, doorknob, and airplane construction. The material played a fundamental role in the development of structural analysis and design and industrial methods of construction, although wood continued to be an essential building material.[111] Ironically, in 1970, the flamingly metallic Britannia, originally constructed on wooden scaffolding, burned in a fire fed by a wooden structure with flammable materials built on its roof to shield its iron from sea air and rainwater.[112]

Historians shouldered the tubular engineers' narrative task. When they composed stories of engineering progress, they wrote as if it were natural to employ historical figures' ironbound perspectives. Perhaps it was, since these storytellers, in many cases trained as engineers, inhabited or inherited a worldview established by the founding fathers of civil engineering. Indeed, a century after the Britannia, an account of the girder bridge in Britain virtually excluded wood and lattice beams.[113]

Historians who invoke the "intractable nature of materials" and "the immutable laws of nature" seem to be reinforcing rather than interrogating ideas put forth by the men who defined "engineering" and its "evolution."[114] Posing questions might have seemed subversive, when the world-altering high-tech network of the day—railroad, *chemin de fer*, *Eisenbahn*—had metal embedded in its name and was used throughout, from rails to stations to warehouses, even if wood was fundamental to its construction.[115] If we were to regard materials as mutable in the hands of technologists, however, we might move our inquiries to the reasons why designers chose materials in specific contexts. During the 1820s and 1830s, builders and manufacturers, including Fairbairn and Hodgkinson, developed I-section members in various materials and configurations, although it was not until the middle of the century that manufacturers produced in iron and on a large scale what today are recognized as I-beams.[116] The lattice was a model in the design, manufacture, and analysis of beams before Stephenson proposed the tube and before builders commonly used wrought iron.[117] But when the lattice does appear in historical accounts, it is as a "bridge." Its transformative nature as a beam went unrecognized. (Who would want to read a book entitled *Beams of America*?) This means: if we discard the evolutionary metaphor, break our addiction to iron-based industrial (r)evolution tales, smash British nationalist shackles, and cease to consider material categories as intractable, then each (wooden) lattice frame appears as a built-up beam, if not a proto-I-beam (figure 6.19). The upper and lower chords place the material where it is needed for compression and bending, respectively.[118] Might we view the original and 1835 lattice bridge patents as seminal I-beams, box beams, and tubes?[119] Suddenly, wood reframes beam history, whose narrative till now was cast in

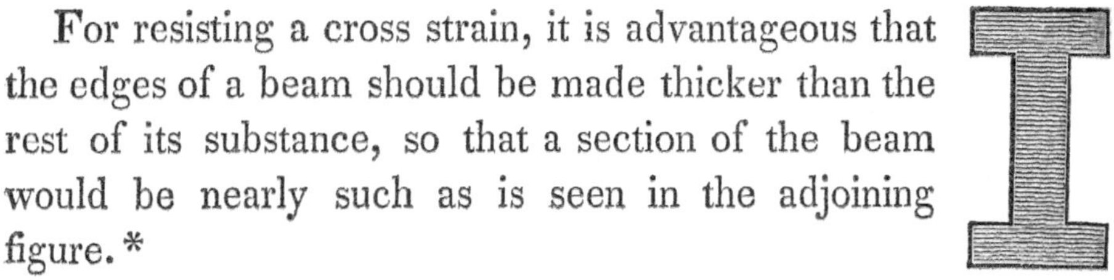
For resisting a cross strain, it is advantageous that the edges of a beam should be made thicker than the rest of its substance, so that a section of the beam would be nearly such as is seen in the adjoining figure.*

Figure 6.19
The I-beam form appeared in a book by Town's business partner's acquaintance, Harvard professor Jacob Bigelow's *Elements of Technology* (Hilliard, Gray, Little, & Wilkins, 1829), 52. UC Berkeley Library, University of California.

evolutionist iron. The metal of industrializing nations needs no longer triumph over the wood of less civilized borrower nations. Instead, we can tell each other about technologists collaborating and competing in a variety of locations and cultures, using a variety of materials and forms as they design, engineer, build.

While lattice and truss immediately eliminated the tube from the market, how did they fare against box or solid plate beam bridges? This is not a simple question to answer, because the specific context of each bridge, including the proposed length of its spans, are key factors in decision-making. Fairbairn built thousands of box girders (hollow beams), which were used in relatively short span bridges, but for longer spans a box girder presented challenges similar to the tubular bridge: it required so much material—a thickness of section and so many stiffeners—that a lighter lattice design was preferred. An analysis of railway bridges built in the German states and Austria concluded that for 10 meters or less, box girders were recommended; longer bridges called for lattice or truss.[120]

BUILDING CATEGORIES AND ENGINEERING NATIONS

Ponts et Chaussées engineer Alexandre Berthault-Ducreux wrote in 1845: "One of the greats of France said: *the style is the man*. Isn't it just as appropriate to say: *the works are the nation?*"[121] Infrastructure does reflect the state of the nation, by demonstrating a government's ability or inability to oversee design, construction, and maintenance of structures and networks that enable individual and nation to function and thrive. Technological objects also allow feelings about what's home-grown and what's foreign to assume tangible form. With the "American" lattice and the "British" tube in mind, let's conclude this chapter by considering the co-construction of bridges and nations.

In the international engineering arena, as iron became common in bridge design; the tubular bridge became an instant curiosity. But unlike most failed inventors, Stephenson had the political clout to "write the book." His 1856 *Encyclopaedia Britannica* article on iron bridges proclaimed:

> It is difficult to conceive even now, with all our subsequent experience, any other means than those adopted of solving the great problem which thus inaugurated a new epoch in the history of bridge construction, and led directly to our present theoretical knowledge of the principles of beams, [as well as] all those numberless elegant and ingenious practical combinations of wrought-iron in bridge construction.[122]

It was, in fact, not that difficult to imagine other means, and many engineers did. Stephenson's hyperbolic account presented the tube as a turning point in theory, practice, and industry. The lattice in this tale becomes an unfit technological form or an adaptation of the tube.[123]

The selection of a bridge type was a local undertaking linked economically, culturally, and socially, far beyond the construction site, with solutions flowing in and out of the ambiguous places we call local, regional, and global.[124] So maintaining national boundaries around design is as complex a task as justifying boundaries around lattice and tube. Great Britain, historian Linda Colley explains, "did not come into being . . . because of an integration and homogenisation of disparate cultures. Instead, Britishness was superimposed . . . in response to contact with the Other." The Other included people who were African, Asian, French, Catholic, and North American. The invention of British identity after the Revolutionary War in the United States was in no small part a response to an equally new American identity. Engineers constructing a tube in opposition to a lattice were building national mythology. According to Colley, "being a patriot was a way of claiming the right to participate in British political life."[125] Similarly, technologists' claims concerning tubes and lattices or the roles of practice and theory, were claims about what it meant to be, inter alia, British and American. These enabled access into cultural and professional engineering communities that existed within and across national borders.

When engineer George Buchanan read a paper concerning bridges and the strength of materials at the Royal Scottish Society of Arts in 1848, a Professor Forbes asked about the resemblance of "these American bridges" to iron tubular bridges. Buchanan replied that "the resemblance in principle to the tubular bridges had struck himself forcibly." He concluded, nevertheless, that despite the strong similarities, "the tubular bridge was, in many respects, a very different structure; and the design of a bridge, of one vast malleable iron tube, was an idea at once happy and original, and was, he considered, due entirely to Mr. Stephenson."[126] Which is to say, the tube was a function of the power, reputation, and business interests of its creators.

Stephenson's status was a national matter. He, along with his father, railroad pioneer George Stephenson, were (and are) figureheads of British engineering and industry. As a national hero, "a member of English parliament and powerfully rich," in the respectful words of prominent railway engineer Auguste Perdonnet, Stephenson stood amongst the technological giants of the world.[127] He also served as president of the Institution of Mechanical Engineers (after his father stepped down as its first president) and the Institution of Civil Engineers. The design of the Britannia Bridge reflected this. The tubular engineers had planned a symbolic manifestation of infrastructure and nation: "it was once contemplated to surmount the centre tower with a figure of Britannia, in stone, sixty feet in height"; in the end, instead of the national goddess, a pair of colossal limestone lions, symbols of British might, guarded each end.[128] The Britannia Bridge was built only a mile from Thomas Telford's Menai suspension bridge (1819, figure 6.1), which had provided a vital transportation link, as Stephenson's bridge was to do for railway traffic. Telford, as Stephenson certainly knew, was the recipient of intensely nationalist accolades.[129] Portraitist John Lucas's renowned painting of the tubular team broadcast Stephenson's status and confirmed the quasi-religious significance of engineering heroes in the culture of the nation-state; the artist apparently patterned the image after depictions of the Last Supper (figure 6.20a, b). Lucas's image of a godlike figure affirmed a technological category (with the tubular bridge, in the background, resembling a cross) and bound disciples into a national-spiritual engineering community.[130] Comparable to Robert Stephenson and his father, George, Jesus and his foster father, Joseph (figure 2.14), are referred to as skilled carpenters, that is, builders, metaphorically if not literally.

Previously, the Christian perspective had affirmed that God, not man, was the force behind creation. The ascension during the eighteenth and nineteenth centuries of inventors and engineers represented a born-again transformation: technologists and industrial techniques were cited as evidence of God's goodness. Stephenson apparently believed that the interior of the tube called to mind "the 'long drawn aisles' of a cathedral"—not the first image most people would invoke inside a dark, dirty, reverberating space, if they could even see it from within a rail car.[131] A depiction of the tubular bridge eventually appeared in the stained glass of Westminster Abbey, the site of Stephenson's funeral and burial (figure 6.21). The technoreligious iconography reminds us of the "civilization" that Great Britain was delivering, facilitated by its engineers, to its far-reaching empire.[132]

As historian Christine MacLeod explains, the rise (and fall) of technologists as heroic avatars in England was the outcome of phenomena including business marketing, efforts to enhance the visibility of the working class, debates about patents, the promotion of science (above technology), and nationalistic pride.[133] The Britishness of Stephenson, however, compels us to remember that heroes seem to represent uncomplicated cultural and technological positions, but they project an image of unity that rarely if ever exists.[134] We've seen that Stephenson's beliefs hardly circumscribed the convictions of all engineers working in Britain. The lattice served, inside and outside

Figure 6.20a
Engineer as national messiah. The Last Supper is the template for one of the most renowned images in engineering history. John Lucas, *Conference of Engineers at Britannia Bridge*, ca. 1850. Institute of Civil Engineers, London / Bridgeman Images. Standing, left to right: Admiral Moorson, Latimer Clark, Edwin Clark, Frank Forster, George Parker Bidder, Hemmingway, Captain Claxton, Alexander Ross; seated, Robert Stephenson, Charles H. Wild, Joseph Locke, Isambard Kingdom Brunel. (Moorson and Claxton were nautical specialists; Hemmingway, a master mason.)

of Britain, as a defense and, sometimes, a weapon, against the elite engineering community personified by Stephenson (figure 6.22) and colleagues.[135] The American label and the prominence of the lattice in Ireland, for example, may have inflamed Stephenson's nationalistic and competitive pride. If he had been able to kill the lattice category, it would have cleared the way to additional wealth and power. That didn't happen. In 1868, an editor of the London-based journal *The Engineer*, when comparing engineering practice in England and France, disapprovingly wrote of his native land, "But ours is a free country, who shall presume to come between the 'great engineer' and 'his board,' or to discuss the consequences of oracles, whether they relate to . . . Britannia Bridges

Figure 6.20b
Juan de Juanes (Vicente Juan Masip), *The Last Supper* (1555–1562). Museo del Prado, Madrid.

or Great Easterns, or the like?" (The Great Eastern, designed by Stephenson's colleague Isambard Kingdom Brunel, was the largest ship ever built, of iron; it had a troubled financial and service history.) The writer contrasted practices in Britain with a process that "enables the great industrial bodies in France seldom to be hoodwinked into works that may be 'engineering monuments,' but more or less financial ruin."[136]

In contrast to Stephenson's glory, Ithiel Town (figure 6.23) is familiar mostly to architecture buffs and covered-bridge enthusiasts. He isn't included on American heroes-of-invention lists such as "Franklin, Whitney, Fulton, and Morse," although his work is of comparable significance; by 1912, a historian had described him as "almost forgotten by posterity."[137] Town was independent; his inventive activity occurred before the founding of a national engineering institution that might have burnished him for future generations. He didn't possess an engineering degree, didn't participate in meetings intended to organize a professional civil engineering organization, and did not establish a business that outlived him. Few of his structures survive and his business records have

Figure 6.21
Robert Stephenson's funeral in Westminster Abbey, London. Bridges portrayed include the Britannia (middle left) and the Victoria (bottom). Stephenson died two months before the first train crossed the Victoria. *Illustrated London News*, October 29, 1859, 423. UC Berkeley Library, University of California.

disappeared. He didn't build in iron, material of progress, or tinker with energy forms synonymous with "future," such as steam and electricity. The ideas crystallized in his patent became so widely employed and adapted that they appear natural rather than invented. Town dispersed his mammoth art and architectural library and there were no plans for the preservation of his monumental home, which was altered beyond recognition and later demolished. His plans for books on architecture and bridges did not materialize (see chapter 7). He did not have a legacy-building heir; there is no evidence of his daughter, Etha, born of unmarried parents, honoring her father's accomplishments, although her husband registered Town's brick patent after his death.[138] Finally, Ithiel Town was American at a time when British promoters were already stoking the British, industrial-revolution history-marketing machine, whose narratives endure. From this perspective, Town is preindustrial, preengineering, and wooden.

In the darkness and far from stained-glass sanctification stands the oldest-known surviving lattice (1832) designed by Ithiel Town, a triangular truss supporting the roof of the First Presbyterian Church in Fayetteville, North Carolina.[139] But nationalist narratives last far longer than most structures. For example, regarding the evidence and analysis offered in this chapter (previously presented in a journal), historian Richard Byrom, author in 2015 of a study of William Fairbairn, asserted: "A claim is made that the Britannia owes its existence to American timber lattice bridges. . . . The timber lattice bridge was a forerunner of the wrought-iron lattice but emphatically not of the Britannia Bridge."[140] These words provide the full extent of the evidence and analysis behind this stark rebuttal. Byrom did not acknowledge the issues and ambiguities examined here. (He does tell us that during "the first half of the nineteenth-century, Britain's engineering leadership was unchallenged."[141]) As we know, even before the first tubular bridge was completed, people noted its connection to the lattice. Stephenson's own account sprung from the lattice, as did Edwin Clark's; in 1848, Lewis Gordon, civil engineering professor at the University of Glasgow, paid fealty to Stephenson while noting that the wooden lattice bridge formed a "box-like girder"; Perdonnet believed the lattice had led to the tube.[142] In Byrom's victoriously presentist interpretation, the Britannia, which ostentatiously failed as a bridge type, transmogrified becomes a remarkable success because "box girder bridges in steel or concrete became a popular choice during the road-building expansion of the 1960s."[143] Byrom conflates the tubular bridge with bridges supported on parallel box beams, which Fairbairn was developing as early as the 1830s and for which he obtained a patent while working on the Britannia. Although his firm was excluded from the construction of the Conway and Britannia bridges, Fairbairn made a fortune with the girders for about seventeen years, during which time designers further developed lattice and truss beams whose financial and material economy put an

Figure 6.22
Robert Stephenson (engraved by Henry Adlard after a photograph by Mayer and Pierson, 1858) M3936, McCord Stewart Museum.

end to Fairbairn's box-girder bridge market.[144] "Emphatically," storytellers who preserve nationalist narratives cannot link lattice to tube.

The goal of two centuries of rhetoric has been the creation of national stories and heroes.[145] We still hear, in reference to Stephenson's iron, but not to Town's wood, "What is not disputed is that the idea of stiffened rectangular tubes through which the trains could run was a brilliant solution."[146] Brilliant, even if engineers rejected it wholesale. Nationalists and evolutionists are affronted by the idea that a source of a vaunted "British" invention might be "American" or that a tube might be made of wood; indeed, that metal is not

Figure 6.23
Ithiel Town (seated, left of center) and family in his only known photograph. The two standing figures are likely his daughter, Etha, and her husband, William T. Peters; the seated woman, perhaps his mother. Probably ca. 1840. New Haven Museum.

superior to wood, but different. In fact, denying Fairbairn's and Stephenson's engagement across borders and disciplines diminishes them as innovators.[147] It is bewildering to imagine Stephenson *not* basing his development of a tube on precedent; that would have been an aberration, counter to inventors' well-known ways, and likely, an impossibility. But as historian Ludmilla Jordanova notes, "The idioms of competition, national heroism, and universal knowledge were never easily integrated."[148] For some, it's a shock to acknowledge that one's perceived worst enemies (economically or politically) may be one's best friends (creatively).

The narrative of struggle among nations as foundational to invention has profoundly shaped understandings of past and future. It surfaced, for example, when nationalist, capitalist forces propelled the race to develop a coronavirus vaccine. Researchers globally collaborated and competed, with American, Chinese, and Indian national pride and business interests driving and threatening the process. Like nineteenth-century bridge-building technology, a successful outcome would have to be safe, affordable, and designed to be mass-produced, transported, distributed, implemented, and monitored; it would shape life around the world. Just as technologists developed diverse structures and strategies for the construction of the transportation infrastructure on which our lives depend, in 2020 researchers were developing ninety types of coronavirus vaccines. Experts believed that several solutions were required.[149]

By the mid-nineteenth century, the tubular engineers may not yet have faced significant competition from industry based in the US—but they were up against American technology.[150] It may have caused consternation: their industrial enterprises were unsuccessfully competing against products and strategies identified with "a new and comparatively poor country," according to a British engineer reporting on "American timber bridges" in 1862.[151] The British engineers and industrialists who failed to obtain contracts for the major long-span bridges described above, in Dirschau, Offenburg, Cologne, and Kehl, however, were not bested by Yankee firms. Europe-based firms wielding technology that might be deemed "American" did the work.[152] Engineer Émile With put into words this nightmare (for some engineering entrepreneurs based in Britain) when he predicted in 1854 that just as wrought iron would replace cast iron, "my innermost conviction is that the tubular system will soon be replaced by lattices of rolled iron, in the manner of the American system."[153]

All the lumber milled from slaughtered American forests, assembled in thousands of "wooden" (that is, wood and iron) railroad bridges, along with all the mythic national characteristics of the supposedly practical, technology-borrowing Yankees, couldn't erase what a reconnaissance team from Britain confirmed during the 1850s: "We shall have to contend with no mean competitors in the Americans." In the United States, iron production was booming.[154] The mass buildability of the lattice bridge concept had made "American bridges" technological, industrial, and military threats to competitors. But no matter how invigorating the nationalist rhetoric and how intense the industrial competition, invention was and remains an outcome of transnational exchange.

The tubular bridge was a colossal construction whose creators positioned it as a symbol of national pride and industrial prominence; it elevated their professional standing and built their wealth. They did not foresee the consequences, though. Bridge chronicler J. G. James believed that the Britannia Bridge steered engineers and manufacturers in Britain to "mass production of box-girder bridges in the 100–200 foot range just at the time when, like other countries, she should have been perfecting cheaper truss forms"—and that this, along with a tradition of cast-iron arch building, led to the "sad decline of British bridge design in the late 19th century."[155] Firms such as Kennard in Britain, however, were successfully focusing on kit-of-parts truss forms; the "sad decline" story might merit revisiting. Still, builders in Britain may have followed the words or business interests of some of their elite engineers too well. Around the world, for a moment, the choice of a lattice over tube in long-span construction conveyed a message about Great Britain's technological-imperial ascendancy: Industrial might did not make technological right. Inventors and entrepreneurs living in an upstart nation, the United States, could participate as leaders. From a longer perspective, in which technological hits become has-beens and local advantages become drawbacks, so-called national activity might more accurately be understood as regional or dispersed and always likely to shift.[156]

Bridges, sensitive links within the networks of roads and rails essential for communication, transportation, business, and defense, were as crucial to the rise of the nation-states as to the lives of the individuals who crossed them. A bridge was a physical and symbolic bond—a structure of political strength. In diverse, unruly places such as England, France, Germany, and the United States, bridges, as structures and icons, helped bound the category, "nation." To local and multinational communities of builders, the success of a bridge type signaled who held the engineering power.

A BORROWER AND A LENDER BE

Historian Nathan Rosenberg asserts: "It cannot be overstressed that America in the first half of the nineteenth century was still primarily a borrower of European technology."[157] During that period, however, builders who lived in the expanding United States were in the forefront of the high-tech field of bridge building. Ithiel Town was one of a network of inventors whose work helped transform the design, construction, and analysis of structures and with it building and manufacturing processes and business practices around the world. Britain's *industrial* might was for a time greater than that of the continental European and North American states. This supported a myth of British *technological* priority that denies the transcultural exchange that is the sine qua non of invention and that was the basis of Britain's "Industrial Revolution."[158] It cannot be overstressed that national competition is only one tale we might tell and that nationalist notions about Europe and America might have tangential relevance to designers working within border-crossing innovative flows. Acknowledging multiple, permeable inventive spaces and the cooperation and competition that binds people within them enables us to alter how we regard our pasts and construct our futures.

Borrowing and lending are quintessential inventive acts. Deeming one nation inventor-lender and another imitator-borrower, however, supercharges a nationalist story of superiority and subservience that inflates claims of technological success and failure. Let's acknowledge this: Nations can levy taxes but nations cannot invent. At specific times and in specific places, individuals, materials, and methods may predominate, or seem to. Engineers and their chroniclers develop and debate ideas about structure, economy, and beauty. As historian Jonson Miller has written, "Practitioners . . . make decisions about what counts as engineering, what engineering is for, what problems it solves, how it defines problems in the first place, and, consequently, who benefits from engineering."[159] They work via experimentation and calculation and acts of denial, celebration, deception, and admiration. Exploring these fundamental technological activities will enable us to overturn rather than reinforce categories and boundaries that camouflage ever-changing relationships.

"Globalization," a word of relatively recent coinage, names a long-standing phenomenon of varying forms and intensities.[160] Questions about national identity smoldered in the nineteenth century and continue to do so. Labels—*Made in America*, *Made in China*—can hide the conceptual, material, and manufacturing origins of technological products.[161] As Adam Smith described in 1776, the fabrication of an article of clothing can be the outcome of a transnational process; that's still true today.[162] As is industrial competition among codependent nations. Still, the extent of twenty-first-century flows does not alter the consequences that national boundaries hold for people denied entry or exit or compelled to live and work in poor conditions.[163] In the nineteenth century, crosscutting inventive networks thrived within explosively expanding webs of transportation, communication, and colonization.[164] As globetrotting technologists engineered the world, making money and history, they built identities along with the bridges, railroads, and ships that sped the flow of people, publications, goods, and capital.[165]

7

GLOBAL POSITIONING

> One of the most extensive kinds of misrepresentation, adopted by European travelers in the United States, is found in the use of the word *American*. This word, when applied to the character, manners, or morals, of the people who inhabit the United States, is scarcely capable of having any meaning. Like *European*, it is an almost merely geographical or political term.
>
> —Yale president Timothy Dwight, ca. 1810

Fairy tales and histories of technological evolution typically feature conjurers (inventors) working within a kingdom (nation), who make the magic (technology) that sparks transformation. These closely allied narrative genres endow objects with life and character.[1] America was a real and imaginary domain that inspired storytellers to portray its technologists as evolutionary spawn whose superpower was "practicality," an endowment of Anglo-Saxon ancestors. Their supposedly unscientific nature (unlike the well-schooled Europeans) would make them builders if not masters of a rising industrial civilization, in contrast to the Indigenous people they were killing and whose land they were taking. As a crossroads and marketplace, America was ideal for the creation and spread of narratives and inventions. Perhaps it's no coincidence that writers gave the fairy tale as a genre its current shape around the beginning of the eighteenth century, also considered the initial phase of industrialization—and the period before the rise of a new devotion: nationalism.[2]

Among the earliest US-based writers to enjoy international fame, and an America-definer who traversed Ithiel Town's social circles, James Fenimore Cooper depicted in 1828 a technological object working the national soil:

> I have seen more beautiful, graceful, and convenient ploughs in positive use here [the United States], than are probably to be found in the whole of Europe united. In this single fact may be traced the history of the character of the people, and the germ of their future greatness.[3]

Viewing technologies as markers of national character and cultural progress was a nineteenth-century allegorical practice that survives into the twenty-first. Observers endowed machines with nativeness and foreignness; they converted the plow and railroad into national symbols of transformation. Cooper's glorification of an agricultural tool was intended to place his American homeland on an equal footing with Europe.[4] While defending the life and culture of the United States (in the 1820s, land east of the Mississippi), Cooper portrayed technology yielding a seed from which national destiny would sprout. Unmentioned was that that seed embodied knowledge conveyed and cultivated through exchange, in the Caribbean and North America, of Europeans with enslaved Africans who contributed their knowledge to the development of iron and agricultural tools.[5]

Ithiel Town's launchpad, New Haven, was located within one of America's most celebrated technological regions—historians once considered Connecticut "the cradle of mass production."[6] The state hosted a dense, ramifying web of people with access to capital and government contracts. The Springfield Armory—about 65 miles from New Haven and near the Connecticut River, site of one of the first lattice bridges (1820) and one of the earliest Howe-lattices (1838–1840)—supported the Connecticut Valley's becoming a place where technologists fostered the fundamentals of management and manufacturing.[7] Yale professor Joseph W. Roe explained in 1934 how the inhabitants of Connecticut had come to possess a highly inventive character: "First, there is race . . . The Connecticut mechanics in the main traced back to the middle class English, the people who brought about the Industrial Revolution and who have led Europe industrially to this day."[8] Roe was drawing on beliefs at least a century old, including the theory that each race or nation had its own specialty. Michel Chevalier, for example, had believed that "ascendancy . . . has passed to the people of Saxon origin." He wondered in 1837, "How and in what form will we manage to appropriate the innovations of the English race?"[9] In 1987, historian Edwin T. Layton Jr.'s quasi-biological observation clarified that "American engineers followed a deep-rooted Anglo-Saxon philosophical bias in favor of an empirical, inductive style. In this respect, then, British and American engineers had much in common. But American engineers carried these shared traits much further."[10] The broad-brush characterization of Americans as inventive Anglo-Saxons helped ensure that, into the twenty-first century, "engineering is constituted of a profoundly raced set of activities."[11] Acknowledging that genes play no verifiable role in invention and that detection of inherited philosophical biases is challenged by the ambiguity and diversity of everyday creative practice, we can avail ourselves of a simpler, verifiable explanation for the making of the innovative American: exclusion and violence.[12]

The mythology of national difference was nurtured in the new and changing landscapes of professionalizing, nation-building engineers; it was about the qualifications, race, social class, and authority to make decisions, and how independent engineers might remain in relation to the organizations that employed them.[13] For example, claims about empiricism or science indicated membership in a specific national or professional community. The real or rhetorical tension between practice and theory was, in any case, long a leitmotif of technological discussion, coloring observations of eighteenth-century Grubenmann bridges and recurring through subsequent centuries.[14] But public declarations tell us little about how inventors and designers actually worked or if there was consistency within nations, technological fields, or epochs.

The co-creation of infrastructure, place, and identity is the triangulation behind community. Keeping in mind the diversity of environments, cultures, and politics contained within and across national borders, we'll explore three lattice-inflected cases that reveal how the term "American," "scarcely capable of having any meaning," can also have so much. We'll take a look at elite creative communities positioning a nation in the world by giving meaning to "America" through symbols, structures, and professional practice.

AMERICA IN THE MAKING

Bridges played a starring role in the construction of "America" and its meanings. Although there is no indication that Town engineered the lattice's symbolic function, he was part of a network of writers, artists, and inventors who crafted pictures, structures, and institutions intended to mold the character of the United States and its citizens.[15]

Town, who thrived in elite society in Boston, New Haven, New York, and several southern states, might be portrayed as the embodiment of white, American male privilege. He might also be seen as an upstart able to cross educational, social, and economic barriers. Unlike some of his friends, Town apparently rose to the upper crust, from carpenter to designer, entrepreneur, and tastemaker. He was not educated within, and apparently not considered a member of, the civil engineering profession, on which he had an outsized impact. Nor was Town a university graduate, though he did receive an honorary Yale University degree. Perhaps the high-class company he kept saw him as a parvenu. In any case, in New Haven, home to Yale and formidable social networks, Town's neighbors and friends included Benjamin Silliman, superstar Yale scientist from a slave-owning family who became an ambivalent abolitionist, and Samuel Morse, anti-Catholic, anti-immigrant, virulently pro-slavery artist-inventor. Town was also connected to religious, slave-owning Yale luminaries such as Timothy Dwight, who like Morse was a Calvinist evangelical Christian who believed that American destiny was preordained, and to abolitionists such as artist Nathaniel Jocelyn.[16] The financial and social ties linking citizens of New Haven to the South, where Yale was raising money, helps explain Town's early development of lattice bridge construction in the Carolinas (see chapter 5).[17] One of numerous

Figure 7.1
Thomas Cole, *The Architect's Dream*,1840. This allegorical landscape highlights historic architecture available for inspiration and appropriation. Dedication: "Painted by T. Cole for I. Town, Archt. 1840." Toledo Museum of Art (Toledo, Ohio). Purchased with funds from the Florence Scott Libbey Bequest in Memory of her Father, Maurice A. Scott, 1949.162.

New Englanders who came to New York City for its opportunities, Town established an office there around 1820. Though evidence is lacking that he participated in the regular lunches that James Cooper enjoyed with Morse, painter Thomas Cole (1801–1848), and members of the intelligentsia, Town and Cole were among the founding members of the National Academy of Design, led by Morse, in 1825.[18]

Projecting an image of the United States and its future was an objective of several members of Town's social group, but there wasn't necessarily agreement about what that image should convey.[19] In 1840, Town rejected *The Architect's Dream* (figure 7.1), a painting he had commissioned from Cole, a founding father of a landscape genre intended to represent "America." Town had expected to receive a picture of verdant American scenery.[20] What Cole produced was an epic architectural panorama of the evolution of civilization, Egyptian to Greco-Roman, with a Gothic church standing near the opposite side of a bridgeless divide: the waterway separates urban and natural, past and present, light and shadow. Between them, the reclining dreamer appears to be Town (though some believe he represents Cole).[21] *Architect's Dream* might be described as a paean to the artist-architect's creative mind, but the image's sweeping, hard-edged blend of architectural historicism and futurism seems to foreshadow construction's and industrialization's environmental destruction. This is something Cole feared, though his and Town's livelihoods depended on the wealth generated by industrial capitalism; American bridges nurtured or destroyed (sometimes both) economies, landscapes, and nations. The painting seems plucked from the artist's renowned five-part *Course of Empire* (1834–36), which portrayed civilization's rise and fall, beginning with *The Savage State*. *Architect's Dream* would have fit before the final two tableaux, *Destruction* and *Desolation*.[22] The series offered an ominous alternative to the techno-utopian narrative of national and human evolution.

Sculpted in marble like a Greek god and festooned with honor cords, Town (figure 7.2) rests on an emblem of his creative energy: Stuart and Revett's *Antiquities of Athens* (published in several volumes, London, 1762–1816). Like the tomes beneath Cole's dreamer, the book underscored the expansive knowledge networks that fed Town's output, all supported by the lattice. His architecture featured tasteful Greek markers intended to instill moral values. The assembling of uniform classical forms was intended to build an ethical, industrial vision of America. Classical architecture (a word with Greek heritage) (figure 7.3) evoked democracy (ditto), the Greek war for independence from the Ottoman Empire (1821–1830), and the slavery of antiquity.[23] Hellenic cachet functioned north as well as south of the Mason-Dixon Line, where enslaved workers, who participated in the construction of classical-style architecture, were essential to Town's and the nation's wealth.[24] Designers in France, Germany, England, and the United States had and

Figure 7.2
Chauncey Bradley Ives, Ithiel Town, 1842. Gift of Dr. William Thompson Peters, BA, 1825, MD, 1830, Yale University Art Gallery.

would draw on a variety of classical styles. It did not matter precisely where the elements originated or the values they once represented. The intent was for citizens of the industrializing nation-states to feel, as we still do, that classical styles reinforce the imposition of order on society by large, venerable institutions.[25]

Town, designer of American bridges and American identity, was a transitional figure who altered the course of his own life, along with architectural and engineering practice. He contributed to the reconception of structure and process and the building of orderly American cityscapes.[26] Intentionally and unintentionally, he helped construct the image of a nation. His proximate cohort positioned America in the world and put its iconography into place.[27] Town's residences, public buildings, and bridges might be called markers of American progress, but, as was apparent to Cole, there was a disconnect between American dreams and the devastation of American environments.

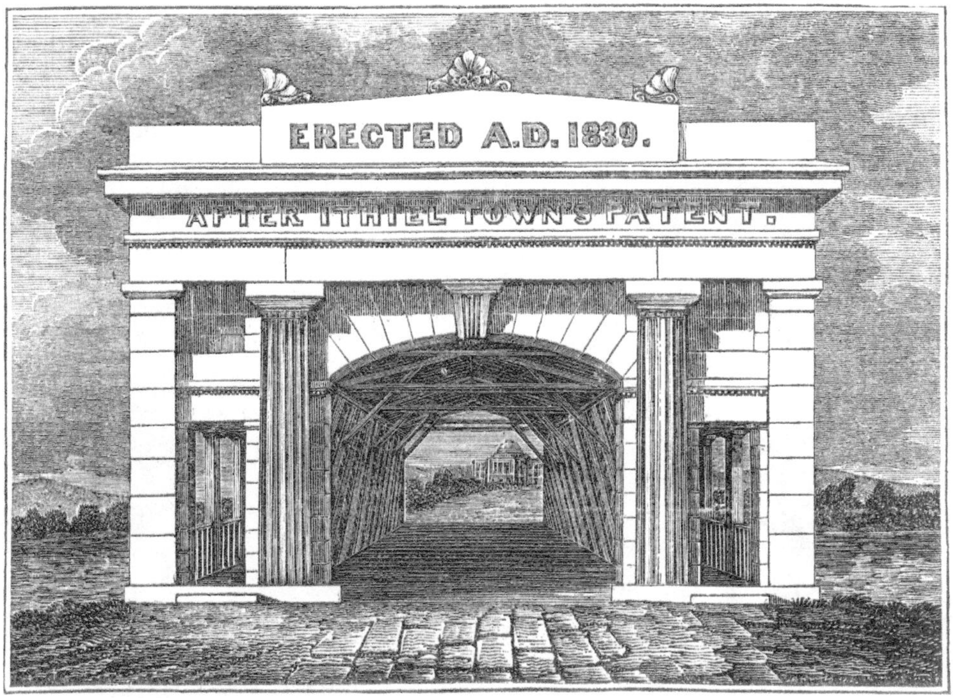

No. 7. A design in perspective, for a front entrance of a roofed bridge, suited to the rich style of No. 5, with side-walks, &c.

Figure 7.3
Bridge on the National Road over the Scioto River, Columbus, Ohio (1832–1824) or over the Potomac, Alexandria, Virginia (1832). A Greek portal masked and reinforced the geometric rigor of the industrialized structure behind it. The bridge stands boldly in the landscape, as if in a Stuart and Revett engraving. Ithiel Town, *A Description of Ithiel Town's Improvement in the Principle, Construction, and Practical Execution of Bridges* (1839), 16. Courtesy of Smithsonian Libraries and Archives.

From the early 1830s, Town and his partner Alexander J. Davis, the latter apparently the more artistically adept of the pair, perhaps inspired by Stuart and Revett, planned a large, luxe multivolume publication, to be printed in England and entitled either *American Views* or *The American Architect*. Beyond promoting their role in the development of architectural archetypes, the tomes would have solidified their vision for the nation.[28] It was never published. Years later, during the summer of 1889, Connecticut demolished Town and Davis's first joint project amid controversy about what the building represented (figure 7.4). Seven men turned a windlass pulling cables that brought down the columns as about 3,000 people watched.[29] We might consider this a small-scale enactment of Cole's *Course of Empire*. In March of that year, the Eiffel Tower and Galerie des Machines had opened at the Paris Exposition. The Grecian style may have seemed outmoded, but

Figure 7.4
Demolition, 1889, of the State House (figure 2.21) (Ithiel Town and Alexander J. Davis, 1831), New Haven. New Haven Museum.

industrial representations of nation rivaled but did not replace—and sometimes, as we'll see in chapter 8, incorporated—classical iconography that affirmed national power.

UN-AMERICAN BRIDGE

In order to understand how bridge designers worked and reworked the lattice over time, should we turn to evolution? Deem materials the masters of men, rather than the reverse? Dive into the souls of nations, on the hunt for engineering traditions?[30] An alternative, exploring the motives of individuals and tracing relationships, is more complex but less magical. Geographer Fred Kniffen, who investigated covered-bridge decision-making, noted that "different [bridge] trusses showed a marked tendency to certain areas, largely because of the activities of ardent disciples or effective salesmen."[31] Covered bridge types built in specific regions of the United States were dependent on government officials, local builders, personal ties.[32] But, as usual, the idea of "local" can mislead: bridges on a rail line, for example, can reflect the thinking of designers working in a railroad or construction company (with sources all over the map) without reflecting the places the railroad runs through.

Place matters, but people define place as they operate across multiple, which is why generalizations and predictions based even on one region can be a challenge.[33] Without deep access to the creation of local knowledge, place-based observations often are merely notional. Still, places are an essential component of borderless local-global webs. In our case, knowledge networks, animated by carpenters, mechanics, engineers, and entrepreneurs working across technological fields, generated the inventive industrial energy of Ithiel Town's base, the Connecticut River Valley in the nineteenth century. This is a condition familiar to observers of Silicon Valley, where, likewise, local social webs formed through practices of inclusion and exclusion plugged into ever-changing networks beyond the region's borders. Before the internet, before the telephone, even before the building of easily passable roads, designers communicated in person, via text (that is, handwritten correspondence), and printed matter.[34]

The wooden lattice railway bridge seems to have been rejected by most railway engineers in the United States by 1850, just as the type was being adapted, in iron, for long spans by engineers based in Europe, as we've seen. Long after metal had supposedly replaced wood, however, railway lattice bridges with members of the latter material continued to be built: there were "at least 100" on the Boston and Maine Railroad in 1900 (figure 7.5). Why? Engineer J. Parker Snow advocated the construction of wooden lattice railroad bridges and had the authority to make it happen. (Snow also applied his knowledge of structures to the iron frames of early high-rise buildings.)[35]

In the United States, some no longer considered the lattice American. The *Journal of the Franklin Institute* in 1868 reported, regarding the design of a bridge by Felician Slataper (figure 7.6), born and educated in Austria, that "the principle of construction is known as the latticed-girder plan with vertical stiffenings, and the structure is believed to be the only one of the kind in the United States, although the principle is used very extensively in modified forms on the German, Russian, and Italian railways."[36] By 1891 in New York State, though, there were about 2,500 "railroad truss bridges," including dozens of riveted iron lattice railroad bridge spans. In the state's largest railroads, the New York Central and the Hudson, thirty-three out of 177 bridges, most built during the thirty previous years, were lattice; on the Delaware and

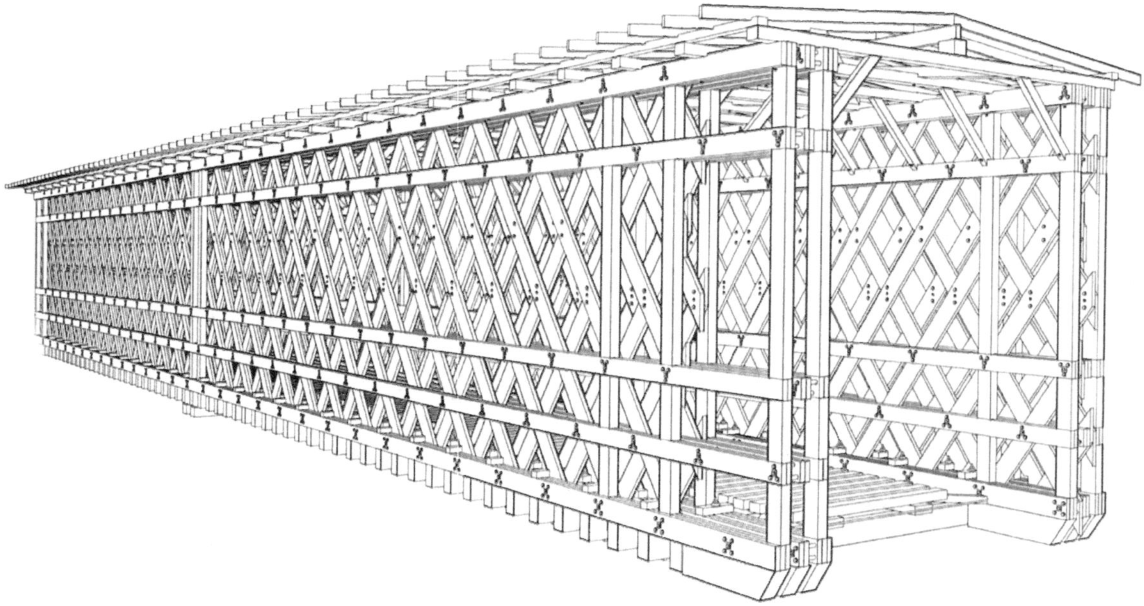

Figure 7.5
The Boston and Maine bridge (J. Parker Snow, 1889; served the railroad till 1962) over the Contoocook River, Contoocook, New Hampshire. The oldest surviving covered railroad bridge in the United States was built the year the Eiffel Tower opened. See fig. 6.15. Amy James, artist; Library of Congress, Prints & Photographs Division, HAER No. NH-38.

Hudson Railroad, thirty-two out of eighty-eight bridges were lattice bridges; none of the 117 bridges on the New York, Lake Erie, and Western Railroad were lattices.[37] Lattice-inspired and -averse designers developed structures reflecting a variety of practices and beliefs.

Engineers of the period believed that three railway companies (through "completely accidental causes," in the words of one), each with its own community of technologists, developed schools of thought representing distinct values, working practices, and specialties.[38] As one engineer expressed it: "In the same way that the Baltimore and Ohio Railroad fostered the Fink type of iron bridge and the Pennsylvania the Whipple and the Pratt trusses, so the New York Central became the promoter of the Town truss and lattice riveted bridges."[39] It's hardly surprising to discover that there were links among these three railroad-company communities. George E. Gray, chief engineer of the New York Central Railroad from 1853 to 1865, referred to Squire Whipple, a founding father of bridge analysis and an early iron bridge inventor and builder associated with the Pennsylvania Railroad, as "my old friend and mentor"; it would not be much of a

Figure 7.6
Alleghany River Bridge, Pittsburgh, Fort Wayne and Chicago Railway (Felician Slataper, built 1864–1868, demolished 1903), Courtesy of Pittsburgh History & Landmarks Foundation,

leap to discern a transrailroad-company link between Whipple's bridge and iron lattice designs—something Whipple himself acknowledged.[40] Gray thus would have had access to a pioneering method of structural analysis before Culmann and Schwedler had published their work (see chapter 4). Whipple, at the beginning of 1847, distributed "some fifty or sixty copies" of his method of determining the stresses in the individual members of truss structures, before printing it later that year within a self-published book; though he did not distribute many copies, he shared his method with engineers that he knew. If we imagine that, during the early days of railway building, only a few began to employ it and each shared it with a colleague (this is a modest scenario for the spread of what Whipple modestly described as "valuable information"), then the mythical "practical" character of the American is due for reappraisal.[41] An understanding of the professional landscape will help. Around 1850, the US census counted 512 engineers, most in states adjacent to New York (where sixty-two lived); this was a select group of professionals who were in communication with each other.[42] They lived in a region of the young United States with a high literacy rate and broad reach of publications—well in advance of such rates in the more populated urbanized areas in Europe.[43] It is possible, though it would be surprising, that someone didn't introduce Whipple's teachings to Culmann when, 1849–1851, he traveled through New York and other states to investigate bridge structures and analysis.[44]

Gray was chief engineer of the newly formed New York Central, which under Cornelius Vanderbilt consolidated multiple lines that eventually stretched into the Midwest.[45] Under his direction, the New York Central appears to be the earliest railroad in the United States to construct large wrought-iron bridges, including lattices, on its lines.[46] When first given responsibility for the rapid design and construction of bridges, he, in his own words, "sought every known publication at home and in Europe." Realizing that he did not have the budget for the materials and labor required to build the iron bridges he read about, Gray determined that he would "cut out what seems unnecessary materials in the European plans."[47] He worked in New York State in 1859–1861, assisted by Howard Carroll, who had "brief experience" assisting iron-lattice bridge pioneer John Macneill in Ireland (see chapter 6).[48] Based on research including lab testing with bridge members assembled from iron shapes obtained from manufacturers, Carroll developed wrought iron lattice bridges, including over the Mohawk River at Schenectady (considered the earliest large all-wrought-iron railway bridge in the United States, 1859; replaced ca. 1900) and the Hamburg Street Canal in Buffalo (ca. 1860), followed by longer spans, including one over the Tonawanda Creek in Batavia, New York (1861).[49]

Engineer Charles Hilton succeeded Carroll (after the latter's enlistment and subsequent death in the Civil War) at New York Central, worked under Gray, and assumed Gray's position when he left. Hilton did not learn directly from Carroll; when he assumed Carroll's position, Hilton had "no experience in iron bridge building," as he had specialized in wooden bridges.[50] Hilton brought that knowledge to the design of the early riveted lattices for the New York Central, at places including Canastota (1864) (figure 7.7), Newark, Rome, and Schenectady (1866); these structures were noted for longevity and safety. Hilton became known for lattice structures and founded his own construction company (figures 7.8 and 7.9).[51]

We've been recounting what may feel like a mundane tale—the aftermath of developments explored in previous chapters of *American Bridge*. Which is to say, we're observing innovators rethinking the lattice as they shifted it from wood to iron. Instead of causing the breasts of American storytellers to swell with metallic pride, however, the transition fell into history's shadows. This may have happened because, at that point, some

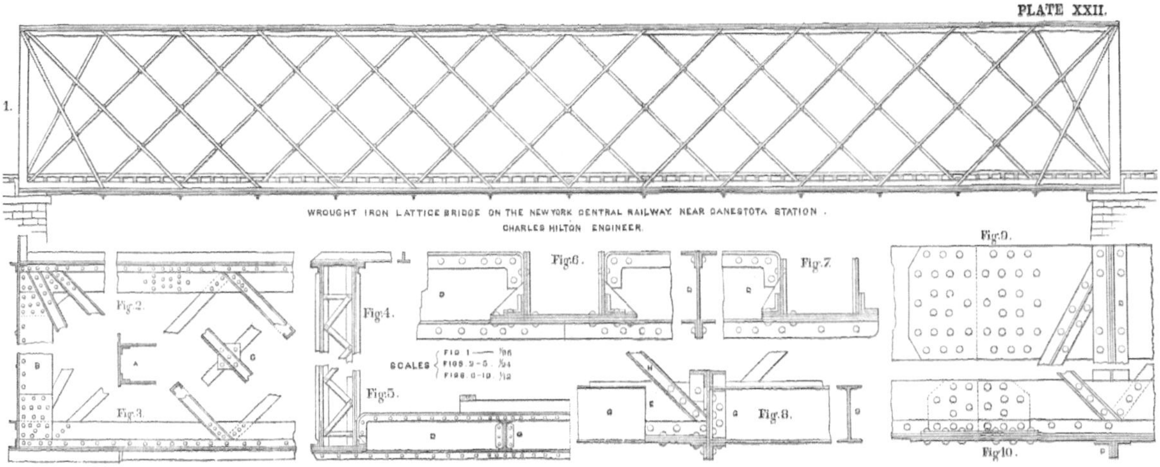

Figure 7.7
Bridge over the Erie Canal at Canastota, New York (Charles Hilton, 1864). In this design for an early iron lattice bridge built in the United States, the parts were riveted (as each little circle indicates), including at each intersection, reflecting Town's patent. George L. Vose, *Manual for Railroad Engineers and Engineering Students* (Lee and Shepard, 1874), pl. 22. UC Berkeley Library, University of California,

engineers in the United States considered riveted iron lattice bridges to be *un*-American. They called them "European" or "English," in reference to places where metal, riveted structures, were becoming common (figure 7.10).[52] In other words, they branded the American bridge a foreigner and attempted to replace it with pin-connected structures.

Pin-connected bridges were lightweight, cheap to manufacture, and quickly buildable—conceived as a kit-of-parts system. Metal "pins," large smooth rounded solid metal cylinders (with nuts tightened on both ends after installation) held eye bars (metal elements with a hole at each end, which could be installed in multiple parallel sets) (figures 7.11 and 7.12). The structures reflected a model that engineers used in truss bridge analysis: they calculated with the assumption that independent members could rotate freely and independently of each other when a load crossed the bridge. This idealized connection would transfer tension or compression without engendering bending or shear forces. It is the typical theoretical condition for a truss bridge (figure 2.9).[53] Theoretical models and reality can diverge, however; in this case, the parts likely did not easily rotate, which meant that the structures performed differently than predicted; the ramifications of the secondary stresses if they did not rotate were not understood for many years. "Strain (i.e., stress) sheets" that indicated the calculations behind the dimensioning of parts dominated the design of pin-connected bridges; in their earliest phase of

Mr. Thomson. Mr. Hilton's first diagonal lattice was the bridge over the Erie Canal near Canastota. The bridge is illustrated in the "Manual for Railroad Engineers," by Professor G. L. Vose, but the attachment of the floor beams to the chords was by a diaphragm distributing to both webs of chords, and not as shown in the book mentioned.

Mr. Hilton was not inclined to eccentric connections, however, and he abandoned panel-point plates after 1862. He always had in mind elastic relations, especially in truss connections, and therefore the connection of a thin angle to a thick plate is not found in his designs. Fig. 3 shows one of his diagonal-system bridges. The horizontal member -*y*- takes both tension and compression; this member is left out in European bridges of this type. He computed trusses from engine

FIG. 1. CARROLL THROUGH LATTICE

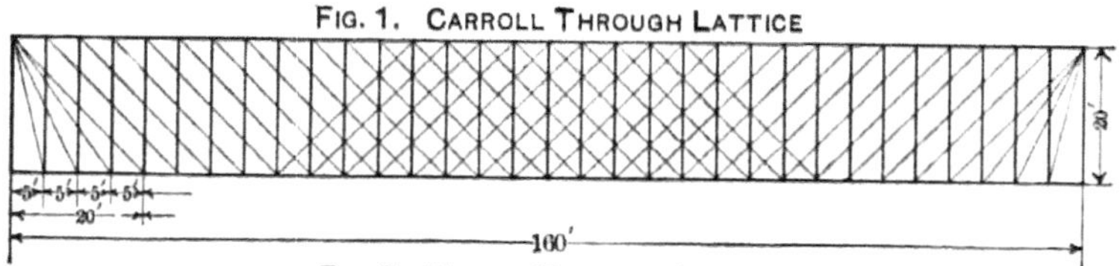

FIG. 2. HILTON THROUGH LATTICE

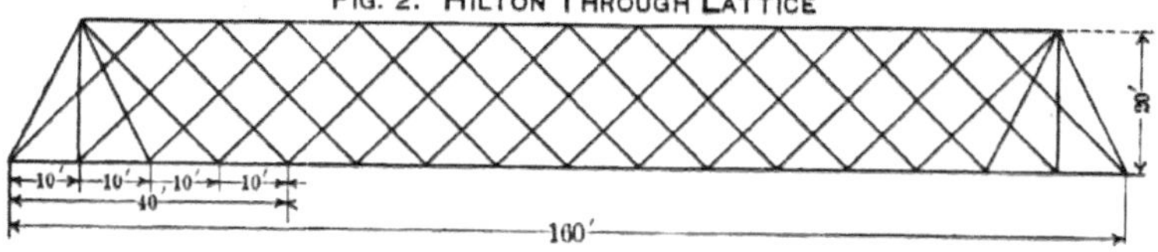

FIG. 3. HILTON LATTICE

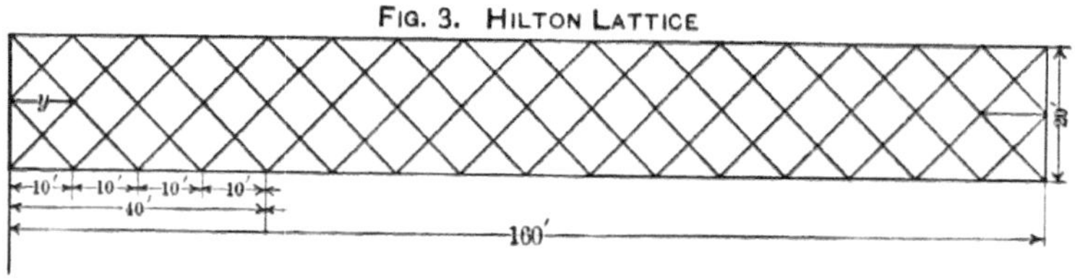

wheels in the year 1862, and, prior to 1878, made specifications for wheel loads such as are now current.

A. S. C. Würtele, M. Am. Soc. C. E., began in 1868 to design lattice bridges following closely Mr. Hilton's practice. In 1869, S. J. Fields, M. Am. Soc. C. E., a pupil of Mr. Hilton, designed lattice bridges, following the same practice with improvements in details of chords, etc. Charles McDonald, M. Am. Soc. C. E., about 1875, also designed lattice bridges with improvements in details. In 1876, John F. Alden, M. Am. Soc. C. E., a pupil of Hilton, further improved details.

As early as 1873, the speaker talked over the matter of short versus long panels with Mr. Hilton, and in 1885 he began to use long panels,

Figure 7.8
Lattice bridge designs by Howard Carroll and Charles Hilton for New York Central Railroad; Design in Figure 2 dates from c. 1864. George E. Gray, "Notes on Early Practice in Bridge Building," *ASCE Transactions* 37 (1897), 6. Princeton University.

Figure 7.9
Bridge (probably Charles Hilton) over Erie Canal at Pittsford, New York, New York Central Railroad. Postcard, Canal Society of New York State.

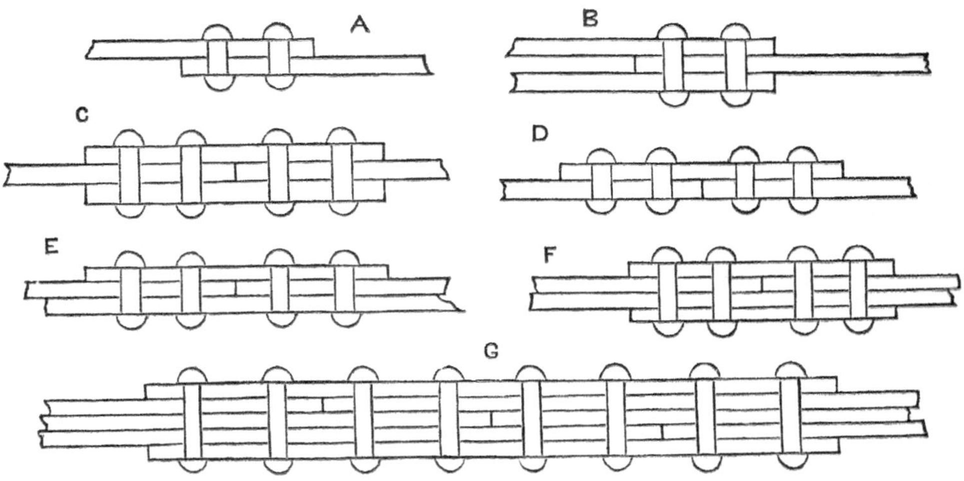

Figure 7.10
Rivet examples (in section). There was a diversity of rivets, riveting configurations, and types of plates. Note: E was used in the Conway and F in the Conway and Britannia bridges. Riveting of metal plates is comparable in some ways to pegging planks of wood. George L. Vose, *Manual for Railroad Engineers and Engineering Students* (Lee and Shepard, 1881), 170. UC Berkeley Library, University of California.

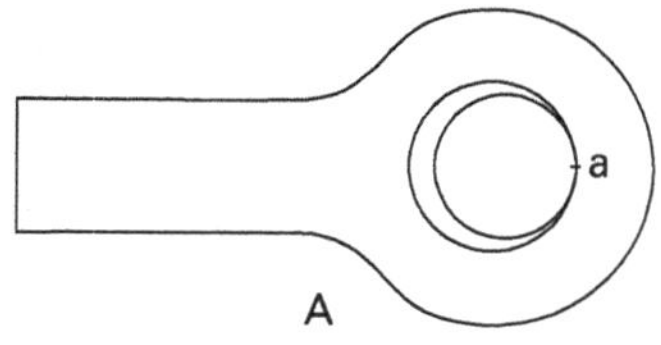

Figure 7.11
The end of an eyebar used in pin-connected bridges. George H. Thomson, "Riveted Lattice for Railroad Bridges of Maximum Span: A Plea for a Return to Rational Design," *Engineering News* 59 (January 23, 1908), 93. Princeton University.

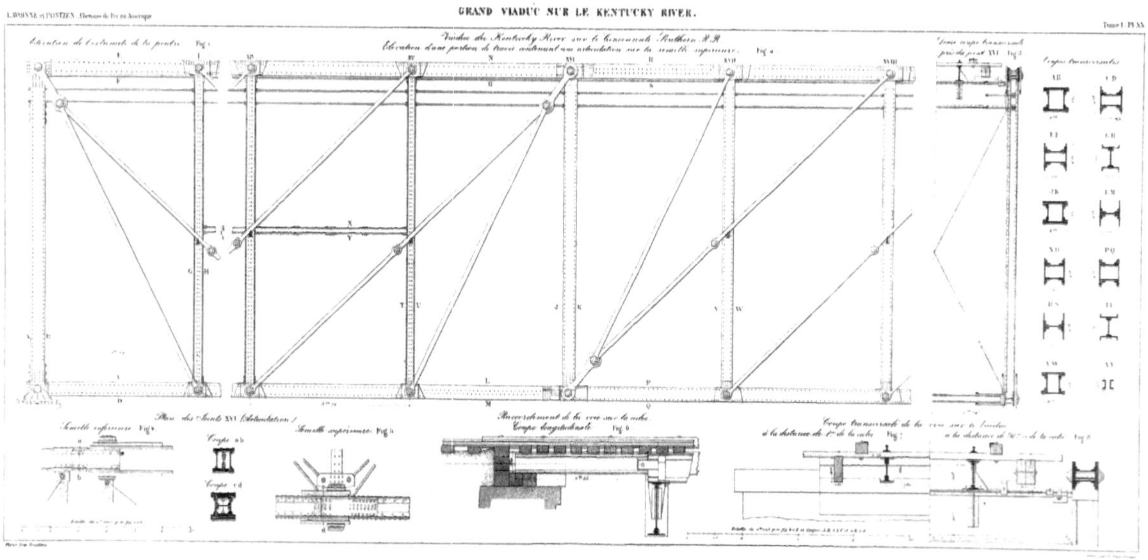

Figure 7.12
Cantilever bridge over the Kentucky River (C. Shaler Smith & Gustave Bouscaren, 1877), Cincinnati Southern Railway. This was a structure with riveted members and pin connections. The latter also served temporary members that supported the bridge during erection. Édouard Lavoinne and Ernest Pontzen, *Les chemins de fer en Amérique* (Dunod, 1880), vol. 1, pl. 20. University of Michigan.

development, engineers appeared to base the design of bridge members primarily on calculations and on ease of construction rather than focusing on in-service conditions.[54]

US bridge-building companies, which became prominent during the 1860s and 1870s, had their specialties.[55] They sold bridge types to customers including railroads that did not have the resources to establish their own bridge design and building departments. Sometimes, joint ownership of bridge building and iron manufacturing facilities helped shape company practices. This led to efficiencies and conflicts of interest that could have life-and-death consequences, as we'll see. Pin-connections enabled firms to minimize the need for skilled riveters at the construction site; workers assembled members from angles and plates and riveted them together in the factory, where work could be more easily implemented. Parts and components were then transported to the site, where workers could rapidly put the bridge together, although the actual ease and speed of construction may have been exaggerated, especially after the first generation of pin-connected bridges.[56] Workers adjusted these bridges during construction, which added to uncertainty about what the stresses actually were and when they would occur.[57] In contrast, riveted lattice bridge truss frames, depending on their size, could be assembled in the factory and brought on rail cars to the site.[58]

A pin-connected-versus-riveted debate continued for years, sometimes couched in America-versus-Europe terms. In Britain, workers were experts in riveting, as exemplified by the Britannia Bridge and based on extensive experience in shipbuilding and boilermaking.[59] Britain-based firms did, however, manufacture pinned bridges from the middle of the century, apparently mostly for export. On the American continent, one bridge type popular in the United States during the 1850s had been developed by Albert Fink of the Baltimore and Ohio Railroad; this American engineer, born and trained in Darmstadt, designed a system that resembled a structure developed by French engineer Polonceau. The Fink spans of the Verrugas Viaduct in Peru (figure 7.13) were an example of a kit-of-parts structure that enabled construction in very challenging conditions. The trusses, shipped in parts from the Baltimore Bridge Company, were pin-connected and bolted. There was no on-site riveting. Experience with export bridges provided training for careers in the United States; engineers involved in this project would be responsible for major bridges in New York.[60] The eyebars employed in pin-connected bridges required an investment in manufacturing precision that took even further the standardization of structural members. Although we've been told that the prioritization of theoretical knowledge was not an American trait, the preference for pin connections appears to be a result of the ability to analyze trusses without theoretical ambiguity, that is, to be "scientific." But the stresses that resulted from both pinned and riveted connections were not precisely understood.[61]

Pin-versus-rivet strife was about more than structural understandings, cost, and the qualifications of construction labor. Riveted lattice bridges appeared to be significantly safer than pin-connected bridges due to redundancy (that is, if one part failed, others could take up the slack). They provided greater rigidity, less vibration (and resulting wear), greater lateral stability, and much better resistance to impact. They functioned more as a unit than as a set of perhaps too easily broken asunder parts that approximated an engineering diagram. Lattices provided greater resistance to phenomena

Figure 7.13
The first Verrugas Viaduct, with four Fink spans (C. Shaler Smith, Charles H. Latrobe, and consultant Walton White Evans, 1873), Oroya Railroad in the Andes, Peru. Ernest Pontzen, *Der Varrugas-Viaduct mit einigen anderen Viaducten verglichen* (Lehmann & Wentzel, 1875). Deutsches Museum Library.

beyond their anticipated capacity, including derailed trains, broken axles and wheels, extra-heavy loads, collisions between trains, objects protruding from trains, and interactions with boat traffic.[62] In other words, lattices were safer, even if no longer considered American.

SAFETY AND "AMERICAN RAILROAD BRIDGES"

Relationships (stretching across regions and nations), place (geographic and political), cost (how much and who should pay), and railroad management (at multiple levels) all collided within debates around the safety of railroad bridges. The lattice, in apparent competition with the new, pin-connected type, became a touchpoint within discussions shaped by evolutionist and nationalist mindsets.[63] Railroad bridges posed a design challenge that was also a moral question about professional responsibility to the public and about who takes the blame. What makes a structure safe? Safe for whom? Should foreseeable accidents be prevented? At what cost? Who's responsible: bridge designers, railroad company managers, insurance companies, government? Beneath these lurk more questions. Who defines the limits of engineering?

Bridge engineer George H. Thomson (1847–1910), who spent twenty-one years (1872-1893) with New York Central, was considered a "violent opponent" of pin-connected bridges.[64] Engineer Benjamin Baker in 1888 read in Bath, England, a paper of Thomson's, entitled "Mechanical Pathology, Considered in Relation to Bridge Design." It was published in the London-based journal *Engineering* under the more sensational title "American Bridge Failures," accompanied by tragic images of collapse and wreckage.[65] One of the pictures presented an incongruous scene. A span of a pin-connected iron railroad bridge, with locomotive and freight cars, had disappeared into the West River in Brattleboro, Vermont, almost without a trace—killing two, severely injuring five (figure 7.14); it was the apparent outcome of railway mismanagement and inadequate design.[66] In the foreground, a couple of cars of the fallen train, including the passenger car, remained on land. Hovering in the middle of the image, damaged slightly by the adjacent collapse, was—could it be?—a wooden lattice bridge. Thomson intended this survivor to impart a lesson. As he later would clarify, "Some of the earliest lattice timber bridges are doing duty to-day and those that have failed have given ample warning of their weakness."[67] Sturdiness and the ability to forewarn made the lattice safe. The spectacle of the Brattleboro collapse, which we might entitle American Bridge vs. American Bridge, brought back into consciousness an evolutionarily passé structure as a model of solidity, dependability, and security.

How different was the lattice from pin-connected structures? In the earlier years of the latter, quite different. There was cause for concern about how well pin-connected bridges performed. But by the 1870s, what Thomson called "an invasion of rivets" revealed that designers of pin-connected bridges increasingly incorporated the lattice bridge design features of Carroll and Hilton, whose "influence [was] pervasive and permanent."[68] Thomson advocated for riveted lattice so that bridges would be able to survive, in his deliciously ironic phrase, "that railroad struggle for existence to come, between structures and wheels."[69]

In January 1908, five months after the horrific collapse of the Quebec Bridge during construction, which we're about to explore, Thomson spoke out again and articulated "lattice dogmas," which were "representative of the 'lattice school' of engineers, as exemplified by their practice for sixty years." The fact that the lattice diagonals were embedded in the top and bottom chords and connected at their intersections created an "interdependence" of members (figure 7.15). They supported each other, serving the members in compression (the pushing-together force that can result in bending or

Fig. 3. West River Bridge Accident; Brattleboro and Whitehall Railroad. August 18, 1886.

Figure 7.14
West River Bridge (A. P. Boller, 1879) collapse, Brattleboro and Whitehall Railroad, August 18, 1886. A span of the pin-connected bridge and a trestle failed, stripping the adjacent covered bridge's siding in two areas. The crowd of onlookers in the photograph does not appear in the published image. George H. Thomson, "American Bridge Failures," *Engineering* 46 (1888): fig. 3; Photograph courtesy of the Vermont Historical Society.

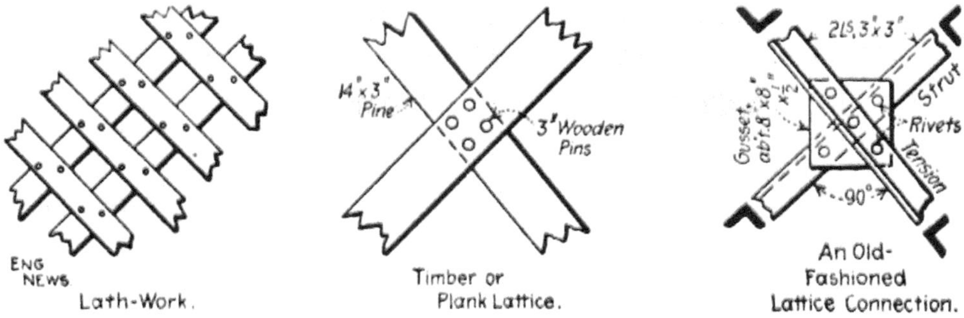

FIG. 3. LATTICE-WORK DETAILS IN TIMBER AND IN STEEL.

Figure 7.15
Thomson's examples of lattice "interdependence." Left to right: wooden lath, a foundation for plaster walls; wooden lattice bridge members; early iron lattice bridge intersection. George H. Thomson, "Riveted Lattice for Railroad Bridges of Maximum Span: A Plea for a Return to Rational Design," *Engineering News* 59 (January 23, 1908): 92. Princeton University.

buckling) and helping to keep the bridge together under impact. Thomson expressed deep skepticism about overdependence on stress calculations and emphasized the importance of knowledge derived from practice, in order to design for actual conditions.[70]

Thomson's approach might be described as wisdom-based. The *Engineering News* editor who published Thomson's 1908 response to the Quebec collapse, however, distanced himself from the author, calling him "frankly and radically reactionary."[71] Almost a century after Town's 1820 patent, the lattice came from the Before Science era; its roots were wooden, and it carried the whiff of craft. Evolutionists saw wood and lattice as low, decaying rungs on the ladder of progress. One engineer insisted that Thomson supply "facts and not mere opinions."[72] Rather than denying facts, Thomson contextualized and interpreted them, while emphasizing engineering as a practice of care. When questioning the endurance of eyebars used in pin-connected structures, for example, Thomson remarked: "It is not opinions, but proofs, that should answer these questions, for they are vital as human life, and as important as the mighty dollar."[73] Some engineers considered Thomson's lack of public obeisance to science apostasy; he was presenting a design philosophy that prioritized the value of human life over "science" (which sometimes seemed to serve as a euphemism for profits). With "34 years' experience with lattice bridges (and nine years prior experience with bridges of stone, wood and pin-connected bridges as well)," Thomson made a "plea" for safety through "rational construction along lines of which we have absolute knowledge" through experience rather than theory and uncertainty.[74]

Engineer Theodore Cooper, chief designer of the Quebec Bridge and judged most responsible for its collapse during construction, was starkly opposed to Thomson's life-safety-first point of view. We know this thanks to a paper Cooper presented and discussed years before, at an 1889 meeting of the American Society of Civil Engineers, and published that July in its *Transactions*, entitled "American Railroad Bridges."[75] The article later became a basic resource for (engineer-)historians because it contains a brief section, without citing sources, that describes bridges built in the early years of the United States. It likely drew on publications such as Howard Douglas's *Military Bridges* (figure 3.5) and helped establish the US structural canon. Rather than focus on this retrospective section, we'll use Cooper's discussion of contemporary bridge building as a jumping-off point for an exploration of the implications of the evolutionist, nationalist mindset.

"American Railroad Bridges." On its face, the title of Cooper's article might today appear unexceptionable. But professional rivalries and debates, nationalist anxieties, and deadly accidents hover behind it. Cooper was not naming a natural category, as one might present a species of tree growing in a specific geographical area. He was attempting to define what "American" practice meant (regarding pin-connected bridges, in particular) and confirm his own significance in an evolution of bridge structures, while leaving a record for posterity in the pages of the primary organ of the American Society of Civil Engineers, which despite its august name could not represent the full diversity of views of engineers working in the United States. Perhaps not coincidentally, Cooper's article appeared during the year that the conspicuously French and remarkably tall Eiffel Tower opened to the public. It almost doubled the height of the formerly tallest in the world, the Washington Monument—a designedly American masonry pile completed about five years earlier, after a forty-year period, and accurately characterized by Gustave Eiffel as "constructed with immense difficulty."[76]

Cooper established in a brief introduction that bridges of the day were "the results of a true evolution:" "We have been able to select the good and reject much that was bad." He explained: "The pioneer life, not only of the earlier settlers, but of each generation to the present day, has developed to a high degree the energies, ingenuity and self-reliance of the American people"; early bridge builders were "wonderful" but unscientific, unlike Cooper's late-nineteenth-century scientifically formed colleagues. The narrative made clear the "advance" from wood to iron and from empiricism to science, while acknowledging "the competitive struggle of the bridge contractors."[77]

In "American Railroad Bridges," which appeared the year following Thomson's 1888 "American Bridge Failures," Cooper evaluates the contemporary state of bridge building in the United States. He provides a lukewarm assessment of lattice bridges relative to pin-connected bridges, because, for example, the lattice required field riveting (which meant potential challenges to quality control) and, supposedly, its nooks and crannies could hold dirt and moisture.[78] As we've noted, what Cooper called "pin-connected" at that point had dramatically changed with the "invasion of rivets" (although Thomson opposed the hybrid nature of this type of design, which it seems would play a role in the Quebec failure).[79] The original pin-connected bridges, such as the Fink (figure 7.13), patented in the United States in the 1850s, had become artifacts of the past. Still pin connections appeared "scientific" because they represented an ideal condition that engineers employed in their calculations.[80] As engineers merged design types, that is, as riveted structures subsumed pin-connected, what was this debate really about? Cooper and his supporters were laying down a story about cost and efficiency, science and safety. They sealed the tale with an evolutionary hierarchy that placed American structures at the top. As we've seen, a genre such as "American bridge" is as much about what people want something to be as its actual characteristics.

From the technological evolutionist perspective, survival is the crux, so it is unsurprising that "American Railroad Bridges" reached its climax with Cooper responding to Thomson (who apparently was present at Cooper's presentation of his paper, yet whose reply was omitted from the ASCE publication).[81] Were pin-connected American bridges survivors or were they headed for extinction? Cooper disputed safety statistics that implicated pin-connections.[82] He asked:

> Does any one advocate the designing and building of bridges to withstand the impact of a railroad train, or the bursting effect of piling two trains on one another inside of the trusses? Are such accidents to be classed as bridge failures or as failures of management?[83]

In 1885, bridge inspector, safety advocate, and engineer Charles Stowell had announced that most pin-connected bridge collapses between 1873 and 1885 occurred due to a train striking the bridge—and that collapse under these circumstances had never occurred with riveted lattice bridges.[84] In the discussion following Cooper's "American Railroad Bridges," Stowell pointed out that Cooper's claim to fame, his bridge specifications, had been predated by another "cover[ing] substantially the same points"; these had been created by Charles Hilton, of lattice-bridge renown.[85] But Cooper harbored a blunt evolutionist perspective: "The intelligent investigator does not decide upon the merits of any developed system by the failures which are necessary steps in its development. Without variations and failures there would be no evolution or the survival of the fittest."[86]

While attempting to separate blame and responsibility, Cooper acknowledged some of the deficiencies of US bridges and some of the reasons behind them.[87] He believed that derailment prevention was required. His view (small comfort to victims) was that it was "a great credit to our style of bridge construction that so few, comparatively, of our bridges have failed." He felt that the safety factor (which takes account of the fact that theoretical model does not accurately anticipate future real-world conditions and which is still in use in the twenty-first century) "cover[ed] our ignorance."[88] Perhaps that's why Cooper did not cite bridges designed during the 1860s and 1870s by industry leaders Joseph M. Wilson and Henry Petit, Pennsylvania Railroad engineers who paid close attention to bridge maintenance and safety.[89]

Cooper quoted from a talk that eminent engineer William Unwin had given to young engineers in England a few months earlier:

> If an engineer builds a structure which breaks, that is a mischief, but one of a limited and isolated kind, and the accident itself forces him to avoid a repetition of the blunder. But an engineer who from deficiency of scientific knowledge builds structures which don't break down, but which stand, and in which the material is clumsily wasted, commits blunders of a most insidious kind.[90]

Ripped from its context (and slightly altered), the quote appears much harsher than the careful, nuanced discussion in which it appeared. Unwin pointed out—it's still relevant today—that the divide between science and engineering might be much narrower than we've frequently been told. Cooper's arrogance appears unbounded because, beyond the question of whether Unwin would have supported Cooper's perspective, in his lecture Unwin had named and praised Thomson! Unwin astutely explained that Thomson, whom he deemed "one of the most distinguished of American bridge builders," was an "adverse critic" whose apparent hyperbole and irony should not obscure his meaningful message.[91]

Benjamin Baker, who had read Thomson's "Failure" paper to an audience of engineers in England, was codesigner, with John Fowler, of the Forth Bridge (figure 7.16), a mammoth, world-renowned cantilever structure of unique design near Edinburgh. (A cantilever is a structure that projects outward from a fixed support, like an arm held outward, or like a lattice bridge span during launching; it enabled construction over extremely wide openings without requiring temporary scaffolding underneath.) The Forth Bridge, like the Eiffel Tower, was completed in 1889, the year Cooper's "American Railroad Bridges" appeared. The structural chutzpah and nationalist aura that characterized these two monuments of engineering and construction impressed themselves onto the minds of civil engineers everywhere.[92] After seven years of construction, the Forth Bridge was the longest cantilever in the world. It was a brash affront to Cooper's conception of what a bridge should be, which is to say, quite different than his image of an American bridge. Cooper deemed the Forth "as clumsy a structure and as awkwardly put together a piece of engineering, from the American point of view, as was

Figure 7.16
Forth Bridge (John Fowler and Benjamin Baker, built 1882–89), near Edinburgh. The renowned bridge is as evocative of Scotland as the nation's traditional garb. Collection of the author.

ever constructed." He believed that "American methods of design, construction, and erection would have saved, in my opinion, 50 per cent. of the cost of workmanship of this structure."[93] An additional factor may have fueled Cooper's pique: Baker had consulted Thomson, with his deep expertise in riveted railroad bridges, on the steel design of the Forth Bridge.[94]

Cooper's criticism of the Forth, a bridge carefully designed with safety in mind, along with his co-optation of Unwin, sprang from an apparently shallow concern about safety, based on relatively shallow bridge-building experience. Unwin had begun his career working with the elderly William Fairbairn, who was responsible for the research and design work behind the Britannia Bridge, which might be considered even more massive and "clumsy" than the Forth. Safety was a paramount concern for the designers of the Britannia (whose development overlapped with the collapse of Stephenson's Dee Bridge, as we saw in chapter 6). Similarly, Baker and Fowler, when they worked on the design of the Forth Bridge, were aware of the notorious collapse of engineer Thomas Bouch's nearby Tay Bridge, whose piers did not possess sufficient resistance to wind. Their design for the Forth Bridge, in fact, replaced Bouch's, which was abandoned after the Tay disaster.[95]

Cooper had advocated for the Quebec Bridge to be a cantilever structure and was wholly responsible, on his own insistence, for its design.[96] He was determined to place an American (built-in-Canada) structure on the world stage. He decided—a fateful choice—to increase the length of the span; this was intended to ease construction of the piers by allowing their placement closer to shore. Cooper knew that this would result in the longest cantilever bridge, longer than the Forth. It would be the longest bridge span in the world.[97] (Today the replacement Quebec Bridge and the Forth remain the two longest cantilever bridges.) Cooper likely did not know, though, that in a scene worthy of wide-screen cinema, John Fowler, codesigner of the Forth Bridge, had said to George Thomson, as the two stood in 1889 on the new structure: "The Americans will exceed this span."[98] Cooper intended to make it happen.

Nineteenth-century engineers' thoughts on a structure's national character, configuration, and size are comparable to scientists' cranial comparisons in relation to race (some scientists considered each race to be its own species). Just as distinctions were made, for example, between Prussian and Teutonic heads, so were bridges categorizable as American and British. Just as craniology confirmed craniologists' status as members of a superior race of white European men, with the largest brains, characteristics of engineering structures were seen as signifiers of who was the most evolved.[99] Today, we might be tempted to dismiss the idea of technological objects as measures of national or racial (evolutionary) status, but these ideas still circulate.

Cooper designed the Quebec Bridge to be lightweight. Using less material saved money, but this meant that the structure had to accomodate high stresses, and there was a lower factor of safety. The "American" construction process compartmentalized design, enabling the fabricator to move ahead quickly in phases; this meant that Cooper did not, at the end of the process, reassess and revise the design of the entire structure, even though it had become heavier and was of unprecedented length. The parts were already in production.[100] This management strategy ultimately enabled Cooper's evolutionist narrative to play out for an international audience: a bridge "variation" did not survive. During construction, it collapsed under its own weight, killing seventy-five workers.

All this points to the undigested and undigestible nature of evolutionist thought in understandings of design. Evolutionists prompt us to imagine bridges "surviving" like plants and animals; concepts such as "economy" and "lightness" and "American" appear to be unambiguous terms with inevitable outcomes rather than ideas subject to ongoing debate and revision. When we investigate decision-making individuals whose actions reflect their values, beliefs, and knowledge, "evolution" can appear to be a mask for carelessness, inexperience, heartlessness, and greed. Despite what evolutionists tell us, bridge failures are not a necessity; the actual contribution of such failures to invention is debatable.[101]

Cooper had stated in "American Railroad Bridges" that "as a rule [the American engineer] has done fully as well as the means given him would admit."[102] This reads as a chilling prophecy, because an excuse offered for the collapse of the Quebec Bridge was that Cooper was not adequately compensated; that is, he did not have the resources to do the necessary analyses.[103] He had acknowledged the unpredictable and changing nature of conditions in which designers work.[104] He recognized the complexity of the idea, "economy." He believed that the experienced designer could design "almost instinctively," and that calculation was only one part of a design process requiring deep understanding of manufacture and construction. But Cooper's instinct was founded on his own and his colleagues' limited experience. He might have learned from the design of the Forth Bridge and the extensive research and testing program behind it, rather than disdaining it.[105] But he was more of a spec writer than experienced bridge designer and his key colleague at the Phoenix Bridge Company did not have the background the project required.[106] Cooper was a different kind of "American" designer than Town and Thomson.

The construction processes of the Forth and Quebec bridges illuminate contrasting approaches. The Forth, much like the Britannia, was "a hand-made production," built onsite by "cheap and abundant labor," while the Quebec Bridge was a "machine-made" product conceived with parts ready to be put together.[107] The Forth design employed riveted connections and on-site assembly that included large components. For the Quebec, much fabrication was done in the factory; components were brought to the site on rail cars, which limited their size. Some parts were riveted but pinned connections and design based on them were integral to its design. The use of mass-construction strategies for a gigantic one-of-a-kind structure may indicate the limit of those strategies; their unproven adoption indicates Cooper's over-confidence.[108]

There was a dramatic difference in the construction strategies of the Forth Bridge, in which workers adjusted parts to fit as they arrived, and the Eiffel Tower, where each piece was precisely manufactured so that workers assembled, but did not adjust, parts on site.[109] But, unlike the Quebec Bridge designers, the tower's builders had the experience and knowledge to successfully build in this way; they had done so already for the viaduct at Busseau d'Ahun over the Creuse (1863–1865) (figure 8.3). As historian Eda Kranakis has noted (unironically, without reference to Cooper's evolutionist Weltanschauung), the design strategy applied to the Quebec Bridge "only made sense within its evolutionary context." She was referring to the designers merely scaling up existing beam designs, employing standard specifications (Cooper's pride) and applying these to the Quebec Bridge—standard actions that contributed to the conditions behind the collapse of a structure of non-standard scale. Cooper's desire to be the structural chief, his disparagement of Thomson, and his publicly articulated evolutionist and nationalist mindset make it difficult to deem him a victim of the (American) organizational practices he advocated and contributed to.[110] During construction, forewarnings had made workers uneasy, but site engineers minimized their observations.[111] Cooper was a leader and enabler of a cocksure "culture" that led to failure.

In 1991, critic Camille Paglia confided, "When I cross . . . any of America's great bridges, I think: *men* have done this."[112] It's a thought to keep in mind when wondering if Cooper's role in the collapse of the longest American-by-design cantilever bridge was the outcome of a toxic mix of evolutionism, nationalism, and machismo. This question might lead us take inspiration from a different biological corner: the construction of sex and gender. What would it mean to neutralize historians' traditional gaze and instead regard technologists and technology as transnational, binational, nationfluid, or anational?[113] A shift in perspective might encourage storytellers to uncover fresh facts for new tales.

Designers' and narrators' masculinity may be a driving force behind the emphasis on firsts and longests, and, not coincidentally, aggressions justified by evolutionism and nationalism. Infrastructure that appeared to be first, longest, or steeliest furnished tangible evidence of the sublime concept of an all-powerful, virile nation advancing world civilization.[114] Within multinational competitive struggles, (American) engineers could view themselves as evolutionary agents of innovation and integrate this hubristic understanding into design. As the industrializing countries remade the world through unbridled construction and corporate competition framed in nationalist terms, the deaths of workers and train passengers became collateral damage in a Darwinian arena. The significance of care and interdependence shriveled next to the exigencies of the struggle to outperform other nations.[115]

We might imagine that an obsession with data analysis might underlie the passion for firsts and longests and, if so, would enhance a quest for safety. Certainly, engineers reported information in a variety of forms and venues; the inquest that followed the collapse of Stephenson's Dee Bridge, as well as Stowell's safety research, indicate that organized collection of information was possible. But beyond sharing information in professional journals, engineers did not establish systematic data collection and analysis in order to promote safe design and construction; conditions today are about the same.[116] They might have. All these years later, though, biological metaphor encourages us to quash our outrage and ignore dysfunction by saying, "Sure it was tragic . . . but it was all part of an evolutionary process."

The popular magazine *Scientific American* commented on the significance of the Quebec Bridge collapse:

> If we were to select out of the many fields of activity which are covered by modern civil engineering, some particular one in which the American engineer has displayed most signally his originality and freedom from tradition, we would choose that of bridge engineering; and if we had been called upon to name some one particular structure which stood as the highest exemplification of his skill in this particular branch of his profession, we would have selected the great cantilever bridge across the St. Lawrence River at Quebec. Not only did it contain the largest and most massive single span of any bridge in the world, but it was constructed upon a system of which the earliest types, on any large scale, were built in this country. Moreover, not only was the bridge American in type, but in the details of its construction also it was essentially American.[117]

The failure of beams assembled of smaller components and called "latticed," employed as compression members beneath the Quebec Bridge's spans, apparently triggered the collapse (figure 7.17).[118] The magazine's editor noted that the failure could not be blamed on flaws in materials or construction. It was about design—American design.

Of the seventy-five workers killed, thirty-three were Kahanwake Mohawks whose families after the tragedy included twenty-four widows and many fatherless children.[119] Before rejoicing in the solidity of the Forth Bridge, though, we might acknowledge that its construction resulted in the death of at least seventy-three workers.[120] Beyond noting their modestly compensated if not life-sacrificing contributions to industrial

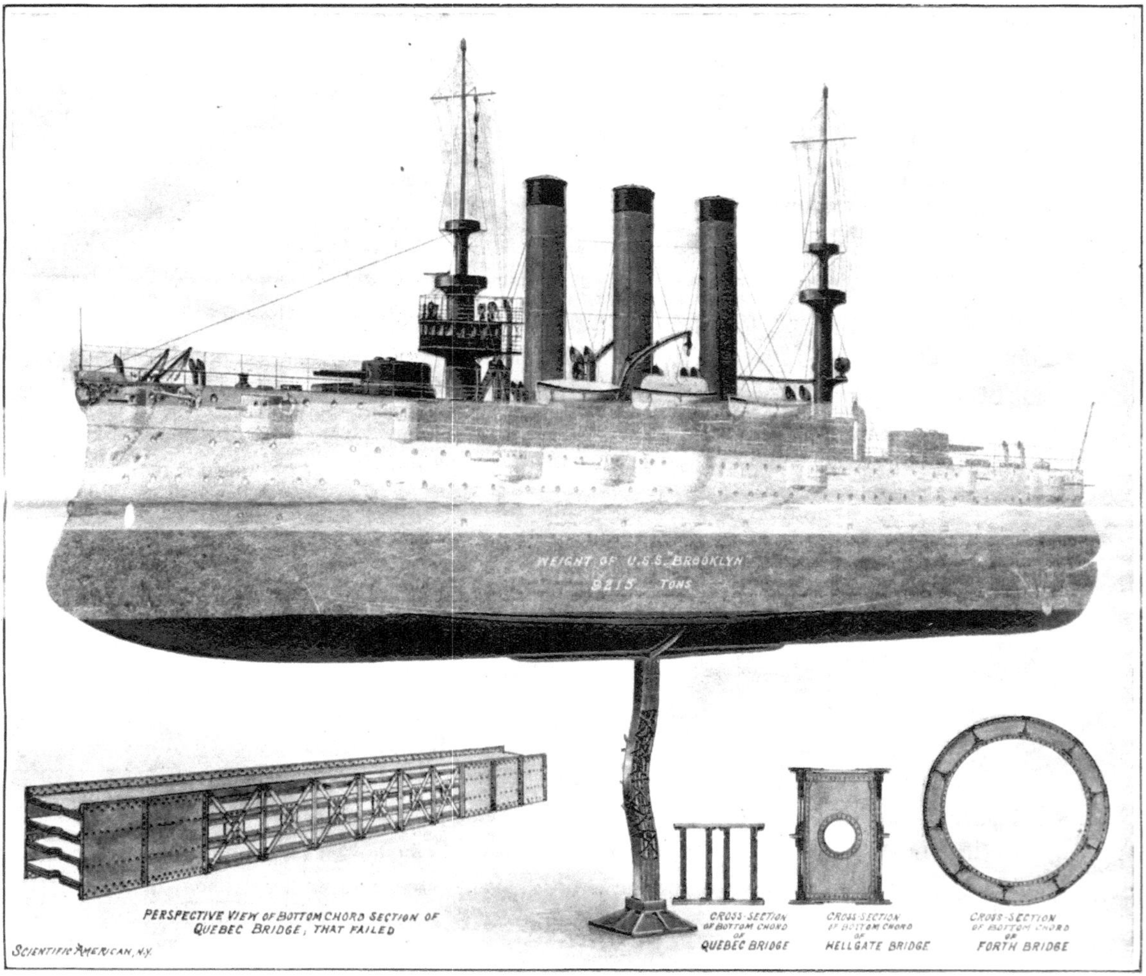

The vertical post, which is shown in the act of breaking down under the load of the cruiser "Brooklyn," is drawn to scale and represents the lower chord member of the Quebec bridge which failed by buckling through the rupture of the latticing. Theoretically this member should have carried 11,820 tons; actually it failed under 8,000 tons.

GRAPHIC REPRESENTATION OF THE ENORMOUS LOAD WHICH THE COLLAPSED QUEBEC BRIDGE BOTTOM CHORD WAS SUPPOSED TO CARRY WITH A MARGIN OF SAFETY.

Figure 7.17
Graphic representing Quebec Bridge load.
This illustration shows (bottom left) the type of "latticed" member (57′ long, and 5′7″ × 4′6″) that failed. The cross section of this member (bottom, near center) is compared to the 12-foot-diameter built-up beam of the Forth Bridge (bottom right corner). The ship represents the 9,000+ ton "safe strength" the beam was designed to carry—which did not include additional loads that the bridge would likely have had to support at various times. "Why the Quebec Bridge Failed," *Scientific American* 97 (October 12, 1907): 257. UC Berkeley Library, University of California.

wealth, the lives of bridge-building workers remain mostly unexplored.[121] If workers had participated in the writing of engineering history, we might have a different understanding of design and maintenance and "national culture" in construction.[122]

Our story doesn't end there. At the dawn of the twenty-first century, when engineer-historian Henry Petroski discussed Cooper's "authoritative paper" on American bridges, he noted:

> Unwin, like Cooper, believed that bridges should be made strong enough to perform their function but not so strong that they are heavier and more expensive than they have to be. This view has always been and always will be shared by the best of engineers, for they recognize that in the final analysis engineering is part of a much larger social enterprise, and money spent unnecessarily for civil-engineering structures becomes unavailable for other civic endeavors, or for initiatives of a humanitarian kind.[123]

Regarding "strength" (which might here translate to "safety"), Unwin, as we've noted, was more aligned with Thomson than Cooper. But how heavy, how expensive, how safe does a bridge "have to be?" Petroski presents design as a zero-sum game: engineers who make structures too safe (in order to save lives) risk squandering resources that otherwise might be applied to aid society or benefit humanity (thus costing lives). This idea deserves further elucidation. In the case of (American) bridges, however, and despite nationalist passions and engineering rivalries, engineers eventually did save lives and benefit humanity, in part by absorbing the lattice dogmas. They designed safer bridges, prompted by factors including the cost of accidents, the wealth and capabilities of railroad companies, and increasing public attention.[124]

If trapped in a burning railroad car that had just plunged through a bridge and into a river, you might agree that rhetoric around evolution, nation, science, and humanitarianism might be less comforting than design based on experience, with humility toward the known unknowns. These would include ever heavier locomotives and freight, poor maintenance, and broken axles. In other words, care and cooperation, rather than survival of the fittest, might be the basis of tales in which designers make and remake technologies that preserve lives.[125] The definition of "unnecessary" spending is worthy of clarification in the United States, where in 2021, 42 percent of bridges have surpassed their service life of fifty years and 7.5 percent are "structurally deficient."[126] Thomson wrote:

> While we expect the physician to be familiar with physiology, without pathology he would be of little use to his fellow-men, and it [is] as much within the province of the engineer to investigate causes, study symptoms, and find remedies for mechanical failures as it is "to direct the sources of power in nature for the use and convenience of man."[127]

When Thomson proclaimed in this way for care, he knew that his would be a lonely voice among his countrymen.[128] The phrase he cited at the end of this passage, however, is more than just another expression of engineering as manly mastership-of-nature. It was and remains the heart of the opening paragraph of the Royal Charter of Britain's Institution of Civil Engineers (1828).[129] We might want to believe that engineering or science made bridges American or made them safe, but it was men given the opportunity to become engineers and scientists and entrepreneurs who determined what engineering, safety, and American might be.

8

GETTING HIGH

The Eiffel Tower's exposed structure spectacularly confirmed the world-changing potential embodied in a reinvention of construction once characterized as "American." When the attraction opened in 1889, office buildings carried by rigid skeletons rather than masonry walls had been buildable for about two decades. The so-called evolution of the tall building, often regarded as an American triumph, was supported by a structural framework erected with uniform parts and connections in a rapid, systematized way.[1] This was not something magically conjured at the moment of the skyscraper's "invention"—a preposterous fairy tale.

Engineers and workers embedded in the skyscraper the structural-constructional strategies of piers that had borne bridges. They worked at the intersection of transnational creative, craft, and industrial practices. But when laborers struggled for better conditions in the United States, they were sometimes labeled "foreigners."[2] For engineers, too, being deemed an outsider rather than held in the national narrative embrace determines who we remember and what we forget.

As we've noted, the lattice and kindred bridges, once they operated as beams, liberated the piers they were placed on. These high-rising towers as a result became objects of structural and constructional systematization. They enabled people to rise and be transported as they surveyed surrounding landscapes. This shared experience of patriotic euphoria helped form national community. Here, we'll touch on industrially-aesthetically-spatially-narratively induced feelings of national belonging as we explore the relationships between building across and building up.

A BRIEF HISTORY OF TALL BUILDING

Historian Joanna Merwood-Salisbury observed that the early skyscraper's "curtain wall façade was an advertisement for an optimistic view of the future of American society under capitalism," within which architects and engineers as managers and knowledge workers, with Ithiel Town as a trailblazer, assumed control that building labor once had.[3] Designers enclosed their high-rise structures with stylings that scaled up classical-national forms that Town had deployed; this allowed businesses and academia to present a traditional, civilized profile to the world. Early skyscrapers' public-facing walls were often themed Greek, Roman, or Gothic, or they evoked natural environments that industrialists were trampling as US citizens pushed "civilization" westward. Engineer William Le Baron Jenney, "inventor of the skyscraper" in popular fantasy, studied, as we've noted, like Eiffel, at the École Centrale; he based his architectural thinking on cultural evolutionism, nationalism, and Anglo-Saxon supremacy.[4]

As load-bearing masonry walls became too thick (that is, took up too much space) and heavy when required to support ever taller buildings, engineers instead employed metal frameworks (figure 8.1). This meant that skyscraper engineers reckoned with some of the same factors as the designers of the tall viaduct piers we saw in chapter 5; that is, in addition to carrying a constant, dead load (the structure itself) and live loads (vehicles or people), the piers had to resist dynamic wind loads, ever-changing forces coming from the side and a variety of directions.[5] As a solution, high-rise engineers assembled in their minds a structural gestalt. Prominent engineer Corydon T. Purdy wrote in 1895, "In the new buildings, the steel frame is a unit—a complete whole—the members of which are closely related to each other and dependent on each other." Purdy viewed the entirety of the skyscraper frame as a giant truss. (The monolithic structure recalled Thomson's principle of "interdependence" exemplified by the lattice, as discussed in chapter 7.) The fact that the walls would no longer bear the load—the frame supported exterior walls rather than being supported by them—inspired Purdy to proclaim, "New conditions in large buildings is a real revolution."[6]

Purdy recognized as revolution the outcomes of innovators who had long been developing the structural and constructional strategies that enabled the high-rise. This they had done literally in the shadow of bridges, by building the tall supporting piers. The work of engineer Wilhelm Nordling (born Stuttgart, 1821–died Paris, 1908), whom we encountered in chapter 5, gives us a means of approaching this as yet undocumented transformation. Nordling studied as an *auditeur externe* in the École Polytechnique and École des Ponts et Chaussées at the beginning of the 1840s, became a French citizen (and changed his name from Nordlinger) in 1858, spent most of his professional career in France, was made Chevalier de la Légion d'honneur in 1864 and later a Grand Officier. He participated in the construction of French railway lines, reached the grade of *ingénieur en chef*, and was involved in a variety of discussions in professional engineering societies, in 1869–1870 as vice-president of the *Societé des Ingénieurs Civils*. He lived in France from 1840 until his death, except for 1870–1879—the years following France's ignominious defeat in the Franco-Prussian war.[7]

Nordling had contributed to the design of the Grandfey Viaduct (1857–1862), built over the Sarine at Freiburg in Switzerland by the French firm Schneider (figure 5.13).

Figure 8.1
"Lattice-girder bracing" in the structural frame of the American Tract Society Building (William W. Crehore, engineer; Robert H. Robertson, architect, 1895; steel by Keystone Bridge Works). This early skyscraper still stands near City Hall in New York. Corydon T. Purdy, "The Use of Steel in Large Buildings," *Journal of Association of Engineering Societies* 14 (1895): 200. UC Berkeley Library, University of California.

Subsequently, as chief engineer on the Orléans Railroad, he was responsible for the design and construction of a series of prominent railroad viaducts (figure 8.2). (A "viaduct" is a bridge that crosses a valley, which might contain a waterway.) The first two, built by the firm Cail/Fives-Lille, crossed the Creuse River at Busseau d'Ahun (1863–1865) (figure 8.3a), with trains passing about 185 feet above land, and over the Cère River near Ribeyrès (1863–1866) (figures 5.15 and 8.12); the great cost of this rail line was justified by the mining and manufacturing industries it was intended to serve. Before and during the construction at Busseau d'Ahun and Ribeyrès, Cail built eight metal viaducts in Spain. Based on the design of these bridges, Nordling gave Cail/Fives-Lille and Gustave Eiffel the freedom to draft designs for four new structures.[8] These would enable railway cars to travel at heights of between 44 and 66 meters on the new Commentry-to-Gannat line of the Orléans Railroad in the department of Allier. Nordling gave Eiffel his

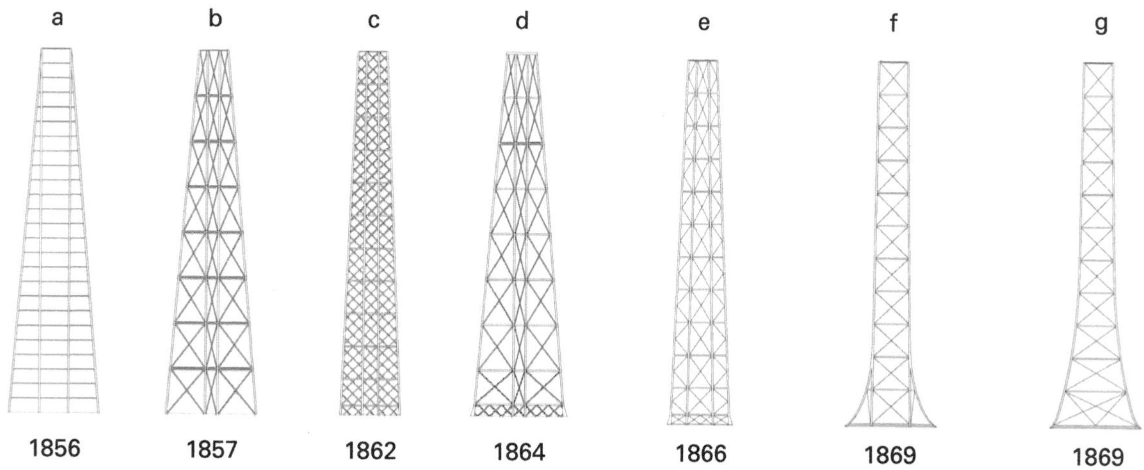

Figure 8.2
Metal viaduct pier construction, 1850s–1860s, known to or built under Nordling.
(a) Sitter (b) Crumlin (c) Grandfey (d) Busseau-sur-Creuse (e) Cère (f) Bouble, Neuvial, Sioule (g) Bellon. Roberto Serrano-López et al., "La adaptación de la forma a las cargas en el arte structural: Desde las pilas y el arco, a la torre / Adaptation of the shape to loads on structural art: from piers and arch, to the tower," *Informes de la Construcción* 66 (2014): 5.

big opportunity; contracts for the viaducts over the Neuvial ravine (1868) and over the Sioule River at Rouzat (1869) (figure 8.3b), enabling him to establish his own company (a month before the completion of the former). Cail built the two other viaducts, over the Bouble River near Echassières (1867–1869) and over the stream at Bellon (1868–70) (figure 8.3c). Assembly and launch of the metal structures took about six months.[9]

Nordling and his colleagues were consciously designing the future of construction.

> Contemporary practice has simultaneously achieved two important advances in the construction of bridges and viaducts. One is the use of compressed air in underwater foundations; the other, the use of metal structures in the piers of viaducts. The first allows us to sink foundations to previously inaccessible depths; the second to reach heights without foreseeable limits.
>
> We are going to deal here only with the second system.

So began the first of Nordling's two landmark papers describing the tall structures that supported his six viaducts of the 1860s. They were written in a "lucid, brief, exact, practical" style and published in 1864 and 1870 in the prestigious *Annales des ponts et chaussées* and as separate offprints.[10] At that time, lattice bridge spans, as Nordling confirmed, were well known.[11] There was nothing remarkable about Eiffel's choice at

Figure 8.3a
Busseau d'Ahun viaduct (1863–65); Sioule and Bellon viaducts on the following pages. Official photographs shared at world expositions offered Nordling's structures as high-tech exemplars, progenitors of a new aesthetic, and hallmarks of national achievement. Jules Duclos, photographer, École Nationale des Ponts et Chaussées.

Neuvial and Sioule of cross-bracing between vertical elements or, under engineer Félix Moreaux, Cail's selection of the lattice at Bellon and Bouble.[12] As they had no need to rethink bridge span design, Nordling and his collaborators took a ninety-degree turn and turned their inventive attention to building up.

The traditional choice, massively heavy stone construction, resisted wind but, as Nordling noted, stone was more expensive, especially for piers of great height, and faced challenges in relation to vibrations and accommodating thermal expansion of the bridge spans.[13] (This did not mean that masonry structures were no longer viable or that stone construction prices were no longer competitive, as the Perrière viaduct, also built on the Commentry-to-Gannat line, indicates.) So they developed piers consisting of columns of hollow cast-iron members stiffened with wrought iron cross-bracing. Some were

Figure 8.3b
Sioule viaduct (1869), Michel Berthaud, photographer, École Nationale des Ponts et Chaussées.

"pyramidal" in form in the transverse (sideways) direction, that is, they curved outward at the base; the curve was the outcome of calculations whose goal was to enhance resistance to wind (figures 8.4, 8.5, and 8.15).[14] This was intended to prevent overturning of the tall narrow structures due to winds and dynamic train loads.[15] While the constructability and light weight of these structures were advantages, wind remained a challenge. It was unpredictable and its impact on light metal structures was not well understood. Nordling noted in 1864 that design for wind resistance required additional study and warned that if there were to be a wind-driven collapse, for any reason, it would cause people to lose faith in metal piers just as failures had resulted in distrust of suspension bridges.[16] He strove to identify the precise manner in which wind impacted pier members; he monitored

Figure 8.3c
Bellon viaduct (1868–70), Michel Berthaud, photographer, École Nationale des Ponts et Chaussées.

the in-service structures in order to determine what worked and if adjustments were required.[17] His team studied in detail the effects of wind through calculation and observation and consideration of various scenarios (figure 8.6). They sought rigid, dependable connections that could sustain thermal expansion, as loosened connections would mean a loss of structural rigidity and accelerated corrosion.

Nordling studied structural precedent. When he noted, "It can be assumed that the example of the great American wooden viaducts, especially the Portage, was not unrelated to the design of the Crumlin and Sitter projects," this was a solid assumption, even though the Portage piers compose a trestle (that is, they're all linked horizontally), rather than independently standing towers (figures 5.21, 8.7, and 8.8).[18] That Nordling, an engineer born in a German state, who studied in France and was responsible for structures in France, Switzerland, and Austria, was familiar with a trestle bridge located 4,000 miles away in rural New York State will not surprise readers of *American*

Figure 8.4
Piers of Sioule and Bouble viaducts. A man standing at the base of each indicates scale. Michel Berthaud, photographer, École Nationale des Ponts et Chaussées.

Bridge. Indeed, upon the Portage Bridge's completion, an early report in a new Canada-based science-and-engineering periodical had unfavorably mentioned its contemporary on the other side of the Atlantic Ocean, Stephenson's tubular bridge (see chapter 6); this account was approvingly reprinted in a London-based journal. Similar to today's flows of information through digital networks, reports of the Portage, many based on the article in *Canadian Journal* (published in Toronto, 1852), spread through organs including *Scientific American* (New York City, 1852), *American Railroad Journal* (New York City, 1852), *Civil Engineer and Architects Journal* (London, 1853), *Allgemeine Bauzeitung* (Vienna,

Figure 8.5
Viaduct pier in the Neuvial ravine. The outward flare provided resistance to large bending forces at the pier base and turned what would have been shear into compressive force; this took advantage of cast iron's compressive strength. Cast-iron columns were employed in early skeleton frames for tall buildings. Michel Berthaud, photographer, École Nationale des Ponts et Chaussées.

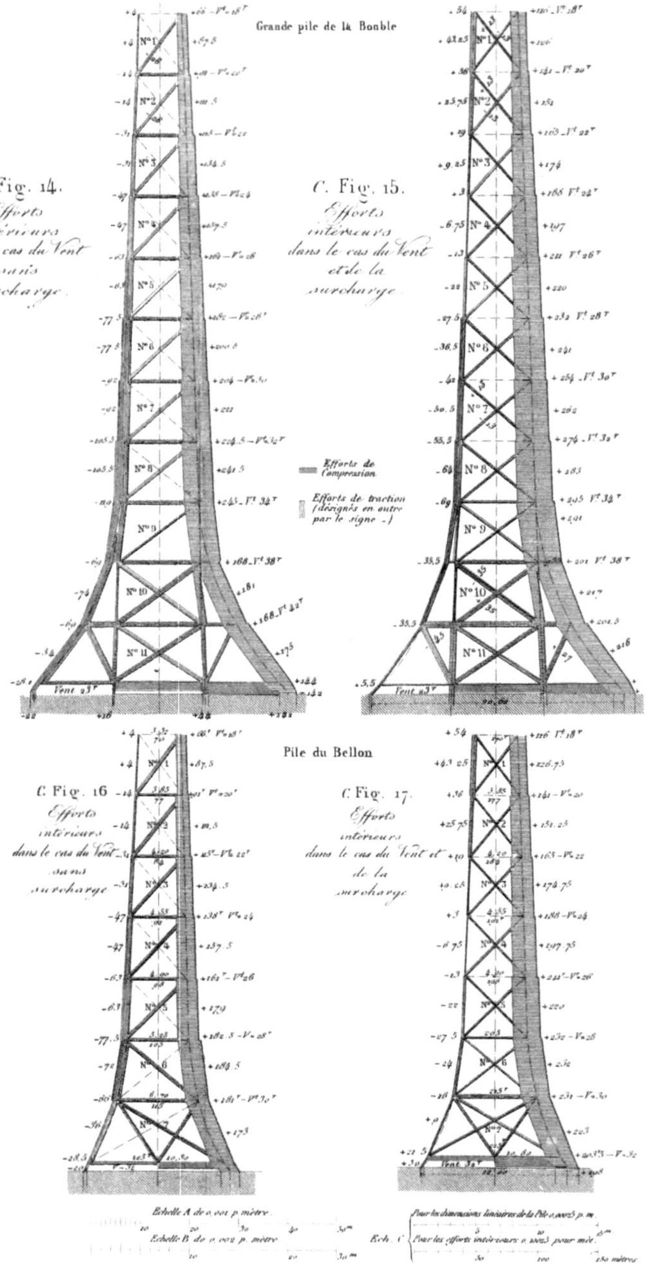

Figure 8.6
Wind loading (without and with overload) studies for tallest and shortest piers of Bouble and Bellon viaducts. Wilhelm Nordling, "Documents relatifs aux viaducs métalliques de la ligne de Commentry à Gannat," *Annales des ponts et chaussées* 4/1 (1870): pl. 215. UC Berkeley Library, University of California.

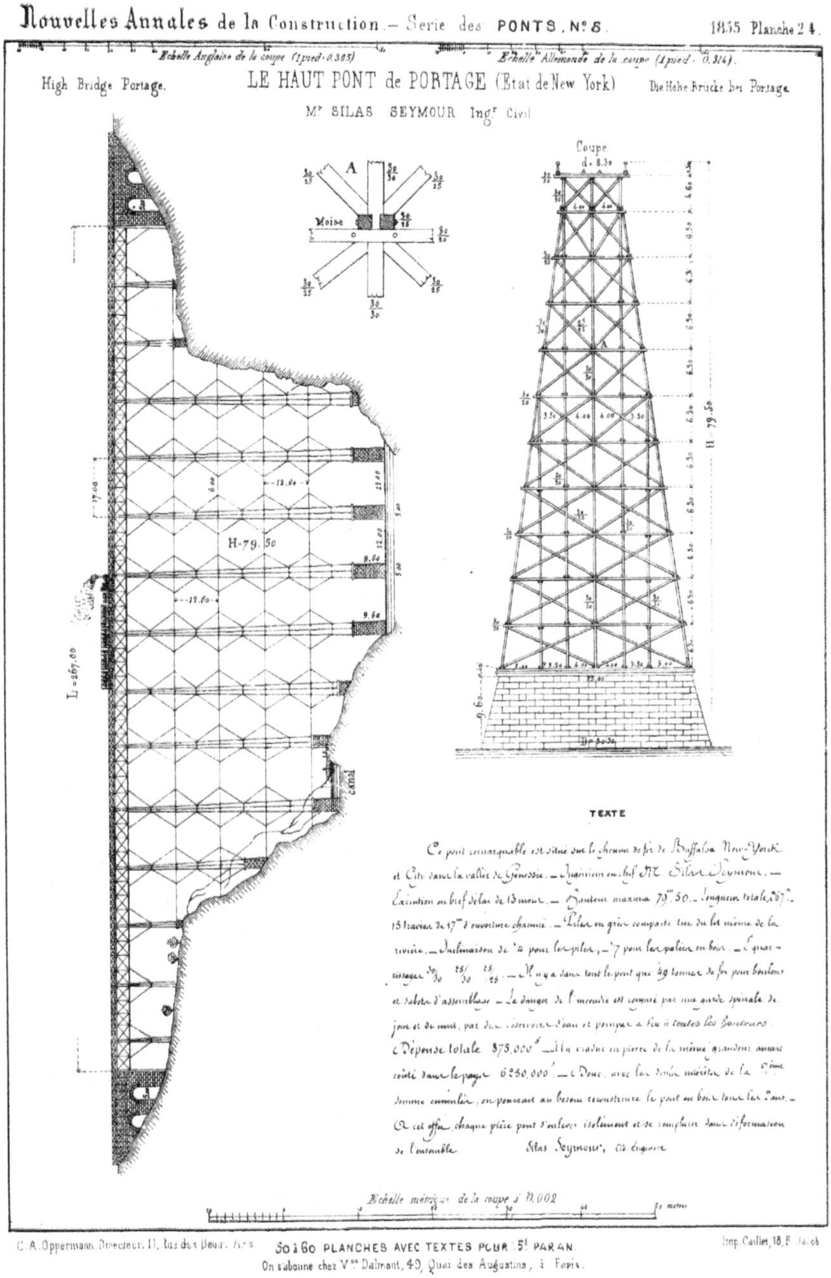

Figure 8.7
Portage Bridge. Elevation drawings of bridge and trestle (1852). This is likely the only surviving drawing to include dimensions and a connection detail. It features scales for French, English, and German readers. The text compares its total cost (75,000 francs) to a stone alternative (6,250,000 francs); the masonry arch shown at the right end was not built. "Haut Pont de Partage [*sic*] dans l'État de New-York," *Nouvelles Annales de la Construction* 1 (June 1855): pl. 24. University of Illinois at Urbana-Champaign.

Figure 8.8
Portage Bridge (1852) pier (center), designed with a base three times as wide as the others, ensured lateral stability. "Pyramidal" design appeared in Nordling's viaducts and the Eiffel Tower. Courtesy of Livingston County Historian's Office, Mt. Morris, NY.

1853), *Mechanics' Magazine* (London, 1853), *Polytechnisches Centralblatt* (Leipzig, 1853), *Zeitschrift des Architecten- und Ingenieur-Vereins für das Königreich Hannover* (Hannover, 1855), *Nouvelles Annales de la Construction* (Paris, 1855), *Dingler's Polytechnisches Journal* (Stuttgart, 1856), and *Zeitschrift des Oesterreichischen Ingenieur-Vereines* (Vienna, 1856).[19]

It's unexceptional that a structure of Portage's scale would attract attention far and wide. We don't know its impact on engineers in its own region, though. There were circumstances worth noting. The Portage design is attributed to engineer Silas Seymour, but more than fifty years later a journalist claimed, without citing his source, that it was determined by a "congress of engineers."[20] He, or they, would have worked on the design soon after engineer Squire Whipple, who lived in New York State, began sharing his pioneering methods of structural analysis; it was reported at the time that "by calculation [the designers] know that it would bear twenty times the weight of any train."[21] Conspicuous also is the proximity in time and space of the Portage Bridge to professionally organizing engineers, who, in the middle of the century lived mostly in New York and nearby states (Massachusetts, Ohio, Pennsylvania, Connecticut, plus further to the west, Wisconsin).[22] Less than three months after its opening, the Portage was a backdrop for a meeting in New York City that became the founding event of the American Society of Civil Engineers.[23]

Innovators who consumed the news from New York State were likely aware of a precedent that had appeared in the pages of *Allgemeine Bauzeitung*: bridge piers built around 1840 for the Little Schuylkill & Susquehanna Railroad in Pennsylvania (figure 8.9).[24] We've noted that a consequence of the ability to construct a bridge such as the lattice as an independent, long, level beam, was that the piers that supported the bridge became independently designed and constructed. They were no longer of a piece with the deck, as they were in an arch bridge, whose vaults dictated where supports had to land, or in an overlaid structure where a variety of systems were intended to work communally (see chapter 2). A pier became a freestanding product. As the Little Schuylkill & Susquehanna Railroad's chief engineer, Solomon W. Roberts wrote, quoting Edward Miller, who preceded him in the position, "The lattice bridges . . . may be supported at any point, indiscriminately."[25] This meant that engineers had the opportunity to design piers that could be built in large numbers, assembled of uniform parts in standard configurations like the bridge spans they supported, and employed in diverse environments.[26]

So why haven't historians narrated the reconception of the pier as an "evolution" that climaxed in high-rise "revolution?" These piers were perceived as wooden so storytellers would have located them in an earlier phase of civilization; wood being seen as a contraindication of industrial revolution (see chapter 1). Plus, the towers, whether standing

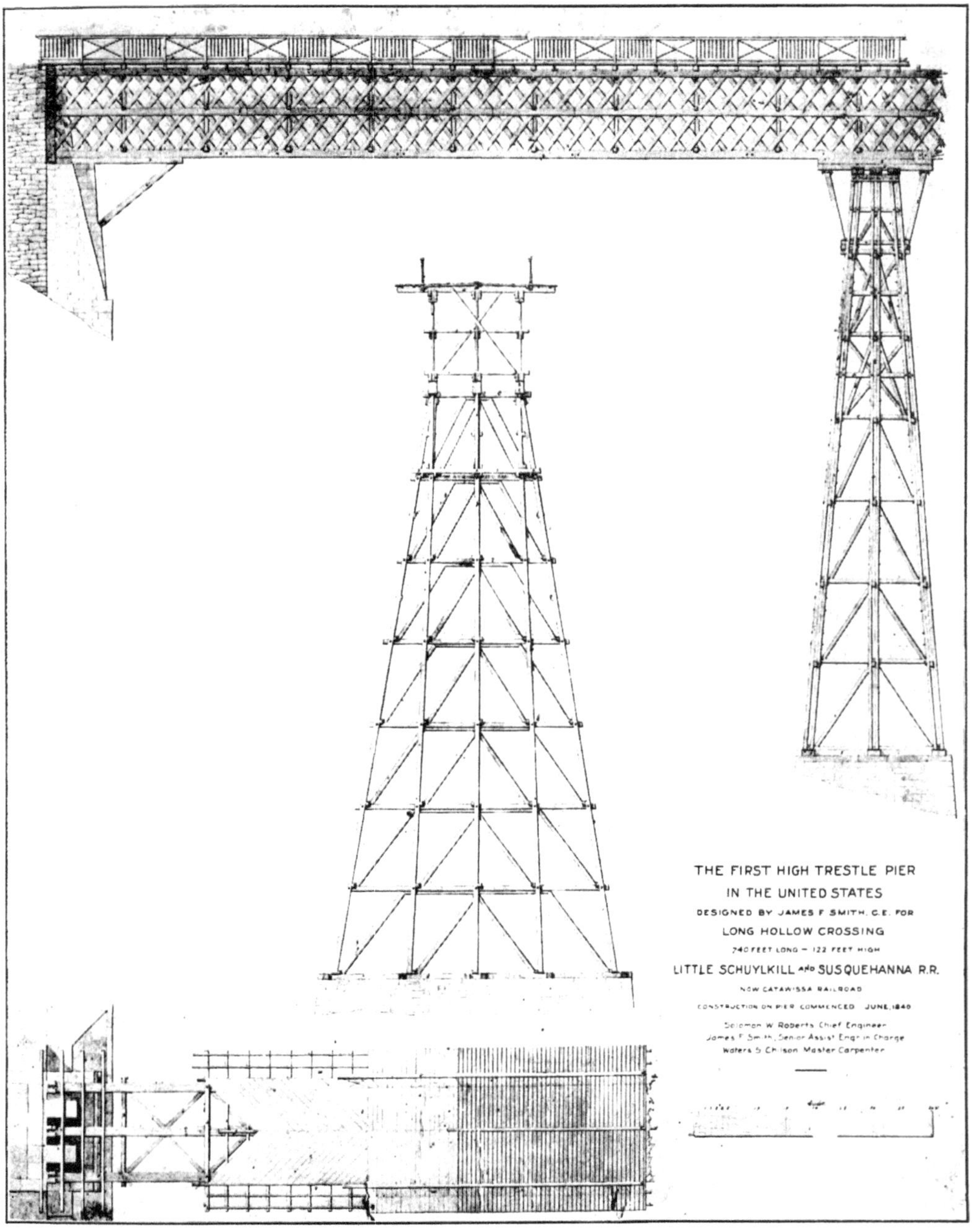

Figure 8.9
Long Hollow Crossing bridge and pier, Little Schuylkill & Susquehanna Railroad (James F. Smith, 1840), Pennsylvania. In 1904, the Engineers' Club of Philadelphia's president presented front and side elevations of this pier, in order to highlight the beginnings of tall, industrialized structures. These pier drawings had been shared in *Allgemeine Bauzeitung* (1845) and George Duggan, *Specimens of the Stone, Iron and Wood Bridges, Viaducts, Tunnels, Culverts, &c. of the United States Rail Roads* (Appleton, 1850). (The lattice bridge drawing, however, was unpublished.) Edwin F. Smith, "Some Early Engineering Works in Pennsylvania." *Engineers and Engineering* 21 (1904): fig. 1, and *Public Works* (London) 4 (1904): 13. Courtesy of Smithsonian Libraries and Archives.

independently or conjoined in a trestle, were rural, rather than urban. So storytellers did not place them into categories with sexy labels, like "skyscraper" or "space truss," though the structures deserved it. With hindsight, we might deem these structures, with skyscrapers about half a century in the future, a world-transforming invention.

Previously, people, animals, and vehicles had walked and rolled up and down the sides of hills and valleys; there was no demand for tall, rigid bridges. An exception were the Roman aqueducts of antiquity, such as the Pont du Gard near Nimes, which channeled water on a descending gradient and so had to be high. Tall medieval masonry bridges were very rare, built in mountainous areas where there was a high risk of flooding.[27] (Compare a tall bridge with a traditional one in figures I.6, 4.1, 5.19, 8.4, or 8.12.) As hauling trains up and down steep slopes was not a viable solution for rail lines, at the end of the initial decade of railway construction in the United States, Little Schuylkill & Susquehanna engineers opted for elevated bridge construction.

Stone piers were initially planned; the line through the mountainous region included seven bridges, with the tallest pier at 114 feet (about as high as a ten-story building). Roberts subsequently recommended as a precaution that the falsework (construction scaffolding) be allowed for a time to remain in place under the completed bridges, and that "timber framed piers" might be built at the midpoints between stone piers (on this line, the longest span between any two piers was 200 feet), just in case additional support was necessary. This strategy was apparently implemented for some of the bridges. Roberts wanted to "err on the side of too much strength"; this was a few years before Squire Whipple and Karl Culmann would share methods for structural analysis that would reduce uncertainty. As a result, the Long Hollow Bridge (1840) had two stone *and* two intermediate wooden piers (figure 8.9).[28] The expense of getting and preparing stone of good quality, however, led the engineers to decide that the bridge over Little Catawissa Creek (ca. 1840) would include only wooden piers.[29] This seems to have been one of the earliest tall railway bridges in which both the spans and supports were lumber frameworks, though already in 1836 bridge inventor Stephen Long had provided a drawing of a typical low-rise pier (figure 4.5).[30] Of course, towers of wood were not new or limited to bridges; the design of a proposed temporary crane for the building of the stone arch High Bridge (or Harlem River Bridge, 1838–1848, a portion of which still stands in New York City) was developed at about the same time (figure 8.10).

The need for an ever-increasing quantity of free-standing, rapidly buildable, rigid, tall piers surely helped catalyze the design of structures made of uniform connectors (bolts) and beams. At Little Catawissa Creek, engineers replaced load-bearing masonry with a skeleton frame; skyscraper engineers would later do the same. This represents the design impulse behind mass-constructable, economical-to-build, tall structural

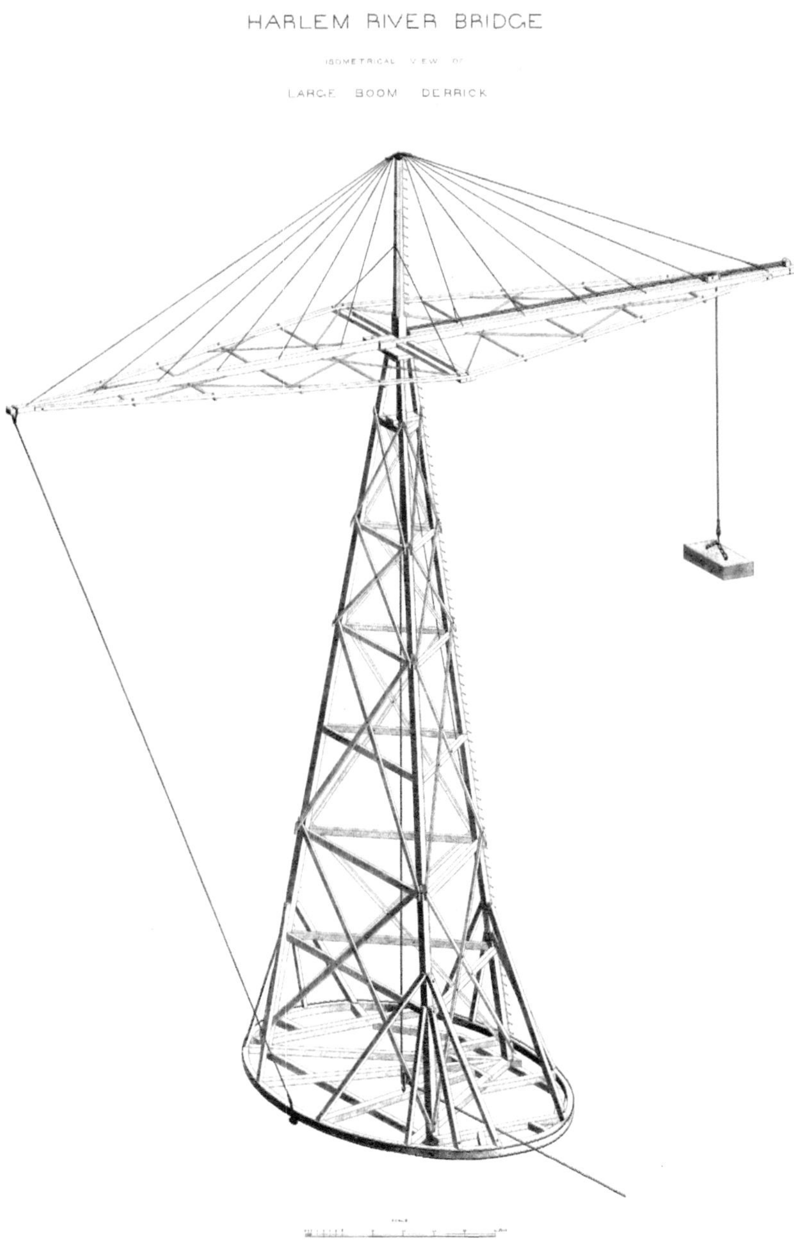

Figure 8.10
Derrick for High Bridge, also known as Harlem River Bridge (John B. Jervis, 1837–1848), New York. Proposed design for a derrick with cable-stayed boom for the construction of a 138-foot-tall stone arch bridge. Courtesy of Jervis Public Library.

frameworks that were customizable per loading and site requirements and made of more-or-less interchangeable parts. It's fitting that engineers designed these earliest towers to support lattice bridges, which shared these constructable characteristics. Portage Bridge designers a dozen years later were likely aware of the learnings provided by the Little Schuylkill & Susquehanna Railroad piers, 200 miles away.[31]

FRENCH, FAMOUS / FOREIGN, FORGOTTEN

Wilhelm Nordling's sharing of research and design, promoted by the French state, highlights the transnational flows that sustain invention. France displayed scale models of the Busseau and Cère viaducts at the Paris Exposition of 1867, which prompted extended discussion in the journal *The Engineer* (London). The fair welcomed more than ten million visitors; diverse aspects of it were depicted in 1,770 images, which appeared in thirty-one newspapers. Drawings and models of Nordling's viaducts (perhaps the ones displayed in Paris), each showing a pier and portion of the deck, were presented at the 1876 Centennial Exposition in Philadelphia, visited also by ten million (figure 8.11). There, an engineer from France was on hand to amplify the detailed descriptions available in an English language publication.[32]

Nordling counted the iron viaducts built in England, Switzerland, Spain, France, Italy, and Austria and exclaimed in 1870, "These 23 viaducts will certainly not be the last!" (He hadn't captured them all.)[33] That year, an image of the Cère Viaduct appeared on the first page of the weekly *Scientific American* (figure 8.12). By 1881, engineers in the United States, discussing the challenges of determining wind loads and safety factors, referred to Nordling's viaducts.[34] In 1882, an engineering professor in Lausanne analyzed Nordling's piers.[35] In 1889, the viaducts were a reference point in a review of structures intended to honor Eiffel and his new tower. Critic J.-K. Huysmans' anti-tower jeremiad included his insulting the structure as "this apotheosis of the viaduct pier." It was an accurate epithet. Eiffel, too, explicitly linked the tower's origins to the piers and acknowledged the significance of Nordling's viaducts in his career: "I have since become attached to this kind of construction." That was evident in his firm's work.[36]

Where do Nordling and the viaducts he was responsible for stand within historians' heroical, national, and structural mental maps? We might think that the iron-flaunting, skyscraping, visually spectacular projects that he led would have inspired technology's storytellers to new heights of narrative effusion. Yet, consistently, leading chroniclers of engineering have barely mentioned or have cancelled Nordling. Historian Eda Kranakis presents Wilhelm Nordeling (*sic*) as an example of someone who left France because of "[frustration] with a system that offered . . . no possibility for advancement." Nordling

Figure 8.11
French Public Works Building, Centennial Exposition, Philadelphia, 1876. Visitors to the pavilion encountered two scale models of Nordling's viaducts (right of center, perpendicular to each other, below the lower left corner of the image of the striped lighthouse). Courtesy of the Free Library of Philadelphia, Print and Picture Collection.

did go to Austria-Hungary, where he remained for nine years, rising to the position of Austrian railway director. Due to political and technological differences, he resigned and returned to France, where he spent most of his life and was apparently highly esteemed. Engineer David P. Billington substitutes Eiffel, a "great artist," for Nordling, and does not cite the latter's publications. He instead observes that "little if anything has been written on Nordling's . . . designs," offering this as proof that Nordling was not a significant designer, even though Nordling's works and observations on "engineering art" (Billington's term) might be considered seminal in the history of (industrial) aesthetics. Historian Antoine Picon, writing of the "influence of the École des Ponts et Chaussées," cites Nordling as an example of a "foreign" student who left France for a "brilliant career"

SCIENTIFIC AMERICAN

A WEEKLY JOURNAL OF PRACTICAL INFORMATION, ART, SCIENCE, MECHANICS, CHEMISTRY, AND MANUFACTURES.

Vol. XXII.—No. 11. [NEW SERIES.] NEW YORK, MARCH 12, 1870. $3 per Annum. [IN ADVANCE.]

The Cere Viaduct.

The viaduct which we illustrate this week crosses the Valley of the Cere, near Ribeyres, and is situate on the line of railway between Figcas and Aurillac, which forms a portion of the central network of the Paris and Orleans Railway, France. The viaduct, which carries the rails at a hight of 181 feet 6 inches above the water level, consists of five spans of lattice girders supported by masonry abutments and by piers formed of clusters of cast-iron columns rising from bases of masonry. The three central spans are of 164 feet each and of the two other spans, that on the one side is 145 feet, and the other 139 feet. The length of each abutment is 118 feet, the total length of the viaduct being 1,012 feet. The width of the viaduct between the rails is 14 feet 9 inches, and it carries a single line of rails. Each of the piers consists of a cluster of eight cast-iron columns united by cross bracings, and fixed to the top of a brickwork base of elliptical shape. Each pier measures 16 feet 5 inches by 8 feet 2½ inches from center to center of columns at the level of the capping, and the columns are disposed at such an inclination that their center lines, if produced upwards, would all meet at a point at 123 feet 1½ inches above the level of the rails. The side columns of each pier have thus a batter in the direction of the line of the viaduct of 1 in 30, and the end columns a transverse batter of 1 in 15, and as all the piers are of the same size at the top, the dimensions of each at the bottom vary with the hight. The brickwork bases of the piers have also their sides built at such inclinations that they form portions of a cone, the apex of which would be at the point of junction of the center lines of the columns before mentioned. The foundations of the viaduct were commenced in June, 1863, the erection of the girders in May, 1865, and by the following October a connection had been formed between the two abutments.

When complete the superstructure was tested by a load of 4,000 kilog. per meter, or about 8,000 lbs. per yard run; and under this test the central piers were compressed 3 millimeters, or about an eighth of an inch. At the same time the central span was deflected 15 millimeters, or three fifths of an inch, and the two spans on each side of it 12 millimeters, or nearly half an inch.

Its total cost was $171,000 in gold. Each of the main piers contains 71 tuns 13 cwt. of cast iron and 51 tuns 10 cwt. of wrought iron, and the cost per yard in hight of the iron portion of the pier was about $400.

The viaduct was constructed under the direction of M. Déglin, engineer-in-chief of the Ponts et Chaussées, and of Mr. Bortoux, acting engineer. The design was by M. Wilhelm Nordling, engineer-in-chief of the northern part of the Reseau-Central.

VIADUCT ON THE PARIS AND ORLEANS RAILWAY, FRANCE.

Railway Economy.

There seems to be a great difference of opinion among railway managers as to the meaning of economy as relating to the roads they operate. By far the greater number apparently holding that it is in cheapness of first cost, without regard to service to be rendered. This has been *the* fault, not only in the working, but in the construction of American railways.

Originally, owing to the scarcity of capital and the necessity of rushing the enterprise through to completion, the roads were hastily graded, insufficiently ballasted, and laid with any iron that could by any possibility be considered as at all suitable, and which cost the least sum per net tun. So, in rolling stock, wheels which have so much to do with the safe transmission of the passengers and freight, were not purchased because of the quality of material used in their composition and the care bestowed upon their casting; but the only question asked was, "What is your price?"

It is no wonder that under such circumstances as these American railways have made themselves illustrious as one of Grim Death's most efficient allies. We believe that an examination of the causes of casualties on railroads will show that of every hundred, ninety-nine have arisen from either a defective wheel, or axle, or rail.

The expenditure of one or two cents a pound more for axles, of three or four dollars additional per wheel, and a few dollars per tun extra for rails, would, we firmly believe, do more to reduce the list of accidents on railways than any other course of action their managers could pursue. Railway supply manufacturers are not slow to discover that it is cheapness, not quality, that secures orders.

Pity 'tis that railway managers have not been as quick to understand that the "penny wise" are frequently the "pound foolish;" that the best is always the cheapest in the long run; that that policy which

Figure 8.12

Cère Viaduct (Wilhelm Nordling, 1863–1866). Designers and builders of tall structures could refer to research and design conducted under Nordling. "Cere Viaduct," *Scientific American* 22 (March 1870): 167; also appeared in *Building News* (London), January 16, 1869, 68, 70. UC Berkeley Library, University of California.

abroad; according to Picon, Nordling was "the Wurtemburger who became director of the Tisza-Pest Railroad."[37] Othering began more than a century earlier: in 1880, two École Nationale des Ponts et Chaussées engineers in a detailed examination of US bridges acknowledged the contributions of "de Nördling" (note the tell-tale addition of an umlaut) and labeled him "European" rather than French.

Nordling appears to have been a brilliant and brilliantly French engineer. So what happened? Perhaps, anxieties generated by French-German competition help explain why historians have almost effaced Nordling's contributions and branded him a foreigner. (He referred to himself as a "Français d'adoption.")[38] Though the four viaducts on the Commentry-to-Gannat line of the Orléans Railroad were completed earlier, the line opened only in 1871, after the six-month Franco-Prussian War had ended and after Nordling had departed, perhaps because of the conflict. (In 1875, though, he was apparently responsible for the Eiffel company receiving its largest project to date, the Pest, now Budapest, train station.)[39] Storytellers could not easily pin an identity on this cosmopolite-technologist who moved among nations. As he was not French enough to be feted by French historians, American historians also did not recognize him. (Nationalist narrative accounts can be the outcome of logrolling confabulations, with each side plumping native and foreign heroes, to mutual benefit.) When Nordling does appear in historical accounts, he's a way station in the ascension of Eiffel, whose tower, designed for collective celebration and international envy, casts a nationalist shadow into the twenty-first century (figure 8.13).[40] In 2018, an engineering paper on the restoration of Nordling's viaducts omits his name and refers to "four historical Eiffel-type rail viaducts."[41] People look through Nordling's work and see a hero, Eiffel, who like Stephenson was a savvy marketer with a national platform who became a figurehead. Nordling's contribution vanishes, his life and work seemingly irreconcilable with the nation-frames-invention narrative. Denationalization is comparable to dehumanization: an innovator not seen to embody the nation is diminished and not a candidate for glorification.[42]

How were the tall viaducts perceived? The draft design of the Grandfey in Freiburg, Switzerland reportedly came from a commission consisting of "a Swiss man, a German, a Frenchman, and an Austrian [Nordling]."[43] Writer and politician Émile Deschanel, in an ecstatic account in a Paris newspaper, called the Grandfey (figures 5.13 and 8.14 and cover of this book) , a "colossal metallurgical masterpiece of our Creusot," a major iron manufacturing enterprise in France. Deschanel conveyed the structure's multinational roots and sheer height by envisioning a ride between landmarks: "Seventy-eight meters! That's exactly the height of the tower of the Freiburg Cathedral. Imagine the majesty

Figure 8.13
"'Great Attraction:' Birds of All Nations Flying to La Tour Eiffel." An American stork is prominent within a flock that has some flying to do to reach 300 meters, which the Chrysler Building in New York did 40 years later. *Punch*, June 29, 1889, 322. Collection of the author.

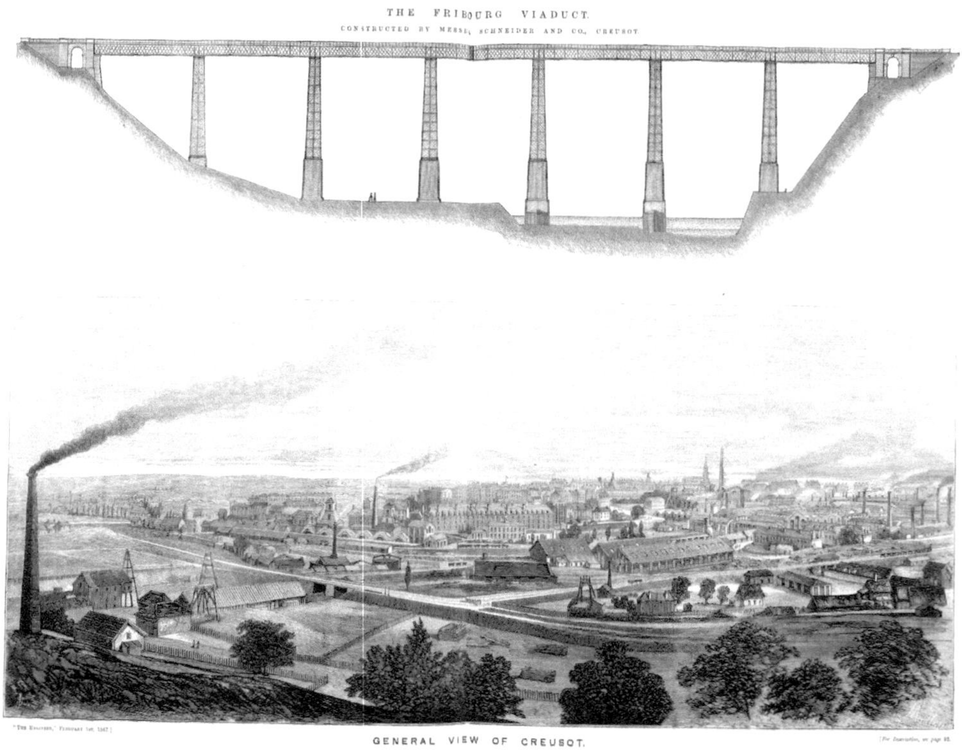

Figure 8.14
Grandfey Viaduct (Wilhelm Nordling and construction engineer C. F. Mathieu, 1857–1862), 78 meters tall and 380 meters long. Le Creusot was the site in France of its manufacture. "Civil Engineering in the Paris Exhibition," *The Engineer* 23 (1867): 102. University of Michigan.

of a train passing [in Paris] over Notre Dame's towers to the Tour St. Jacques, linked by an iron roadway: that's the Grandfey Viaduct." This sacrilegious reverie of church tower as bridge pier contrasts with the anxieties of artists who years later asserted that the Eiffel Tower would "humiliate" the city's monuments. That apprehension was as much about the tower's form and size as its being a product of the industrializations transforming Paris.[44] In the case of Grandfey, however, the classical gods had the last laugh, when in 1925–1926 engineers entombed the "longest, highest, and most important railroad bridge in Switzerland" in concrete, transforming the Grandfey into a ponderous Roman-aqueduct-inspired arch bridge that "merge[d] more harmoniously into the landscape."[45] Rather than harmony, photos show heavy-handed canonical architecture restoring a traditional articulation of cultural identity.

Ithiel Town's boxy lattice bridges were also at one time foreign figures in the landscape (figure 4.1). The classical cladding (figure 7.3), which helped people accept the new technology, might be seen as a reactionary carapace on a revolutionary skeleton. But even Nordling's viaducts' exposed structures featured classical touches. Subsumed in the complexity and impossible for train passengers to spot, these details must have been intended for consumers of the drawings and photographs, that is, conservative tastemakers of the art-historical establishment and engineering officials whose imprimatur was required. The details nudged these transnational (one might say, stateless) products and their radical aesthetics back into a national narrative of civilizing evolution (figure 8.15a, b).

Contemporary photographs of the Orléans Railroad viaducts help us imagine how mind-blowing it must have been to see these aliens looming in the countryside. Nordling understood that they diverged from traditional expectations of how structures should visually perform for the public. As "everyone knows" from seeing metal roof structures that looked like a "thicket," he wrote, simplicity should be a design goal. He recognized that "ordinary rules of symmetry and uniformity do not apply to these large works; witness the two viaducts over the Sioule and Neuvial, whose irregularities aren't noticed by anyone." In any case, few people would see these "industrial structures" after the opening reception.[46] In 1860s Paris, however, architectural designers and critics were engaging in debates that would continue through the century, about sustaining or shedding artistic traditions, the roles of engineers and architects and the educational institutions that trained them, what counted as architecture and how industry might alter it, and what exposed iron was actually saying about (French) society. At that time, new or under-construction public buildings marked by the prominence of visible iron (including Les Halles, Gare du Nord, and the Bibliothèque Nationale reading room) were reproducing or expanding on traditional arched and trussed design.[47] In comparison, the railway viaducts stand outside of conventional architecture, representing a new way of seeing. Nordling explained:

> There is no more surface; nothing but lines. All these lines, whether they belong to the front façade or the side or rear or even to the interior structure, appear all at once; none escapes the eye, if you only choose the right point of view. In fact, it's a transparent architecture.[48]

Enthusiastic commentators compared the Grandfey and Cère to lace and spider webs.[49] As people looked at, they looked through the structures. Transparency enabled multiple vantage points and surprising perspectives, which photographers and painters began to capture (figures I.7 and 8.15). In 1869, Eiffel was so taken with the beauty of

Figure 8.15a
Bellon Viaduct. Catwalk through bridge span and bridge span support. Nordling wrote in 1864 of "a new example of the frequent coincidence of the laws of statics and the requirements of taste." Michel Berthaud, photographer, École Nationale des Ponts et Chaussées.

Figure 8.15b
Bellon Viaduct. Lower portion of pier. Michel Berthaud, photographer, École Nationale des Ponts et Chaussées.

the Sioule Viaduct (figures 8.3 and 8.4), which his firm had just built, that he sent two photos of it to his mother.[50]

The constructive vanguard, located far from big cities and mainstream media, had made a conspicuous statement that would eventually reach the urban masses and overtake the old guard: two decades after the launch of the Sioule, the Eiffel Tower (initial design, 1884; built 1887–1889), much taller than a mere viaduct pier nestled in the landscape, offered a rush to the public. Its structure had been enabled by lattices, industrial building strategies, and soaring piers, along with graphic statics, a visual method of analysis employed by Maurice Koechlin, who (with Émile Nouguier) was the initial designer of the tower and a student of Karl Culmann, author of the well-known reports on bridge building in the United States (see chapter 4). Still, they overlaid the classical: a "triumphal arch," an ancient architectural form, was grafted onto each side of the tower's base. It softened the silhouette and, according to Eiffel, provided a tribute to "the honor of France in the fields of science and industry." An arcade at the perimeter of Eiffel Tower's first level also featured architectural ornamentations (designed by Stephen Sauvestre and erased in a 1927 refurbishment). These substantially altered the tower's appearance from a distance and up close.[51] Painted a vibrant red and rising over the capital, the Eiffel Tower, as the Portage Bridge had (figures 1.3, 1.6, and 8.16) inspired visitors to experience engineering feats as intoxicating national achievements that heightened and nationalized the experience of landscape. At its base, forty-four full-scale replica houses hosted by "native" guides and arrayed along the Seine provided an architectural experience organized by racial and national-cultural categories. It portrayed Romans and Aryans as the source of the cultures of Europe. Visitors to the exposition could wander through an evolutionary "history of human habitation" and then glance up at an overwhelming representation of the next phase.[52]

Gustave Eiffel, unlike Nordling, was not able to attend the École Polytechnique; he studied at the École Centrale. (Like Nordling, Eiffel's family changed its name, from Boenickhausen.)[53] He majored in chemistry; early on he made a name for himself as a construction manager; he networked with prominent engineers such as Nordling, who had a decisive impact on his career; he attracted talented and ambitious designers to the company he named after himself.[54] Eiffel's story adheres to a stereotypical "American" trajectory of a self-promoting, driven, entrepreneurial manager who achieves high-tech success thanks to a team of expert collaborators, thinks big, and late in his career reinvents himself. Perhaps the Eiffel Tower's entrepreneurial provenance, as well as its latticed roots, encouraged some to view it as a foreign intrusion.[55] No French engineer comes close to receiving the attention Eiffel has received, despite the numerous successful creative technologists who worked in France during the nineteenth century. Engineer-entrepreneurs such as Ernest Gouin, Jean-François Cail, Pierre Schaken, and

Figure 8.16
Portage Bridge (1852–1875) and Eiffel Tower, (France, 1889–present). Erie railway trains stopped on the Portage Bridge, allowing passengers to descend into the structure and stroll along pedestrian walkways. "Sectional view of Portage High Bridge," Library of Congress. Eiffel Tower, 2019 Jack Young—Places/Alamy Stock Photo.

Eugène Schneider, along with Eiffel, whether or not they attended prestigious schools, established businesses that built around the world.[56]

Builders eventually erected transportation and communication networks that linked nations across the planet. When Mahatma Gandhi suggested in 1931 that "America should say: 'We know how to build bridges; we won't keep it a secret, but we shall say to the whole world, "We will teach you to build bridges and we will charge you nothing,"'" Gandhi likely did not know that "America" had "said" as much a century earlier.[57] Technologists in industrializing nations had listened to and taught each other—despite inventors' profit-seeking goals—how to build. Manufacturers in Britain and France, for example, shipped thousands of bridges worldwide. When colonizing nations such as Britain built in India, however, this was not about creative exchange. "In most semi-civilized places the cost of [unskilled] labor was much less than in civilized places," and workers lacked riveting expertise; engineers designed accordingly.[58] In some locations, not just skilled labor, but labor of any kind could be scarce. Lattice bridge construction set the standard; it functioned in the less-than-civilized slave states of the US and everywhere workers labored under what Saint-Simon called "industrial liberty."[59]

When Michel Chevalier rhapsodized about Americans' "practical liberty," he did not know that fifty years later French citizens would initiate in New York Harbor the erection of a monument dedicated to this ever-morphing idea.[60] The project was conceived after the Union victory in the Civil War in 1865, which was followed by a period of explosive industrial growth in the United States. In 1879, Eiffel was called on to design the Statue of Liberty's supporting skeleton. Intense exposure to wind, water, and sun would differentially affect skeleton and copper skin—a condition typical of skyscrapers. Eiffel's firm had the know-how, seeded by Nordling's viaducts, to devise a solution; Koechlin designed the pier-like internal structure of Liberty, which opened two and a half years before the Eiffel Tower.[61] This Brobdignagian symbol stood for a mélange of ideas about individual, corporate, and national liberties tempered by survival-of-the-fittest ideology, at a time when only select Americans enjoyed the guarantees of freedom and Black Codes had replaced the Slave Codes. The statue embodied techno-utopian ideals of international interconnection that ran through its decades of development.[62] Its structure supported a monument to liberty that invited people to come in, climb up, and gaze at a borderland crisscrossed by ships whose wakes formed a network map in the water.

AFTERWORD

Restoring the Story

Imagine if I, the author of *American Bridge*, declared, "I'm going to design and build a bridge. I've been researching the history of construction for years. I have some knowledge about structures. Okay, I'll do it."

And why not? Engineers have done the reverse, for years. With curiosity sparked by the complexities of design and construction, or seeking to promote themselves, some undertake the writing of history.

Stories about the past are always about the future. Robert Stephenson developed a bridge type that was a failure, yet he and his colleagues succeeded in pushing their work into future history, helping structure how we think about iron, engineering, and Britain. If engineers can write history, why can't a historian design a bridge?

"Wait a minute!" you'll exclaim. "People would die if a historian designed a bridge!" The implication is that lives are at stake if a historian designs a bridge, but if an engineer writes history, no lives are at stake. Let's explore this assumption.

As we stated at the start: stories are our bridges to the future. We entrust our lives to the bridges we cross; the same is true of the stories we believe. Narratives enable nations to become "imagined communities," whose social, political, legal, and physical infrastructure enfolds the profoundly different mindsets that exist within such communities.[1] Often, the consequences of national narratives have been deadly.

If an engineer unthinkingly adopts evolutionist or nationalist attitudes—for example, assumes the Atlantic Ocean was a divider rather than a connector, or deems Americans "practical," or asserts that innovation arises within rather than across national borders—that person supports ways of thinking that have contributed to uncountable deaths. Many more than have resulted from bridge collapses.

"Hold on," you might say, "It's not just engineers! Many storytellers uncritically adopt nationalistic perspectives." You'd be right. History as a discipline has been a nationalist project whose mission is to construct beliefs about national destiny. Historians, engineers, and many others have supported it. Their key partner is you. Broad public support gives these tales life-defining potential.

Figure 9.1
Bridge (destroyed 1915) over the Thur, near Thann, Alsace, France. Postcard, collection of the author.

For example, an idea long supported by Chicagoans asserts that engineer Willliam Le Baron Jenny "invented the skyscraper." Even if it were possible to define "skyscraper," how could a gigantic assembly of diverse technologies emerge from the brain of one person? Interconnected networks of engineers and workers and companies over many years enabled the tall building. Still, the great-man-invents-skyscraper legend, as credible as a stork delivering a baby, thrives in popular media and many minds.[2] Why? In Chicago, it's civic pride—nationalism at a local scale. Something deemed first is first here, in this place, my city, my home.

By reverse engineering history, that is, by dismantling stories in order to understand their construction and maintenance, we can uncover why a structure or person might be nationalized, called "first," or forgotten. But it's a challenging task. The autopsy conducted at the dismantling of Jenney's Home Insurance Building (1885), along with subsequent documentation confirming that the building did not have a skeleton frame independent of its walls, has not put an end to myths about the origins of the skyscraper.[3]

Figure 9.2
Lattice bridge over the Vistula at Dirschau (today Tczew, Poland) (Karl Lentze, 1850–1857, still partially exists). At the very start of World War II, Polish defenders blew up the bridges in an attempt to halt the German army invasion. J. Gollnow und Sohn Stahlbau-Werk, *Dirschau, die Eisenbahnbrücke über die Weichsel* (Presentation album for Adolf Hitler, [1940]). Third Reich Collection, Library of Congress.

If stories are our bridges to the future, the many bridges destroyed in war remind us how often our stories fail to carry us forward.[4] Can we design bridges that will? We'll need to demolish the old, simple tales, embrace latticelike complexity, and establish a declaration of interdependence that replaces independence as the supposed foundation of liberty. This will enable us to retrieve lost evidence, recover ignored histories, and remix them to create pathways to shared futures. It's worth exploring the narratives that bring us together and tear us apart because the construction of history is a matter of life and death. The ability of stories to bridge gives us the power to transform.

LE MONDE ILLUSTRÉ

JOURNAL HEBDOMADAIRE

ABONNEMENTS POUR PARIS ET LES DÉPARTEMENTS
Un an, 21 francs ; — Six mois, 11 francs ; — Trois mois, 6 francs.
Le numéro : 35 c. à Paris — 40 c. dans les gares de chemins de fer.
Tout numéro demandé quatre semaines après son apparition sera vendu 40 c.
Le volume semestriel : 11 fr. broché. — 16 fr. relié et doré sur tranche.
LA COLLECTION DES 25 VOLUMES : 261 FRANCS.
Adresser tout ce qui concerne la partie littéraire et artistique à M. PAUL DALLOZ, *directeur*.

BUREAUX DE VENTE ET D'ABONNEMENT
9, RUE DROUOT, OU 13, QUAI VOLTAIRE

14e Année. N° 694. — 30 Juillet 1870

DIRECTION ET ADMINISTRATION
13, QUAI VOLTAIRE
Toute demande d'abonnement non accompagnée d'un bon sur Paris ou sur la poste, toute demande de numéro à laquelle ne sera pas joint le montant en timbres-poste, sera considérée comme non avenue. — Toute réclamation, toute demande de changement d'adresse doit être accompagnée d'une bande imprimée. — On ne répond pas des manuscrits envoyés.
Adresser tout ce qui concerne les abonnements et l'administration à M. BOURDILLIAT, *administrateur*.

Destruction du pont de Kehl par les Badois. — D'après les croquis de MM. Mirland et Broutta. — (Voir la *Correspondance*, page 73.)

Figure 9.3
"Destruction of the Kehl bridge by the Badois." Located between France and Germany, this lattice bridge (Fleur Saint-Denis and Vuigner, 1858–1861), linking Strasbourg and Kehl over the Rhine, was strategically designed with a revolving drawbridge on each end. This allowed the link to be severed as needed (figure 6.9). Still, the eastern drawbridge was blown up at the beginning of the Franco-Prussian War in 1870. Subsequently, the bridge became part of Germany, where it remained until 1919. France blew up part of the bridge in 1940 and Germany did the same in 1944. *Le Monde Illustré*, July 30, 1870, 65. Collection of the author.

Figure 9.4
Railroad Bridge over the Seine, Argenteuil, France. This bridge was bombed during the Franco-Prussian War, 1870–1871. Its surviving piers are visible in Claude Monet's 1873 paintings of the bridge's replacement. From *Disasters of the War*, Jean Andrieu, photographer. Paris Musées, Musée Carnavalet.

NOTES

PREFACE

1. Thomas Smits, *The European Illustrated Press and the Emergence of a Transnational Visual Culture of the News, 1842–1870* (Routledge, 2020), 13.

2. Within the large literature that introduces structural concepts, see the classic Robert A. Heller and Deborah J. Oakley, *Salvadori's Structure in Architecture: The Building of Buildings*, 4th ed. (Pearson, 2017).

INTRODUCTION

1. Neil Smith and Ward Dennis, "Restructuring of Geographical Scale: Coalescence and Fragmentation of the Northern Core Region," *Economic Geography* 63 (1987): 160–182.

2. Marjorie Nice Boyer, *Medieval French Bridges* (Medieval Academy of America, 1976), xi. "The railroad bridge now occupies a place in connection with the economy and facilities of land transportation, second in importance only to that of the steam engine itself"; W. P. Trowbridge, "Town's Truss," [Columbia College] *School of Mines Quarterly* 9 (1888): 317.

3. Re fig. I.1, Kathryn Elizabeth Sackett, "Organized Nostalgia: G. H. Durrie's Winter Landscapes and Nineteenth-Century American Art Associations' Promotion of a National Identity" (master's thesis, Michigan State University, 1999), 4, 26–27.

4. David Frisby and Mike Featherstone, eds., *Simmel on Culture* (Sage, 1997), 171–173. In *American Bridge*, "invention" refers to something perceived as new. I use "innovation" and "design" almost interchangeably.

5. Mimi Sheller and John Urry, "New Mobilities Paradigm," *Environment and Planning A* 38 (2006): 210.

6. Artha Tuladhar, *Trail Bridge Building in the Himalayas* (Helvetas, 2007).

7. David A. Hounshell, *From American System to Mass Production, 1800–1932* (Johns Hopkins University Press, 1984), 15–17, 303–315, 331–336.

8. I have adapted Eleanor Byrne and Martins McQuillan's question about Disney films, cited in Tracey Mollet, "'With a smile and a song . . .' Walt Disney and the Birth of the American Fairy Tale," *Marvels & Tales* 27 (2013): 110.

9. Dorothy Ross, "Grand Narrative in American Historical Writing: From Romance to Uncertainty," *American Historical Review* 100 (1995): 652–653. Ross attributes "romance" to Northrop Frye. For comparable use of "evolutionism," see David Graeber and David Wengrow, *The Dawn of Everything: A New History of Humanity* (Farrar, Straus and Giroux, 2021).

10. Stephen K. Sanderson, *Social Evolutionism: A Critical History* (Basil Blackwell, 1990). Mike Hawkins, *Social Darwinism in European and American Thought, 1860–1945* (Cambridge University Press, 1997), 184. John Z. Langrish, "Darwinian Design: Memetic Evolution of Design Ideas," *Design Issues* 20 (2004): 8–9.

11. Anthony Leeds, "Darwinian and 'Darwinian' Evolutionism in the Study of Society and Culture," in *The Comparative Reception of Darwinism*, ed. Thomas F. Glick (University of Chicago Press, 1974), 437–477; Robert M. Young, *Darwin's Metaphor: Nature's Place in Victorian Culture* (Cambridge University Press, 1986), 80–81; Bernard Lightman, "Darwin and the Popularization of Evolution," *Notes and Records of the Royal Society of London* 64 (2010): 5–24.

12. Sanderson, *Social Evolutionism*, 10–12. On Spencer, see Richard Hofstadter, *Social Darwinism in American Thought*, rev. ed (Beacon, 1955), 31–50.

13. Derived from research of Henry Schmidt (PhD diss., University of California Berkeley, forthcoming).

14. Siep Stuurman, "Dehumanization before the Columbian Exchange," in *The Routledge Handbook of Dehumanization*, ed. Maria Kronfeldner (Routledge, 2021), 39–40; John P. Jackson Jr. and Nadine M. Weidman, *Race, Racism, and Science: Social Impact and Interraction* (ABC-CLIO, 2004), 80–88. Fritz Machlup and Edith Penrose, "Patent Controversy in the Nineteenth Century," *Journal of Economic History* 10 (1950): 1–29.

15. Georges Canguilhem, "The Role of Analogies and Models in Biological Discovery," in *Scientific Change*, ed. A. C. Crombie (Basic, 1963), 507–520; Darwin apparently used technological analogy to enhance his arguments; John F. Cornell, "Analogy and Technology in Darwin's Vision of Nature," *Journal of History of Biology* 17 (1984): 308–312, 320–322. Stephen M. Downes, "Evolutionary Psychology," in *Stanford Encyclopedia of Philosophy*, ed. Edward N. Zalta (Spring 2020), https://plato.stanford.edu/archives/spr2020/entries/evolutionary-psychology/. Re "ecosystem" and "meme," see Kevin N. Laland and Gillian R. Brown, *Sense and Nonsense: Evolutionary Perspectives on Human Behavior* (Oxford University Press, 2002), 225, and Merlin Donald, "Material Culture and Cognition," in *Cognition and Material Culture: The Archaeology of Symbolic Storage*, ed. Colin Renfrew and Chris Scarre (McDonald Institute, 1998), 186–187; Matthew Rampley, *The Seductions of Darwin: Art, Evolution, Neuroscience* (Pennsylvania State University Press, 2017), 5–58; Langrish, "Darwinian Design," 16–18. Also, Frank B. Golley, *A History of the Ecosystem Concept in Ecology: More than the Sum of the Parts* (Yale University Press, 1993).

16. Young, *Darwin's Metaphor*, 80–125; Gregory K. Dreicer, "The Long Span. Intercultural Exchange in Building Technology: Development and Industrialization of the Framed Beam in Western Europe and the United States, 1820–1870" (PhD diss., Cornell University, 1993), 49–50; Giuliano Pancaldi, "Darwin's Technology of Life," *Isis* 110 (2019): 680–700. Charles Darwin, *On the Origin of Species by Means of Natural Selection* (Murray, 1859), 31.

17. Langrish, "Darwinian Design," 4; Peter J. Bowler, "The Changing Meaning of 'Evolution,'" *Journal of History of Ideas* 36 (1975): 99, 114.

18. Charles Davis Jameson, "The Evolution of the Modern Railway Bridge," *Popular Science Monthly* 36 (February 1890): 461–481, with G. Lindenthal letter, 844–845; J. E. Greiner, "The American Railroad Viaduct: Its Origin and Evolution," *ASCE Transactions* 25 (1891): 349–372; Charles C. Schneider, "Evolution of the Practice of American Bridge Building," *ASCE Transactions* 54 (1905): 213–234; Llewellyn Nathaniel Edwards, "The Evolution of Early American Bridges," *Transactions of the Newcomen Society* 13 (1933): 95–116; James K. Finch, "Wind Failures of Suspension Bridges or Evolution and Decay of the Stiffening Truss," *Engineering News Record*, March 27, 1941, 74–79; Llewellyn Nathaniel Edwards, *A Record of History and Evolution of Early American Bridges* (University Press, 1959); Carl W. Condit, "Two Centuries of Technical Evolution Underlying the Skyscraper," in *Second Century of the Skyscraper*, ed. Lynn S. Beedle (Van Nostrand Reinhold, 1988), 11–24; William Addis, *Structural Engineering* (Ellis Howard, 1990), 205; J. G. James: "The Evolution of Wooden Bridge Trusses to 1850," *Journal of Institute of Wood Science* 9 (1982): 116–135, 168–193, and "The Evolution of Iron Bridge Trusses to 1850," *Transactions of the Newcomen Society* 52 (1980–81): 67–101, and "Some Steps in the Evolution of Early Iron Arched Bridge Design," *Transactions of the Newcomen Society* 59, no. 1 (1987–1988): 153–187. James, "Evolution of Iron Bridge Trusses," 84. Engineers used theories of evolution before *Origin of Species*, e.g. Jos. Langer, "Ueber R. Schifkorn's priv. balken- und bogenförmige Trägerkonstruktion," *Zeitschrift des Oesterreichischen Ingenieur-Vereines* 6 (1854): 256–259.

S. C. Gilfillan, *The Sociology of Invention: An Essay in the Social* (Follet, 1935; reprint, MIT, 1970), 14, equated natural and technological forms. On George Basalla, *The Evolution of Technology* (Cambridge University Press, 1988), see W. Bernard Carlson and Michael E. Gorman, "Understanding Invention as a Cognitive Process: The Case of Thomas Edison and Early Motion Pictures, 1881–91," *Social Studies of Science* 20 (1990): 419–420, and Colleen Dunlavy's review, "Technological Change," *Science* 245 (1989): 991. Also Joel Mokyr, *The Lever of Riches: Technological Creativity and Economic Progress* (Oxford University Press, 1990); Gregory Dreicer, "Building Myths: The 'Evolution' from Wood to Iron in the Construction of Bridges and Nations," *Perspecta* 31 (2000): 130–141.

19. Jules Gaudard, *De l'évolution dans la construction des grands ponts* (Viret-Genton, 1892). Henry Petroski, *Engineers of Dreams: Great Bridge Builders and the Spanning of America* (Vintage, 1995), 36, 79, 80, and *Invention by Design: How Engineers Get from Thought to Thing* (Harvard University Press, 1996), 204–205. George Michael Danko, "The Evolution of the Simple Truss Bridge" (PhD diss., University of Pennsylvania, 1979), 72, and Stephen F. Eisenman, "Design in the Age of Darwin," in *Design in the Age of Darwin: From William Morris to Frank Lloyd Wright*, ed. Stephen F. Eisenman and Corinne Granof (Northwestern University Press, 2008), 5–10.

20. Cornell, "Analogy and Technology," 303–344. Darwin was likely aware of developments in civil engineering; he read the *Edinburgh Philosophical Journal*, which contained early reports of bridges in the United States; Peter J. Vorzimmer, "The Darwin Reading Notebooks (1838–1860)," *Journal of History of Biology* 10 (1971): 107–153. Darwin recognized that "selection" led to misunderstandings; Young, *Darwin's Metaphor*, 87–99. Bernard, cited in Canguilhem, "Role

of Analogies," 509. On "mechanical," see J. Walter Wilson, "Biology Attains Maturity in the Nineteenth Century," in *Critical Problems in the History of Science*, ed. Marshall Clagett (University of Wisconsin Press, 1959), 407ff. Karl Marx, from *Capital* (Modern Library), 1: 406n2, cited in Nathan Rosenberg, *Inside the Black Box: Technology and Economics* (Cambridge University Press, 1982), 34–35; thanks to Judith Reppy.

21. Robert J. Richards, *Darwin and the Emergence of Evolutionary Theories of Mind and Behavior* (University of Chicago Press, 1987), 590–593. On evolutionary models of thought, see Donald Campbell, "Evolutionary Epistemology," in *Evolutionary Epistemology, Rationality, and the Sociology of Knowledge*, ed. Gerard Radnitzky and W. W. Bartley III (Open Court, 1987), 47–89. Steven Vogel, *Cats' Paws and Catapults: Mechanical Worlds of Nature and People* (Norton, 1998), 31–38, 298–302.

22. Charles Darwin, *On the Origin of Species by Means of Natural Selection* (Murray, 1859), 402. Passages in Darwin's writing appear progressivist; Sanderson, *Social Evolutionism*, 30–33, 196–203; George Holton, "Science and Progress Revisited," in *Progress: Fact or Illusion?*, ed. Leo Marx and Bruce Mazlish (University of Michigan Press, 1996), 31; Adam Kuper, "Anthropology," in *Cambridge History of the Social Sciences*, vol. 7, *The Modern Social Sciences*, ed. Theodore M. Porter and Dorothy Ross (Cambridge University Press, 2003), 359. "Progress towards perfection" was "uncharacteristic" of Darwin, according to Holton, "Science and Progress Revisited," 31.

23. Henry Petroski, *To Engineer Is Human: The Role of Failure in Successful Design* (St. Martin's, 1982), 67; Otis T. Mason, *Origins of Invention: A Study of Industry among Primitive Peoples* (Scribner's, 1895), 349.

24. Henry Petroski, *The Pencil: A History of Design and Circumstance* (Knopf, 1989), 24.

25. Andrea Palladio, *Four Books of Andrea Palladio's Architecture* (Ware, 1738), 3:65. Richard Shelton Kirby et al., eds., *Engineering in History* (McGraw-Hill, 1956), 138, 227. Condit, "Building and Building Construction," in *Technology in Western Civilization* (Oxford University Press, 1967), 1:369, believed that builders in the United States "adapted" trusses depicted by Palladio. Gennaro Tampone and Francesca Funis, "Palladio's Timber Bridges," *Proceedings of the First International Congress on Construction History*, ed. S. Huerta (2003), 1909–1919; Rowland J. Mainstone, "Structural Analysis, Structural Insights, and Historical Interpretation," *Journal of Society of Architectural Historians* 56 (1997): 335–336; Jacques Heyman, "Palladio's Wooden Bridges," *Architecture Research Quarterly* 4 (2000): 81–86; Howard Burns, "From Julius Caesar to the Grubenmann Brothers: Soane and the History of Wooden Bridges," in *John Soane and the Wooden Bridges of Switzerland*, ed. Angelo Maggi and Nicola Navone (Archivio del Moderno, Accademia di Architettura, Università della Svizzera italiana; John Soane's Museum, 2003), 19–29. David Yeomans, "Soane and Swiss Bridges," *Construction History* 19 (2003): 49–54; Mario Rinke, *Vom Konstrukt zum Typus: Der Wandel der strukturellen Form von Tragwerken im 18. and 19. Jahrhundert* (Deutscher Kunstverlag, 2016), 24–25. Before Palladio, see for example, Leonardo da Vinci; Carla Ceraldi and Ennio Russo Ermolli, "The Swiss Covered Bridges of Eighteenth Century. A Special Case: The Bridge at Schaffhausen," in Huerta, *Proceedings*, 570–571.

26. Augustus Henry Lane-Fox Pitt-Rivers, *The Evolution of Culture* (Clarendon, 1906), 44; Franz Boas, *Race, Language and Culture* (Macmillan, 1940), 290–294.

27. Caroline Winterer, *The Culture of Classicism: Ancient Greece and Rome in American Intellectual Life, 1789–1910* (Johns Hopkins University Press, 2002); Raymond Quek, "Nationalism and Architecture: An Introduction," in *Nationalism and Architecture*, ed. Raymond Quek et al. (Ashgate, 2012), 5–9.

28. Steven Sloman and Philip Fernbach, *The Knowledge Illusion: Why We Never Think Alone* (Riverhead, 2017), 133–134. Steven Johnson, *Where Good Ideas Come From: Natural History of Innovation* (Riverhead, 2010), 22, 198. Vaclav Smil, *Invention and Innovation: A Brief History of Hype and Failure* (MIT Press, 2023), 1–3.

29. Bill Addis, *Building: 3000 Years of Design Engineering and Construction* (Phaidon, 2007).

30. David E. Nye, *America as Second Creation: Technology and Narratives of New Beginnings* (MIT, 2003), 18–19, 200–203. Ann Norton Greene, "War Horses: Equine Technology in the American Civil War," in *Industrializing Organisms: Introducing Evolutionary History*, ed. Susan R. Schrepfer and Philip Scranton (Routledge, 2004), 143–165, and *Horses at Work: Harnessing Power in Industrial America* (Harvard University Press, 2008).

31. Leda Cosmides, John Tooby, and Jerome H. Barkow, "Introduction: Evolutionary Psychology and Conceptual Integration," in *The Adapted Mind: Evolutionary Psychology and the Generation of Culture*, ed. Jerome H. Barkow et al. (Oxford University Press, 1992), 5.

32. David Albert, "Explaining It All: How We Became the Center of the Universe," *New York Times Book Review*, August 12, 2011, 16, and Robert Foley and Marta Mirazón Lahr, "On Stony Ground: Lithic Technology, Human Evolution, and the Emergence of Culture," *Evolutionary Anthropology* 12 (2003): 109–122. Anthropologists link increased brain size and intelligence to development of stone tools—over 2.5 million years; Kathy D. Schick and Nicholas Toth, *Making Silent Stones Speak: Human Evolution and the Dawn of Technology* (Simon & Schuster, 1993), 17–18; 219. Re brain and self-control, see Brian Hare, "Survival of the Friendliest: *Homo sapiens* Evolved via Selection for Prosociality," *Annual Review of Psychology* 68 (2017): 172–173. Re social cooperation, see Peter J. Richerson and Robert Boyd, *Not by Genes Alone: How Culture Transformed Human Evolution* (University of Chicago Press, 2005), 196–197, 235–237, 55, 51–52. Also Martin A. Nowak, "Five Rules for the Evolution of Cooperation," *Science* 314 (2006): 1560–1563; Coren L. Apicella et al., "Social Networks and Cooperation in Hunter-Gatherers," *Nature* 481 (2012): 497–501.

33. Vogel, *Cats' Paws*, 21–22. "There is no demonstrable, direct, genetic basis for most human behavior"; C. Michael Barton and G. A. Clark, "Evolutionary Theory in Archaeological Explanation," *Archeological Papers of the American Anthropological Association* 7 (1997): 12.

34. Downes, "Evolutionary Psychology." Also, Stefan Linquist et al., "Exploring the Folkbiological Conception of Human Nature," *Philosophical Transactions of the Royal Society B* 366 (2011): 444–453.

35. Mike Hansell, *Bird Nests and Construction Behaviour* (Cambridge University Press, 2000), 1. Chris de Bresson, "The Evolutionary Paradigm and the Economics of Technological Change," *Journal of Economic Issues* 21 (1987): 754–757.

36. Re biologists, see J. Toulmin, in Discussion in Crombie, *Scientific Change*, 612. Re engineers, see Kees Gispen, *New Profession, Old Order: Engineers and German Society, 1815–1914* (Cambridge

University Press, 1989). Ross, "Grand Narrative," 651; re evolution as a scientific explanation of history, see Dorothy Ross, "Historical Consciousness in Nineteenth-Century America," *American Historical Review* 89 (1984): 924–925. Karl E. Weick, "Small Wins: Redefining the Scale of Social Problems," *American Psychologist* 39 (1984): 43.

37. Claire Kloppmann-Lambert, "Specialized Aspects of Architectural Discourse: Metaphors in the British Magazine 'The Architectural Review,'" *ASp: La revue de GERAS* 73 (2018): 16–17, http://journals.openedition.org/asp/5111. Elizabeth Kolbert, "That's What You Think," *New Yorker*, February 27, 2017, 66–71.

38. Judith Schueler, *Materialising Identity: The Co-construction of the Gotthard Railway and Swiss National Identity* (Foundation for History of Technology & Aksant Academic, 2008), 15–16, 79–82; Michael Adas, *Machines as the Measure of Men: Science, Technology, and Ideologies of Western Dominance* (Cornell University Press, 1989), 221–224.

39. Jerome Bruner, "The Narrative Construction of Reality," *Critical Inquiry* 18 (1991): 4–5, 13–15. Daniel Kahneman, "Don't Blink! The Hazards of Confidence," *New York Times Magazine*, October 23, 2011, 32; Daniel Kahneman, *Thinking, Fast and Slow* (Farrar, Straus and Giroux, 2011), 212, 218. Ross, "Grand Narrative," 666–668. Re biologists, see M. Elizabeth Barnes, "The Spandrels of San Marco and the Panglossian Paradigm: A Critique of the Adaptationist Programme," by Stephen J. Gould and Richard C. Lewontin, in *Embryo Project Encyclopedia* (2014-11-14), ISSN: 1940-5030, https://hdl.handle.net/10776/8239.

40. "Great divides" comes from from Latour, who cites Jack Goody's "great dichotomy"; Bruno Latour, "Drawing Things Together," in *Representation in Scientific Practice*, ed. Michael Lynch and Steve Woolgar (MIT Press, 1990), 19–20. Addis, *Structural Engineering*, 14–23.

41. Nora Barlow, ed., *Darwin and Henslow: The Growth of an Idea* (Murray, 1917), 80–81.

42. Harris Mylonas and Maya Tudor, "Nationalism: What We Know and What Still Need to Know," *Annual Review of Political Science* 24 (2021): 110–112. Peter J. Taylor, *World City Network: A Global Urban Analysis* (Routledge, 2004), 27.

43. Trajan, letter to editor Philo White, dated September 12, 1825, *Western Carolinian* (Salisbury), October 11, 1825, 3; John L. Sanders, "Town, Ithiel," in *North Carolina Architects and Builders: A Biographical Dictionary* (2009): http://ncarchitects.lib.ncsu.edu/people/P000032. In the 1820s and 1830s, concern about the republic inflamed nationalism in the United States; Ross, "Historical Consciousness," 913. On geometric forms and state character, see Amir Alexander, *Proof! How the World Became Geometrical* (Scientific American/Farrar, Straus and Giroux, 2019): 227–257.

44. In 1840s France, some observers interpreted iron structures as symbols of democracy; Sang-hun Lee, "Technology and Form: Iron Construction and Transformation of Architectural Ideals in Nineteenth Century France, 1830–1889" (PhD diss., MIT, 1996), 245–249. Also, Tom F. Peters, "American Culture of Construction," *Perspecta* 25 (1989): 160n19; David Monteyne, "Framing the American Dream," *Journal of Architectural Education* 58 (2004): 24–33.

45. James A. Secord, *Victorian Sensation: The Extraordinary Publication, Reception, and Secret Authorship of "Vestiges of the Natural History of Creation"* (University of Chicago Press, 2000), 5.

46. "Sedentarism treats as normal stability, meaning, and place, and treats as abnormal distance, change, and placelessness"; Mimi Sheller and John Urry, "The New Mobilities Paradigm," *Environment and Planning A* 38 (2006): 208. J. M. Blaut, "Diffusionism: A Uniformitarian Critique," *Annals of Association of American Geographers* 77 (1987): 30–47, and Jennifer Tann, "Space, Time and Innovation Characteristics: Contribution of Diffusion Process Theory to History of Technology," *History of Technology* 17 (1998): 143–163.

47. Darwin H. Stapleton, "Moncure Robinson: Railroad Engineer, 1828–1840," in *Benjamin Henry Latrobe and Moncure Robinson: The Engineer as Agent of Technological Transfer*, ed. Barbara E. Benson (Hagley Foundation, 1975), 33; Howard Mumford Jones, *America and French Culture, 1750–1848* (University of North Carolina Press, 1927), 4.

48. Darwin H. Stapleton, "Benjamin Henry Latrobe and the Transfer of Technology," in *Technology in America: A History of Individuals and Ideas*, ed. Carrol W. Pursell Jr. (MIT Press, 1986), 28, 34. John M. Staudenmaier, *Technology's Storytellers: Reweaving the Human Fabric* (MIT, 1985), 123.

49. Quotation in James C. Burke, "North Carolina's First Railroads: A Study in Historical Geography" (PhD diss., University of North Carolina, 2008), 9, from North Carolina Constitutional Convention, "Proceedings and Debates of the Convention of North-Carolina, Called to Amend the Constitution of the State, which Assembled at Raleigh, June 4, 1835" (Joseph Gales, 1836).

50. Burke, "North Carolina's First Railroads," 9–10.

51. Peter J. Taylor and Ben Derudder, *World City Network: A Global Urban Analysis*, 2nd ed. (Routledge, 2016), 181 and throughout.

52. Burke, "North Carolina's First Railroads," 22–30. I intend no criticism of Burke's work.

53. Margaret T. Hodgen, *Change and History: A Study of the Dated Distributions of Technological Innovations in England* (Wenner-Gren Foundation for Anthropological Research, 1952), 189–190, 200; Christine MacLeod, "European Origins of British Technological Predominance," in *Exceptionalism and Industrialisation: Britain and Its European Rivals, 1688–1815*, ed. Leandro Prados de La Escosura (Cambridge University Press, 2004), 111–126.

54. Eugene S. Ferguson, "On the Origin and Development of American Mechanical Know How," *Midcontinent American Studies Journal* 3 (1962), 15; D. L. Burn, "The Genesis of American Engineering Competition, 1850–1870," *Economic History* 2 (1931), 309. A. G. Kenwood and A. L. Lougheed, *Technological Diffusion and Industrialisation before 1914* (St. Martin's, 1982), 19–22; Karl Erich Born, "The Role of Social Factors in the Acquisition and Development of Technology in Germany," in *L'acquisition des techniques par les pays non-initiateurs* (CNRS, 1973), 282.

55. Hodgen, *Change and History*, 192.

56. Maxine Berg and Pat Hudson, *Slavery, Capitalism and the Industrial Revolution* (Polity, 2023), and Maxine Berg and Pat Hudson, "Rehabilitating the Industrial Revolution," *Economic History Review* 45 (1992): 24–50.

57. Anna Bezanson, "The Early Use of the Term Industrial Revolution," *Quarterly of Journal of Economics* 36 (1922): 343–349; Emmanuel Fureix and François Jarrige, *La modernité désenchantée* (La Découverte, 2015), 68–69.

58. All of J. M. Blaut, *The Colonizer's Model of the World: Geographical Diffusionism and Eurocentric History* (Guilford, 1993); quote, 11–12. Biological or cultural factors are usually offered for (European) inventiveness.

59. Berg and Hudson, *Slavery, Capitalism*, 131–136.

60. Arnold Pacey, *Technology in World Civilization* (MIT Press, 1990), vii, 101–102, 117–120. Sven Beckert, *Empire of Cotton: A Global History* (Knopf, 2015), 420–422; Kanakalatha Mukund, "Indian Textile Industry in 17th and 18th Centuries: Structure, Organisation and Responses," *Economic and Political Weekly* 27 (1992): 2057–2065. Also, Edward A. Gargan, "As Looms Fall Silent, India's Social Fabric Frays," *New York Times*, October 7, 1992, A4.

61. Paul L. Knox, "World Cities and the Organization of Global Space," in *Geographies of Global Change*, ed. R. J. Johnston et al. (Blackwell, 1995), 245.

62. Ross, "Grand Narrative," 652, 656–657; Ross, "Historical Consciousness," 917; and Nye, *America as Second Creation*, 9–12, 40–42. Re impacts, see Leo Marx, "Domination and the Redefinition of Progress," in Marx and Mazlish, *Progress*, 208–210. "Elite time colonized and helped construct the boundaries of territories that we have come to call nations"; John R. Gillis, "Memory and Identity: History of a Relationship," in *Commemorations: Politics of National Identity*, ed. John R. Gillis (Princeton University Press, 1994), 6. Ben Marsden and Crosbie Smith, *Engineering Empires* (Palgrave Macmillan, 2005), 16–21; 159–160.

63. John Tresch, *Romantic Machine: Utopian Science and Technology after Napoleon* (University of Chicago Press, 2012), 4–5.

64. Juliet Wilson-Barreau, *Manet, Monet and the Gare Saint-Lazare* (Yale University Press, 1998), 65–101; Cecilia E. Ramos, "Caillebotte's 'On the Pont de l'Europe': A Transversal Vista of Modernity," *Thresholds* 31 (2006): 32–39; Pietro Galfi della Bagliva and Claude P. J. Ghez, "Deconstructing Gustave Caillebotte's 'Le Pont de l'Europe' (1876)," in *Gustave Caillebotte: Impressionism in Modern Paris*, ed. Yasuhide Shimbata (Bridgestone Museum of Art, Ishibashi Foundation, 2013), 230–241; Isabel Cabezas, "Creating a Scene to Make an Impression: How Gustave Caillbeotte and his Street Scenes of Haussmannian Modernity Support Impressionism Without Subscription," *Bowdoin Journal of Art* 4 (2018): 13–18, 23–26; J. Kirk T. Varnedoe, "Caillebotte's Pont de l'Europe: A New Slant," in *Gustave Caillebotte and the Fashioning of Identity in Impressionist Paris*, ed. Norma Broude (Rutgers University Press, 2002), 8–20; Tyler Edward Ostergaard, "Best Within: Contested Image of the Railroad in French Visual Culture, 1837–1877" (PhD diss., University of Iowa, 2014), 203–271, esp. 233–238; Peter Soppelsa, "How Haussmann's Hegemony Haunted the Early Third Republic," Hollis Clayson and André Dombrowski, *Is Paris Still the Capital of the Nineteenth Century? Essays on Art and Modernity, 1850-1900* (Routledge, 2016), 35–51.

65. David E. Nye, *American Technological Sublime* (MIT, 1994), 43. Nye takes a national approach (xiv, 43). See also, Richard Haw, *Brooklyn Bridge: A Cultural History* (Rutgers University Press, 2005), 509.

66. Frederick Brown, *For the Soul of France: Culture Wars in the Age of Dreyfus* (Knopf, 2010), 151–154. Priscilla Parkhurst Ferguson, *Paris as Revolution: Writing the Nineteenth-Century City* (University of California Press, 1994), 220–224. A contemporary account of the opening:

"L'achèvement de la tour Eiffel," *Le Temps*, April 7,1889, 2. T. M. Charlton, *History of the Theory of Structures in the Nineteenth Century* (Cambridge University Press, 1982), 171. Patrick Kingsley, "Bias Shared by Extremes of Right, Left and Islam," *New York Times*, April 5, 2019, A1, A8. Lee, "Technology and Form," 265–269.

CHAPTER 1

1. Darwin, *On the Origin of Species by Means of Natural Selection* (Murray, 1859), 486. Michael Adas, *Machines as the Measure of Men: Science, Technology, and Ideologies of Western Dominance* (Cornell University Press, 1989), 308.

2. George W. Stocking Jr., *Race, Culture, and Evolution* (Free Press, 1968), 131–132. The barbarian concept was employed in a variety of cultures; see, for example, Magnus Fiskesjö, "On the 'Raw' and the 'Cooked' Barbarians of Imperial China," *Inner Asia* 1 (1999): 139–142; 162, n. 68. It was used, too, in ancient Greece.

3. Alfred Russel Wallace, *The Wonderful Century: Its Successes and Failures* (Dodd, Mead, 1898), 357–360, 370.

4. Theodore W. Pietsch, *Trees of Life: A Visual History of Evolution* (Johns Hopkins University Press, 2012), 85–122; Manuel Lima, *The Book of Trees: Visualizing Branches of Knowledge* (Princeton University Press, 2014), 38–40, 46–47. David A. Baum and Stacey D. Smith, *Tree Thinking: An Introduction to Phylogenetic Biology* (Roberts, 2013), 9–30. "Misinterpretations and Intuitive Ideas About Evolutionary 'Progress' or 'Advancement,'" *The Tree Room*, https://evolution.berkeley.edu/evolibrary/article/0_0_0/evotrees_interpretations_02#a2.

5. See bridge tree in Christian Schädlich, "Der Baustoff Eisen als Grundlage für die Herausbildung qualitative neuer Baukonstruktionen im 19. Jahrhundert," in *Zur Geschichte des Konstruierens*, ed. Rainer Graefe (Deutsche-Verlags-Anstalt, 1989), 138–139; tree portraying "timber construction" in Mathias Seraphin, "On the Origin of Modern Timber Engineering," in *Proceedings of the First International Congress on Construction History*, ed. S. Huerta, (Juan de Herrera, Benvenuto, F. Dragados, 2003), 1846; on Banister Fletcher's "Tree of Architecture" (1921), see Carla Yanni, "Development and Display: Progressive Evolution in British Victorian Architecture and Architectural Theory," in *Evolution and Victorian Culture*, ed. Bernard Lightman and Bennett Zon (Cambridge University Press, 2014), 242.

6. For the standard perspective, see Bill Addis, *Building* (Phaidon, 2007), 610; William Addis, *Structural Engineering* (Ellis Howard, 1990), provides insightful discussion on engineering historiography. Re wood's role, see Carl Condit, *American Building Art: The Nineteenth Century* (Oxford University Press, 1960), 16, 4, 8. Re coexisting "races in all stages of culture," see Henry Balfour, "Introduction," in A. Lane-Fox Pitt-Rivers, *Evolution of Culture* (Clarendon, 1906), xiv–xv.

7. Conversation with Henry Schmidt, January 23, 2024.

8. Iñaki Iriarte-Goñi and María Isabel Ayuda, "Wood and Industrialization: Evidence and Hypotheses from the Case of Spain, 1860–1935," *Ecological Economics* 65 (2008): 177–186. Nicholas A. Bill, "Timber Bridge Construction on British and Irish Railways, 1840–1870: The Scale of Construction and Factors Influencing Material Selection," *Construction History* 31 (2016):

75–98. Sven-Erik Åström, "English Timber Imports from Northern Europe in the Eighteenth Century," *Scandinavian Economic History Review* 18 (1970): 12–32. Seraphin, "On the Origin of Modern Timber Engineering," Huerta, *Proceedings*, 1845–1854 and Mathias Seraphin, "The Influence of the German Railways on the Birth of Modern Timber Engineering," *Proceedings of the Second International Congress on Construction History*, ed. James Campbell et al. (Construction History Society, 2006), 2869–2889.

9. G. A. Breymann and H. Lang, *Allgemeine Bau-Konstsruktions-Lehre, mit besondere Beziehung auf das Hochbauwesen*, vol. 2: *Die Constructionen in Holz* (Weise, 1870), 1, cited in Mario Rinke, "Konstruktive Metamorphosen: Holz als Immerwährendes Surrogat," in *Holz: Stoff oder Form*, ed. Mario Rinke and Joseph Schwartz (Niggli, 2014), 264. Seraphin, "Influence of the German Railways," 2869, 2876, 2878, 2884.

10. Lee H. Nelson, "Early Wooden Truss Connections vs. Wood Shrinkage: From Mortise-and-Tenon Joints to Bolted Connections," *Association for Preservation Technology Bulletin* 27 (1996): 11–23; David T. Yeomans, "Early Carpenters' Manuals, 1592–1820," *Construction History* 2 (1986): 28.

11. "'High Bridge,' Portage, New York," *Canadian Journal* 1 (1852): 68–69, is the source of many reports, including "'High Bridge,' Portage, New York," *Civil Engineer and Architects Journal* 16 (February 1853): 65–66, pl. 8. "1,600,000 feet of timber and 106,820 lbs. of iron," as stated on a stereograph in "Sectional view of Portage High Bridge—Among the Timbers," Library of Congress, https://www.loc.gov/item/2018650267/. There is a description in "Railroad Bridge at Portageville," *Wyoming County Mirror* (Warsaw, NY), September 16, 1851, 3 (typescript from Madden Collection, Canal Society of New York State). Mike Kierstead, "Addendum to Erie Railway, Buffalo Division, Bridge 361.66 (Portage Viaduct)," Written Historical and Descriptive Data, HAER, National Park Service, US Department of Interior, 2018, from Prints and Photographs Division, Library of Congress (HAER NY-54), 6–7. Engineers Preston Lincoln and H. C. Brundage apparently worked with Seymour, "First Crossing of Portage Bridge," *Hornellsville Weekly Tribune* (New York), August 21, 1852, 2.

12. Stefan M. Holzer, "Structural Iron Elements in German Timber Roofs (1600–1800)," *Construction History Society* 23 (2008): 33–57. Jacob Leupold, *Theatrum Pontificiale, oder Schau-Platz der Brücken und Brücken-Baues* (Zunkel, 1726). Helmut Hilz, "Heinrich Gerber (1832–1912): Eisenbrückenbau und Unternehmertätigkeit in Süddeutschland im 19. Jahrhundert" (PhD diss., Ludwig-Maximilians-Universität, Munich, 1992), 33. The well-known "wooden" arch bridge over the Vils near Wilshofen (1809) was made up of 210.7 cubic meters of wood and 2256.8 kg of iron; Carl Friedrich von Wiebeking, *Theoretisch-practische Wasserbaukunst*, vol. 3, new ed. (Zängl, 1814), 326 ff.

13. Re casting, see Rainer Fremdling, "Continental Responses to British Innovations in the Iron Industry during the Eighteenth and Early Nineteenth Centuries," in *Exceptionalism and Industrialisation: Britain and its European Rivals, 1688–1815*, ed. Leandro Prados de la Escosura (Cambridge University Press, 2004): 153–154. Re construction and advertising, see David de Haan, "Iron Bridge—How Was It Built?," *BBC History*, February 17, 2011, http://www.bbc.co.uk/history/british/victorians/iron_bridge_01.shtml; Ironbridge Gorge Museum Trust, "The Iron Bridge," https://www.ironbridge.org.uk/about-us/the-iron-bridge/, and Wikipedia, "The Iron Bridge," https://en.wikipedia.org/wiki/The_Iron_Bridge#cite_note-UNESCO1-71. Re Coalbrookdale's modest impact on bridge design, see J.G. James, "Some Steps in the Evolution of Early Iron

Arched Bridge Design," *Transactions of Newcomen Society* 59 (1987–88): 154–156. Barrie Trinder, "The First Iron Bridges," *Industrial Archaeology Review* 3 (1979): 120, concludes that the bridge's form and techniques were not adopted by others. Re the doorknob, see Steven Lubar, "New, Useful, and Nonobvious," *American Heritage's Invention and Technology* 6 (1990): 14.

14. The slave trade was a source of wealth for Thomas Goldney and Thomas Goldney III, investors and participants in the Darby business; see Maxine Berg and Pat Hudson, *Slavery, Capitalism and the Industrial Revolution* (Polity, 2023), 132.

15. Bernard Marrey, *Les ponts modernes: 18e-19e siècles* (Picard, 1990), 105, reports the year as 1755; Georg Christoph Mehrtens, *Vorlesungen über Ingenieur-Wissenschaften*, Teil 2, *Eisenbrückenbau* (Engelmann, 1908), 268, gives the year 1719; Garrin was the engineer.

16. Amy E. Slaton, "Introduction," in *New Materials*, ed. Amy E. Slaton (Lever, 2020), 11–14; 18–20.

17. Addis, *Structural Engineering*, 1–13.

18. Gregory K. Dreicer, "The Long Span. Intercultural Exchange in Building Technology: Development and Industrialization of the Framed Beam in Western Europe and the United States, 1820–1870" (PhD diss., Cornell University, 1993), 27–29; Lauren Beck, "Introduction: Firsting in the Architecture of Decolonizing Scholarship in the Early-Modern Atlantic World," in *Firsting in the Early-Modern Atlantic World*, ed. Lauren Beck (Routledge, 2020), 1–6. Interest in firsts dates to antiquity; see Pamela O. Long, *Openness, Secrecy, Authorship: Technical Arts and the Culture of Knowledge from Antiquity to the Renaissance* (Johns Hopkins University Press, 2001), 12–13.

19. Forest History Society, "Fueling the Fires of Industrialization," https://foresthistory.org/education/trees-talk-curriculum/fueling-fires-industrialization/essay-fueling-the-fires-of-industrialization/; Sarah Hurtes, "European Union Signals a Move Away from Wood Energy," *New York Times*, September 15, 2022, https://www.nytimes.com/2022/09/15/world/europe/europe-wood-energy-deforestatation.html.

20. Town stated in his brochures that the bridge could be built in iron.

21. Michel Chevalier, *Les voies de communication aux États-Unis* (excerpted from *Journal de l'Industriel et du Capitaliste*, and Chevalier's *Lettres sur l'Amérique du Nord*) (F. G. Levrault, 1837), 48, referring to bridges of Burr, Town, and Long.

22. William Petit Trowbridge, a military, hydrological, and civil engineer who became professor at Yale and Columbia: "The [lattice] thus introduced for timber bridges not only required no modification when wrought iron took the place of timber, but rather rendered the use of iron and steel possible, as these materials are almost universally employed at the present day in bridge building." William Petit Trowbridge, "Town's Truss," [Columbia College] *School of Mines Quarterly* 9 (1887–1888): 317.

23. Charles Goschler, "Note sur les chemins de fer suisses," *Mémoires et compte rendu des travaux de la société des ingénieurs civils* 9 (1856): 172.

24. Franz Anton Gerstner, *Die innern Communicationen der vereinigten Staaten von Nordamerika*, ed. L. Klein (Förster, 1842–1843), 2:106–107. "Wonderful plenty of wood," from Brooke Hindle,

"The Artisan during America's Wooden Age," in *Technology in America: A History of Individuals and Ideas,* ed. Carrol W. Pursell Jr. (MIT Press, 1986), 16.

25. Forest G. Hill, *Roads, Rails and Waterways: The Army Engineers and Early Transportation* (University of Oklahoma Press, 1957), 104, and Daniel Hovey Calhoun, *The American Civil Engineer: Origins and Conflicts* (Technology Press/MIT Press, 1960), 126–132. "Remarks Upon a Paper 'On the Building Materials of the United States of North America' by David Stevenson C. E., Edinburgh," *American Railroad Journal* 7 or 13 (1841): 262; re Stevenson, see chapter 3.

26. Justin M. Spivey, "Philadelphia & Reading Railroad, Peacock's Lock Viaduct," Written Historical and Descriptive Data, HAER, National Park Service, US Department of Interior, 2000, Prints and Photographs Division, Library of Congress (HAER PA-118), 2. Re construction speed, see Carl Ghega, *Über nordamerikanischen Brückenbau und Berechnung des Tragungsvermögens der howe'schen Brücken* (Kaulfuss Witwe, Prandel, 1845), 4–5.

27. W. R. C., "Notes on Practical Engineering: Bridges," *American Railroad Journal* 17 (1844): 10. Edwin Layton, "Mirror-Image Twins: The Communities of Science and Technology in 19th-Century America," *Technology and Culture* 12 (1971), 568. According to Layton, "The intractable nature of materials constituted one of the most important barriers to the development of technology"—removed thanks to "science."

28. Julian Young and Kenneth Haynes, "Translators' Preface," in *Martin Heidegger, Off the Beaten Track* (Cambridge University Press, 2002), ix.

29. "Ueber Eisenbahnen in Nordamerika," *Dingler* 36 (1830): 322. Also "Nordamerikanische Literatur (Eisenbahnen)," *Dingler* 35 (1830): 155–160.

30. "Remarks upon a Paper by David Stevenson," 226–227.

31. There is a contemporary account in *First Annual Report of S. W. Roberts, Chief Engineer of the Little Schuylkill and Susquehanna Rail Road Company* (Clark, 1839), 8.

32. Forest History Society, "The Railroad and the Tie," https://foresthistory.org/education/trees-talk-curriculum/fueling-fires-industrialization/the-railroad-and-the-tie/; and "Fueling the Fires of Industrialization."

33. Steven W. Usselman, *Regulating Railroad Innovation: Business, Technology, and Politics in America, 1840–1920* (Cambridge University Press, 2002), 75–96; Iriarte-Goñi and Ayuda, "Wood and Industrialization," 182.

34. Squire Whipple and John Roebling discussed the Rider Bridge in *American Railroad Journal* 23 (1850); Victor C. Darnell, "The Pioneering Iron Trusses of Nathaniel Rider," *Construction History* 7 (1991): 69–81.

35. Engineer James O. McClure's account points to arson: Lockwood R. Doty, ed., *History of Livingston County, New York* (Van Deusen, 1905); information about the pine lands of 300 acres comes from McClure, 511. Re the watchman, see "Haut pont de Partage [*sic*] dans l'État de New-York," *Nouvelles Annales de la Construction* 1 (June 1855): 3, pl. 24. Re the rain, see "Burning of a Bridge: Portage Bridge on Erie Railway Totally Destroyed by Fire: Supposed Incendiarism," *New York Times* May 7, 1875, 2; "Portage Bridge: Its Destruction Described by an Eye-Witness," *Buffalo*

Courier, May 7, 1875, http://www.letchworthparkhistory.com/burningbridge.html. The description is based on an interview with contractor: "Railroad Bridge at Portageville," *Wyoming County Mirror*, September 16, 1851, Madden Collection, Canal Society of New York State. On the replacement bridge, see George S. Morison, "The New Portage Bridge," *ASCE Transactions* 5 (1876): 1–8; Charles Macdonald and George S. Morison, "On the New Portage Bridge," 235–239, pls. 1–3; M. E. Pontzen, "Chronique," *Annales des ponts et chaussées* 5–2 (1876): 171–173. Also Gleim, "Eiserne Brückenbauten in den Vereinigten Staaten von Nord-Amerika," *Deutsche Bauzeitung* 10 (1876): 139. Macdonald believed a launched, riveted lattice structure would have taken half the time; the new structure's engineer disagreed; Morison, "New Portage Bridge," 235–239. Five years later, the bridge was rebuilt: Simine Short, *Locomotive to Aeromotive: Octave Chanute and the Transportation Revolution* (University of Illinois Press, 2011), 88–91. On marketing, see Herbert Gottfried, *Erie Railway Tourist, 1854–1886: Transporting Visual Culture* (Lehigh University Press, 2018), 1–23.

36. "Haut pont de Partage," 3, plate 24.

37. Town's four advertising brochures will be referred to as: Town (year). Town (1821), 9; Town (1831), 8; Town (1839), 12, taken from Robert Fulton, *Treatise on the Improvement of Canal Navigation* (Taylor, 1796), 126.

38. Mansfield Merriman and Henry S. Jacoby, *A Text-Book on Roofs and Bridges*, vol. 3: *Bridge Design* (Wiley, 1894), 10. Specific bridge types were recommended for specific span lengths.

39. J. Parker Snow, "Wooden Bridge Construction on the Boston and Maine Railroad," *Journal of Association of Engineering Societies* 15 (1895): 32; also "Life of Different Kinds of Timber in Bridges of Various Kinds and Advisibility of Protecting Same from Weather," *Proceedings of the Ninth Annual Convention of the Association of Railway Superintendents of Bridges and Buildings*, October 17–18, 1899 (Rumford, 1899): 50–65.

40. Nicholas Bill, "The Safety Record of Timber Railway Bridges in the UK: 1840–1870," *Proceedings of the Institution of Civil Engineers, Engineering History and Heritage* 167 (2014): 34–46.

41. George L. Vose, "Safety in Railway Travel," *North American Review* 135 (1882): 383. Frank Griggs Jr., "Tariffville Bridge Disaster," *Structure*, February 2021, https://www.structuremag.org/article/tariffville-bridge-disaster/.

42. Gordon MacLeod, "Place, Politics and 'Scale Dependence': Exploring Structuration of Euro-Regionalism," *European Urban and Regional Studies* 6 (1999): 248.

43. Squire Whipple, *A Work on Bridge Building: Consisting of Two Essays, the One Elementary and General, the Other Giving Original Plans and Practical Details for Iron and Wooden Bridges* (Curtiss, 1847), 39–46.

44. Georges Ribeill, "Vie et mort des ouvrages d'art: L'exemple des ponts de chemins de fer" (unpublished manuscript, 1992), 7. Wood rotted and bolts had become *Wasserleiter* (water pipes); "Ueber die Dauer hölzerner Brücken," *Allgemeine Bauzeitung* 27 (1862): 81–82. The bridge was built in 1836 and opened 1837, though the pine had to be replaced within six years and in 1847 major repairs were undertaken; Eugène Flachat in Auguste Perdonnet, *Nouveau portefeuille de l'ingénieur des chemins de fer* (E. Lacroix, 1857), 185. Also, "Eugene Flachat," *Science* 8 (1898): 14–15. The provisional bridge design was based on American models: Ernest Deligny, "Notice

sur la chute du pont du chemin de fer de Paris à Saint-Germain à Asnières, incendié le 25 février 1848, et sur le pont provisoire en charpente établis par M. Eugène Flachat, ingénieur en chef," *Memoires et compte rendu des travaux de la Société des ingénieurs civils* 1 (1848): 61–62. See chapter 5.

45. Charles Couche, "Travaux d'art des chemins de fer d'Allemagne" *Annales des mines* 5 (1854): 311–312; Robert Crawford, "The Railway System of Germany," *Institution of Civil Engineers: Transactions* 22 (1862–1863): 13; "Cast Iron Bridges," *The Engineer* 23 (1867): 126.

46. J. G. James, "Overseas Railways and the Spread of Iron Bridges, c. 1850–1870," (Author, 1987), 27; Snow, "Wooden Bridge Construction," 31–43. Re the Civil War, see, for example, Wikipedia, "Richmond and Petersburg Railroad Bridge," https://en.wikipedia.org/wiki/Richmond_and_Petersburg_Railroad_Bridge, and Bull Run Bridge in chapter 4. Re New York Central, see George E. Gray in Greiner, "American Railroad Viaduct," 369–371, and chapter 7. Engineer Thomas C. Clarke estimated there were 300 miles of iron and 700 miles of wooden railroad bridges in the US; John Alexander Low Waddell, *A System of Iron Railroad Bridges for Japan* (Tokyo University Press, 1885), 2.

47. Phyllis Deane, *The First Industrial Revolution*, 2nd ed. (Cambridge University Press, 1979), 137–141.

48. Condit, "Building and Building Construction," 1: 373; Condit, *American Building Art*, 4, 16. Also Sanghun Lee, "Technology and Form: Iron Construction and Transformation of Architectural Ideals in Nineteenth Century France, 1830–1889," (Ph.D. diss., MIT, 1996), 295–301. Scott Gabriel Knowles and Jose Torero, "Plyscrapers, Gluescrapers, and Mother Nature's Fingerprints," in Slaton, *New Materials*, 151–173; Bettina and Steffen Franke, eds., *Proceedings of the 4th International Conference on Timber Bridges* (Bern University of Applied Sciences/Empa, 2002).

49. John A. Roebling, *Report of John A. Roebling, Civil Engineer, to the Presidents and Directors of the Niagara Falls Suspension and Niagara Falls International Bridge Companies, on the Condition of the Niagara Railway Suspension Bridge* (Murphy & Bechtel, 1860), 20.

50. The wood in the deck structure was eventually replaced with metal, due to deterioration in the damp environment; William Irwin, *The New Niagara: Tourism, Technology, and the Landscape of Niagara Falls, 1776–1917* (Pennsylvania State University Press, 1996), 40.

51. This preceded by five years the publication of Darwin's *Origin of Species*. "Niagara Suspension Bridge," *American Railroad Journal* 28 (1855): 188.

52. William F. Ogburn and Dorothy Thomas, "Are Inventions Inevitable? A Note on Social Evolution," *Political Science Quarterly* 37 (1922): 83–98.

53. Carl Condit, *American Building: Materials and Techniques from the First Colonial Settlements to the Present* (University of Chicago Press, 1968), 94. Jennifer Whyte, "Evolutionary Theories and Design Practices," *Design Issues* 23 (2007): 53–54.

54. Engineer Benjamin Wilder Guppy, B & M Railroad in Boston, continued, "In steel designing, the sections of the main members are determined first and then the details designed to transmit the stress properly. In straight timber work it is the design of the connection that generally determines the dimensions of the main members. In metal work the unit stresses to be considered are those of tension, compression, shear, and bearing. In timber the unit stresses

to be accounted for are tension, compression, bearing on end grain, bearing across grain, shear parallel to fiber, and shear normal to fiber on planes parallel to and normal to direction of fiber"; Robert Fletcher and J. Parker Snow, "A History of the Development of Wooden Bridges," *ASCE Transactions* 99 (1934): 365.

55. Mario Rinke, "The Infinitely Shapable Structure: Structural Iron and the Decontextualization of Construction," In *Before Steel: The Introduction of Structural Iron and Its Consequences*, ed. Mario Rinke and Joseph Schwartz (Niggli, 2010), throughout and 75–76, 84; Rinke, "Konstruktive Metamorphosen," 271.

56. Milton S. Graton, *The Last of the Covered Bridge Builders* (Clifford-Nicol, 1978), 1–5, 9.

57. Bill, "Safety Record of Timber Railway Bridges," 34–46. On wooden lattice bridges surviving train collisions, see Benjamin W. Guppy in Snow, "Wooden Bridge Construction," 41.

58. Guppy in Snow, "Wooden Bridge Construction," 39, says, "Neither is it safe to entrust light iron bridges wholly to the care of workmen not technically educated; the timber parts of wooden ones may be." Snow described working methods of wood bridge builders and challenges of dealing with metals: "It has come within the observation of the writer many times that when an intelligent master-carpenter has had the care for a term of years of a line of wooden bridges . . . he gradually brings their parts, when building new ones, to almost the exact size called for by scientific analysis when actual loads are used in calculation. He will use iron rods that are too small for they show him no distress unless they break, but the timber parts guide him to right results"; Snow, "Wooden Bridge Construction," 39. Re camber, see Snow, "Wooden Bridge Construction," 39, and John S. Lupold and Thomas L. French Jr., *Bridging Deep South Rivers: The Life and Legend of Horace King* (University of Georgia Press, 2004), 32, 56.

59. Rinke, "Konstruktive Metamorphosen," 276; Karl Culmann, "Der Bau der eisernen Brücken in England und Amerika," *Allgemeine Bauzeitung* 17 (1852): 163.

60. Joachim Radkau, *Technik in Deutschland: Vom 18. Jahrhundert bis heute* (Suhrkamp, 1989), 63.

61. Robert Thorne, "Introduction," in *Iron Revolution: Architects, Engineers and Structural Innovation, 1780–1880*, ed. Robert Thorne (RIBA Heinz Gallery, 1990), 7–8.

62. Corydon T. Purdy, "The Use of Steel in Large Buildings," *Journal of Association of Engineering Societies* 14 (1895): 186; Snow, "Wooden Bridge Construction," 31. C. Shaler Smith, *Comparative Analysis of the Fink, Murphy, Bollman & Triangular Trusses* (Woods, 1865), 3. "Railroad Bridges," *Western Railroad Gazette*, December 1, 1860, 2, provides a list of wooden bridge fires and notes that replacement iron bridges "will become a heavy and very profitable branch of business."

63. Stephen F. Eisenman, "Design in the Age of Darwin," in *Design in the Age of Darwin: From William Morris to Frank Lloyd Wright*, ed. Stephen F. Eisenman and Corinne Granof, (Northwestern University Press, 2008), 5. Spencer shaped the thinking of Louis Sullivan and Daniel Burnham; Joanna Merwood-Salisbury, *Chicago 1890: The Skyscraper and the Modern City* (University of Chicago Press, 2009), 22–24. Dankmar Adler, "The Influence of Steel Construction and Plate Glass upon Style," *Proceedings of the Thirtieth Annual Convention of the American Institute of Architects* (1896): 244.

64. W. R. C., "Notes on Practical Engineering," 8. Darwin noted this in 1871; Langrish, "Darwinian Design," 13.

65. Evolutionist indoctrination continues, e.g., John F. Unsworth, *Design of Modern Steel Railway Bridges*, 2nd ed. (CRC, 2018), 2.

66. J. W. Powell, "From Barbarism to Civilization," *American Anthropologist* 1 (1888): 121; Otis T. Mason, *The Origins of Invention: A Study of Industry among Primitive People* (Scribner, 1895), 22; also, Merwood-Salisbury, *Chicago*, 25–26. E.g., Jonathan Clarke, "From Rational to Structurally Ornamental: Exported English Iron Architecture of the Mid-Nineteenth Century," in *Function and Fantasy: Iron Architecture in the Long Nineteenth Century*, ed. Paul Dobraszczyk and Peter Sealy (Routledge, 2016), 113–140. Robert Thorne, "Introduction," in *Structural Iron and Steel, 1850–1900*, ed. Robert Thorne (Ashgate, 2000), xiv, frames metal and engineering in national terms.

67. Bill Cooke, "The Denial of Slavery in Management Studies," *Journal of Management Studies* 40, no. 8 (2004): 1895–1918; see chapter 5. Re "larger, more complex, more advanced societies," see Edward T. Layton Jr., "European Origins of the America Engineering Style of the Nineteenth Century," in Reingold and Rothenberg, *Scientific Colonialism*.

68. William M. Taylor, "Iron, Engineering and Architectural History in Crisis: Following the Case of the River Dee Bridge Disaster, 1847," *Architectural Histories* 1 (2013): 2, notes that the terms "spirit of the age" and "practical thinking" dehistoricize and minimize failure and crisis.

69. Andrew Carnegie, *The Gospel of Wealth* (Century, 1901), 4. Ali A. Mazrui, "From Social Darwinism to Current Theories of Modernization: A Tradition of Analysis," *World Politics* 21 (1968): 69–83; Hofstadter, *Social Darwinism*, 4–12. 45–46; Daniel Becquemont, "Social Darwinism: From Reality to Myth and from Myth to Reality," *Studies in History and Philosophy of Biological and Biomedical Sciences* 42 (2011): 12–19; and Thomas C. Leonard, "Origins of the Myth of Social Darwinism: The Ambiguous Legacy of Richard Hofstadter's *Social Darwinism in American Thought*," *Journal of Economic Behavior and Organization* 71 (2009): 37–51. Mike Hawkins, *Social Darwinism in European and American Thought, 1860–1945* (Cambridge University Press, 1997).

70. Jilin Xu, "Social Darwinism in Modern China," *Journal of Modern Chinese History* 6 (2012): 182–197.

71. John Seabrook, "The Flash of Genius," *New Yorker*, January 3, 1993, 44. Re rockets, see previous chapter.

72. Mazrui, "From Social Darwinism," 69–83.

73. Brian Hare and Vanessa Woods, *Survival of the Friendliest: Understanding Our Origins and Rediscovering Our Common Humanity* (Random House, 2020), xvi.

74. Nineteenth-century scientist Alexander von Humboldt, for example, believed in a unity of nature that rejected racist views and presented a more harmonious picture of humanity and nature. This perspective did not align with views of economy and nation that justified the oppression and violence inherent in a struggle for survival. Eleanor Jones Harvey, *Alexander von Humboldt and the United States: Art, Nature, and Culture* (Princeton University Press, 2020); re other theorists, see Hawkins, *Social Darwinism*, 177–182.

75. Brian Hare, "Survival of the Friendliest: Homo sapiens Evolved via Selection for Prosociality," Annual Review of Psychology 68 (2017): 170–172; Andreas Nieder, *A Brain for Numbers: The Biology of the Number Instinct* (MIT Press, 2019), 63–75.

76. Ernie Smith, "Cheese American Style," *Tedium*, July 4, 2017; https://tedium.co/2017/07/04/american-cheese-culture-history/; Rohit Kapoor and Lloyd E. Metzger, "Process Cheese: Scientific and Technological Aspects—A Review," *Comprehensive Reviews in Food Science and Food Safety* 7 (2008): 194–214; Jenny Gross, "Gruyère is Still Gruyère, even if It's Made in America, a Federal Judge Rules," *New York Times*, January 13, 2022: B5; Neil Vigdor, "The F.D.A.'s New Approach to French Dressing? Laissez-Faire," *New York Times*, January 15, 2022, B3; Jennifer Calfas and Talal Ansari, "The U.S. Federal Government Deregulates French Dressing," *Wall Street Journal*, January 12, 2022, https://www.wsj.com/articles/the-u-s-federal-governmentderegulates-french-dressing-11642036962.

77. Swiss Platform for Sustainable Cocoa, "Produced in the South—Consumed in the North," https://www.kakaoplattform.ch/about-cocoa/cocoa-facts-and-figures; Patricia Cohen, "How West Africa Can Reap More Profit from the Global Chocolate Market," *New York Times*, September, 28, 2023.

78. Elian Peltier, "How Africans Are Changing French—One Joke, Rap and Book at a Time," *New York Times*, December 12, 2023. Knox, "World Cities," 242–243.

79. David Krakauer, "Home Improvers Are a Natural Fact," *Times Higher Education Supplement*, October 4, 2007, https://www.timeshighereducation.com/books/home-improvers-are-a-natural-fact/196647.article.

CHAPTER 2

1. Fulton, *Treatise on the Improvement of Canal Navigation* (Taylor, 1796), x, 130. He proposed a design for a cast iron aqueduct with uniform parts, pl. 13. Rebecca Solnit, *A Field Guide to Getting Lost* (Penguin, 2005), 4–5.

2. "In America, building construction was always the chief form of capital investment and the chief occupation of skilled craftsmen. . . . Investment in fixed capital for manufacturing equipment before 1825 may be seen as a spin-off from profits made in trade, real estate, and building"; Thomas C. Cochran, *Frontiers of Change: Early Industrialization in America* (Oxford University Press, 1981), 51. An estimated million houses (100 per day) were built between 1800 and 1825; Eugene S. Ferguson, "The American-ness of American Technology," *Technology and Culture* 20 (1979): 7. Daniel Hovey Calhoun, *The Intelligence of a People* (Princeton University Press, 1973), 292–294; Kevin H. O'Rourke and Jeffrey G. Williamson, *Globalization and History: The Evolution of a Nineteenth-Century Atlantic Economy* (MIT Press, 1999), 35.

3. Town's friend Samuel Morse foresaw the significance of the Erie Canal for New York's economy and his own future; Thomas Bender, *New York Intellect: A History of Intellectual Life in New York City from 1750 to the Beginnings of Our Own Time* (Knopf, 1987), 119–121.

4. André Guillerme, "Network: Birth of a Category in Engineering Though[t] during the French Restoration," *History and Technology* 8 (1992): 155–156 and throughout. G[uillau]me

Tell Poussin, *Travaux d'améliorations intérieures projetés ou exécutés par le gouvernement général des États-Unis d'Amérique, de 1824 à 1831* (Anselin et Carilian-Goeury, 1834–1836), xxi.

5. Town (1821), 4. Town's four advertising brochures will be referred to as: Town (year). Fulton, *Treatise*, 125: "By what means can a given quantity of materials be arranged, so as to incur the least expence [*sic*] in erecting, and be rendered most permanent; and by which mode will the least materials answer the purpose?" Fulton is quoted by Town (1821, 1831, 1839).

6. The letter was endorsed after "slight examination" by Yale president and mathematician Jeremiah Day; Yale scientist Benjamin Silliman; Yale graduate, lawyer, politician, and businessman Simeon Baldwin; and Yale mathematics professor Alexander M. Fisher; Letter, Eli Whitney to Ithiel Town, New Haven, December 26, 1820, New Haven Museum MSS 105, series 1, box 1, folder d. Endorsements also included in Town (1825), 0–1 and Town (1831), 5. Letter, Simeon Baldwin, New Haven, to Town, April 12, 1823, in Whitney Library of New Haven Museum, contains a request for "your plan" of bridge over the pond on the Whitney property; Baldwin asks Town not to charge for patent rights and says it will be a good way to "introduce" the bridge to the public. It was reportedly built in 1823, moved upstream in 1860, and demolished in 1891; James Wickham Roe, "Early New Haven Inventors," in *Inventors and Engineers in Old New Haven*, ed. Richard Shelton Kirby (Books for Libraries, 1969), 13. David A. Hounshell, *From American System to Mass Production 1800–1932* (Johns Hopkins University Press, 1984), 3, 29–32. It would have been complex to remove a plank pinned at many points, and the bridge would have had to be supported during the process; conversation with bridge builder Will Truax, January 15, 2024.

7. Mary N. Woods, *From Craft to Profession: Practice of Architecture in Nineteenth-Century America* (University of California Press, 1999), and Julia M. Truettner, *Aspirations for Excellence: Alexander Jackson Davis* (University of Michigan Press, 2003), 48. Re Town and Davis in New York, see Morrison H. Heckscher, "Building the Empire City: Architects and Architecture," in *Art and the Empire City: New York, 1925–61*, ed. Catherine Hoover Voorsanger and John K. Howat (Metropolitan Museum of Art; Yale University Press, 2000), 168–187. Re Town's church building in New Haven, see Gretchen Carol Townsend, "Protestant Material Culture and Community in Connecticut, 1785–1840" (PhD diss., Yale University Press, 1995), e.g., 20–21, 35–36, 132–133, and John Donoghue, "Alexander Jackson Davis, Romantic Architect, 1803–1892" (PhD diss., New York University, 1979), 93. Town partnered with Davis in 1829–1835 and 1842–1843. The State House was apparently occupied in 1830, before completion; "Gone . . . But Not Forgotten," *Nineteenth Century* (Autumn 1976): 4–5. The fourth building on the Green was United (originally North) Church (David Hoadley, 1815).

8. Ian K. Steele, *The English Atlantic, 1675–1740: An Exploration of Communication and Community* (Oxford University Press, 1986), 266. Reportedly, engineering books were hard to find in the United States and were imported "at great expense"; *North American Review* 8 (1818): 25. Town may have had access to books of architect Charles Bulfinch at Boston Athenaeum; Sarah Allaback, "Louisa Tuthill, Ithiel Town, and the Beginnings of the Architectural History Writing in America," in *American Architects and Their Books to 1848*, ed. Kenneth Hafertepe and James F. O'Gorman (University of Massachusetts Press, 2001), 201. Michael C. Quinn, "Library of Ithiel Town: Illustrated Publications and Early Nineteenth Century American Architecture"

(master's thesis, Yale University, 1976), 19, asserts that the libraries would not have been rich resources. Re Town's Boston milieu, see James F. O'Gorman, *Edward Shaw of Boston: Antebellum Architect and Author—An Introduction* (American Philosophical Society, 2016), 2–16.

9. A surviving notebook contains geometrical and calculus problems Town worked on while building one of his earliest bridges. Town Papers, "Mathematical Exercises of Ithiel Town," MS, Yale University, Ithiel Town Papers, group 499, folder 3.

10. Roger Hale Newton, *Town and Davis, Architects: Pioneers in American Revivalist Architecture, 1812–1870* (Columbia University Press, 1942), 22–23. Lydia H. Sigourney, "Residence and Library of Ithiel Town, Esq.," *Ladies' Companion and Literary Expositor* 10 (1839): 123.

11. His pose mirrors that of the instructor featured in George Fisher, *The Instructor, or Young Man's Best Companion* (Booksellers, 1786), frontispiece. Kimberly Orcutt, "Unintended Consequences: The American Art-Union and the Rise of a National Landscape School, *Nineteenth-Century Art Worldwide* 18, no. 1 (2019), http://www.19thc-artworldwide.org/orcutt-omeka/items/show/93709. "Among the pieces which may now be seen at his studio, 115 Canal Street, is a large portrait of the celebrated Ithiel Town"; "Sketches of Artists in New York," *Brother Jonathan*, 8 July 1843, 290. Town's daughter gave the painting to Yale, which sold it to Center Church in the 1940s, according to *Pearl Street, New York*, https://pearlstreetrevival.typepad.com/photos/ithiel_town_images/town_portrait2.html. Ludmilla Jordanova, *Defining Features: Scientific and Medical Portraits* (Reaktion, 2000), 41–43.

12. Henrietta Frances Silliman Dana, *Hillhouse Avenue from 1809 to 1900* (privately printed, 1900; Tuttle, Morehouse and Taylor, 1907). Re the date Town moved into his New Haven villa, see R. W. Liscombe, "A 'New Era in My Life:' Ithiel Town Abroad," *Journal of Society of Architectural Historians* 50 (1991): 13, and Charles E. Peterson [based on H. Allen Brooks], "Town's New Haven Villa, 1836–37," *Journal of Society of Architectural Historians* 13 (1954): 27–28. Re the date Town moved in (June 1837), "book mania," and Town calling Silliman "my nearest neighbor," see Ithal [*sic*] Town, "America" [letter to John Weale], *Civil Engineers and Architects Journal* 1 (1837/1838): 52. (Town owned a copy of Thomas Frognall Dibden's *Bibliomania; or Book Madness* (1809)); "Literary Gossip," *New York Times*, May 28, 1864, 3. Dibden inscribed a book for Town; *Catalogue of Library of Late Francis B. Hayes of Boston* (Bangs, 1898), 56.) Re Jefferson, "I labour grievously under the malady of Bibliomanie"; see "Thomas Jefferson to Lucy Ludwell Paradise, 1 June 1789," *Founders Online,* National Archives, https://founders.archives.gov/documents/Jefferson/01-15-02-0166. Re collecting relationships: Town to publisher John Weale: "It will be difficult for you to send me any new work unless just published, which I have not already obtained on those subjects, and our importers now bring out all common books as soon as published and more especially all practical works on Architecture & Engineering"; letter, Ithiel Town, New York to John Weale, London, [Sept?] 17, 1836, British Architectural Library, Royal Institute of British Architects, Manuscripts & Archives Collection, Weale Papers. Re Town's library, see Sigourney, "Residence of Ithiel Town," 123–127; also in *Baltimore Literary Monument* (April 1839), 245; *Ladies' Garland and Family Wreath* (June 1841); "Architects and Architecture of New York," *Brother Jonathan*, July 15, 1843, 301–303; William Dunlap, *History of the Rise and Progress of the Arts of Design in the United States* (Scott, 1834), 2:299–300, 410; William H. Pierson, *American Buildings and Their Architects*,

vol 2, *Technology and the Picturesque: The Corporate and the Early Gothic Styles* (Oxford University Press, 1978), 125–134; Liscombe, "New Era," 5–17; Sarah Allaback, "Louisa Tuthill," 199–215, and Sarah Allaback, "Writings of Louisa Tuthill: Cultivating Architectural Taste in Nineteenth-Century America" (PhD diss., MIT, 1993), 60–61; Newton, *Town and Davis*, 19–23, 66–72. Also, W. B. Carnochan, "Education as Architecture: The Yale Report of 1828," *Journal of Aesthetic Education* 36 (2002): 8–12. Town's former residence, altered beyond recognition and employed as a science lab at Yale, was demolished in 1957 for an expansion of Dunham Laboratory, which had been built on an adjacent lot in 1912; https://buildingsofnewengland.com/tag/ithiel-town. Re Town's print collection and its auction, see Helena E. Wright, *The First Smithsonian Collection: The European Engravings of George Perkins Marsh and the Role of Prints in the U.S. National Museum* (Smithsonian, 2015), 39–45; A. J. Davis shared with Town a love of books, 52–58. Nathaniel Jocelyn and Thomas Cole were self-taught artists who relied on books, including Town's; Toby Maria Chieffo-Reidway, "Nathaniel Jocelyn: In the Service of Art and Abolition" (PhD diss., College of William and Mary, 2005), 19. Also Robin L. Thomas, "Architects' Libraries," in *Companions to the History of Architecture*, vol. 2, *Eighteenth-Century Architecture*, ed. Caroline van Eck and Sigrid de Jong (Wiley, 2017), 1–21.

Little is known about Ithiel Town. Newton, *Town and Davis*, did not meet scholarly standards at time of publication; Edna Donnell, "Roger Hall Newton, Town & Davis, Architects," *Art Bulletin* 25 (1943): 164–166. Quinn, "Library of Ithiel Town," is essential for understanding Town's library; re network of people connected to it, 35–41; re Town's wealth, from his bridge-building practice, 20; re his New York library, 22. Town letters are conserved at Yale, New Haven Colony Historical Society, Metropolitan Museum, New York Historical Society, and Royal Institute of British Architects. In "From a Correspondent, New York, July 26, 1839," *North Carolinian* (Fayetteville), August 17, 1939, 2, the writer, apparently a North Carolina resident, mentions "a substantial viaduct constructed upon the principle which has made the fortune of our acquaintance Ithiel Town, Esq." There is a brief biography in Joan Marter, ed., *Grove Encyclopedia of American Art* (Oxford University Press, 2011), 1:56–58. Death announcements are in "Death's Doings," *New World: Weekly Family Journal*, June 22, 1844, 794, and *Times Picayune* (New Orleans), June 25, 1944, 2. Re Town, Davis, and the library, see Amelia Peck, ed., *Alexander Jackson Davis, American Architect, 1803–1892* (Rizzoli/Metropolitan Museum of Art, 1992), 18, 41–42, 62; J. Marshall Bullock and Catherine W. Bishir, "Davis, Alexander Jackson (1803–1892)," *North Carolina Architects & Builders: A Bibliograhical Dictionary* (2009), http://ncarchitects.lib.ncsu.edu/people/P000003. Re Town's and Davis's office gallery, see Carrie Rebora, "Alexander Jackson Davis and the Arts of Design," in Peck, *Alexander Jackson Davis*, 37.

When Town died in 1844, his estate included fifty shares in the Clarendon Bridge Company, valued at $15,000 ($500,000 in 2020); the remaining five-years value of the patent was determined to be $1000 ($35,000 in 2020). His possessions were evaluated at $14,055 and his real estate at $14,000. It is not clear if these amounts include the value of his library and home and the land it was on. Quinn, "Library of Ithiel Town," 20, and New Haven Probate Court, No. 10,604, Last will and testament of Ithiel Town, Inventory (July 6, 1844). An offer to rent Town's New Haven home appeared in "For Sale—The House and Lot," *New York Daily Herald*, May 7, 1850, 3.

13. Letter, Ithiel Town to daughter Etha L., November 17, 1829, New Haven Colony Historical Society, MSS 105, box I, folder G; re Town's travels, see note 13. Morse's trip was underwritten by patrons; Bender, *New York Intellect*, 123.

14. Newton, *Town and Davis*, 15–75, 26–29. Town's training took place from about 1804. Asher Benjamin apparently wanted to start a school of architecture in 1802; [Talbot F. Hamlin], "Asher Benjamin," *Dictionary of American Biography* 1:179–180. Town's relationship to Benjamin has not been documented; Town and contractor Seth Geer advertised Benjamin's builder's guide, *Practice of Architecture*; "New Work of Architecture," *Evening Post* (New York), January 18, 1835, 1. See also Yeomans, "Early Carpenters' Manuals," 13–33; Rudolf Wittkower, "English Literature on Architecture," *Palladio and English Palladianism* (Thames and Hudson, 1974), 93–112.

15. Newton, *Town and Davis*, 40, 43, 93; Richard Sanders Allen, *Covered Bridges of the Northeast*, rev. ed. (Greene, 1983), 15, 66, 74; Historic Northampton, "Isaac Damon," http://www.historic-northampton.org/highlights/damon.html. See also David Merrill, "Isaac Damon and the Southwick Column Papers," *Old-Time New England* 54 (1963): 48–58.

16. Re Austin, see Henry Howe, letters in *New Haven Journal and Courier* (1883–1884), cited in George Dudley Seymour, "History of the State House," manuscript, 6, Yale University Library, George Dudley Seymour Papers, group 442, series 5, box 105, folder 1504; see also the biography by Seymour, "Ithiel Town: Architect, Engineer, Art-Collector, 1784–1844," manuscript, 13, 23–24, 41, Yale University Library, George Dudley Seymour Papers, group 442, series 5, box 105, folder 1504. See also James F. O'Gorman, *Henry Austin: In Every Variety of Architectural Style* (Wesleyan University Press, 2008), 1–6.

17. Letter re shopfronts in Paris, see Ithiel Town, Paris, to A. J. Davis, January 12, 1830, Metropolitan Museum, New York. Within five years, Town designed a metal storefront; plan and elevation in Metropolitan Museum, Misc. Portfolio, in Public Buildings folder, 24.66.392; folder 24.66.1401: sketch of "Store front with show-case, Lyceum Nat. Hist"; [Ithiel Town], New York, Board of Assistant Alderman, "Communication from Ithiel Town, Esq. dated January 28th, 1836, at Cincinnati (Ohio) Containing Suggestions for Improving the Style and Manner of Building in the City of New-York, &c. &c. Presented by Mr. Cleaveland," document no. 65 (February 15, 1836). [New York:] Board of Assistant Aldermen, February 15th, 1836; Ellen W. Kramer, "Contemporary Descriptions of New York City and Its Public Architecture ca. 1850," *Journal of Society of Architectural Historians* 27 (1968): 269. David T. Yeomans, "British and American Solutions to a Roofing Problem," *Journal of Society of Architectural Historians* 50 (1991): 266–272; Ithiel Town, "Octagonal Plan of School-House," in [Henry Barnard], *School-House Architecture* (Case, Tiffany and Burnham, 1842), 46–48; Joseph Masheck, "Meaning of Town and Davis' Octagonal Schoolhouse Design," *Journal of Society of Architectural Historians* 25 (1966): 302–304; US Patent Office, patent 3762, "William T. Peters of New Haven, Connecticut, Executor of Ithiel Town, Deceased. Molding of Bricks" (September 27, 1844); Seymour, "Ithiel Town: Architect, Engineer," 15.

18. Town (1821), 3.

19. Ronald Kline, "Charles P. Steinmetz and Development of Electrical Engineering Science" (PhD diss., University of Wisconsin, 1983), 13; Staudenmaier, *Technology's Storytellers*, 103–119;

William Addis, *Structural Engineering* (Ellis Howard, 1990), 51–59. A design rule Town used related lattice frame height to span length. "Mathematical Exercises of Ithiel Town," MS, Ithiel Town Papers, group 499, folder 3, Yale University, show Town's skill in math and geometry. Also, A. Hertwig, "Die Entwicklung der Statik der Baukonstruktionen im 19. Jahrhundert," *Technikgeschichte* 30 (1941): 82–98; Stefan M. Holzer, "The Polonceau Roof and Its Analysis," *International Journal for the History of Engineering and Technology* 80 (2013): 22–54.

20. Town (1821, 1825, 1831), brochure covers.

21. James Utterback et al., *Design-Inspired Innovation* (World Scientific, 2006), 92–94, 98–106; "Indispensable," according to Bent Flyvbjerg and Dan Gardner, *How Big Things Get Done* (Currency, 2023), 157–184.

22. Town (1821), 4.

23. Town, Letters Patent (1820), 401. "Brace" may be understood as "strut." Capt. Trumbull requested from Town two cost estimates for a Potomac aqueduct of wood and of iron; W. Trumbull, Washington, to Town, November 3, 1832, Whitney Library of the New Haven Colony Historical Society, Ithiel Town Collection, MSS # 105, box I, folder C.

24. White pine was considered optimum for structures requiring long members subject to tension. Town (1821), 4: "The diagonal bearing of these trusses, is composed of sawed plank ten or eleven inches wide, and from three to three and half inches thick."

25. Iron bolts were suggested for an iron bridge; Town (1821), 6. Town recommended that oaken treenails be cut from a die; Town (1821), 5. Later, he stated, "the best and most economical is, to saw them out square from plank, with a circular saw, and then turn them with a small lathe, attached to some water or other machinery" (1839), 14. William Hosking, *Essay and Treatise on the Practice and Architecture of Bridges* (John Weale, [1843]), xxix, and Michel Chevalier, "Pont sur le James-River, à Richmond, en Virginie; par M. Moncure Robinson," *Revue générale de l'architecture et des travaux publics* 1 (1840): 33; and Michel Chevalier, *Histoire et description des voies de communication aux Etats-Unis et des travaux d'art qui en dépendent* (Charles Gosselin, 1840–1841), 2:565. Also, Eric Sloan, *A Reverence for Wood* (Ballentine, 1973), 37. Re hole drilling, see Chevalier, *Histoire* 2:565. A double lattice bridge required twice the number of treenails; Allen, *Covered Bridges of the Northeast*, 27.

26. Town, Patent (1820) and Town (1821), 4–5. For a report on one of oldest existing lattices, see Joseph D. Conwill et al., "Bath-Haverhill Bridge," Written Historical and Descriptive Data, HAER, National Park Service, US Department of Interior, 2002, from Prints and Photographs Division, Library of Congress (HAER NH-33). For engineering analysis, see Todd Craig Radford, "Town Lattice Truss: An Appropriate Bridge Technology for Developing Countries" (PhD diss., MIT, 2010); re chords, 129–130, 107–110, 130–131. Also J. Parker Snow, "Wooden Bridge Construction on the Boston and Maine Railroad" *Journal of Association of Engineering Societies* 15 (1895): 31–43.

27. The deficiencies are described by Town (1839), 4. Moncure Robinson (chapter 3) tried to address issues in the design of the nineteen-span James River bridge, Richmond (1836–1838), and the "Peacock Bridge" over the Schuylkill Canal, Reading (1839–1840). He included vertical members along the length of the bridge, as "reinforcement" (this became common); Chevalier,

Histoire, 2:567, 571–579. Re patenting, see Steven Lubar, "The Transformation of Antebellum Patent Law," *Technology and Culture* 32 (1991): 932–959.

28. "Chord Terminations/Butt Joints in the Chords," in *Covered Bridge Manual*, ed. Philip C. Pierce et al. (FHWA-HRT-04-098), prepared for US Department of Transportation, Federal Highway Administration, Turner-Fairbank Highway Research Center (2005), 160.

29. Town (1839), 6. Eli Whitney's similar analysis in a testimonial, Town (1825), 1. Long used similar terms, but for his bridge he claimed no tensile stress in cross braces; Dario Gasparini and David Simmons, "American Truss Bridge Connections in the 19th Century I: 1829–1850," *Journal of Performance of Constructed Facilities* 11 (1997): 121. Town (1839) includes the terms "lattice-braces" (4), "truss-braces," "lattices," and "trusses" (5), "sides" (6), side-trusses (11); also "diamond," per note below.

30. Addis, *Structural Engineering*, 69–72, 157, and Bill Addis, *Building* (Phaidon, 2007), 311, 315. Navier (1820s) noted that a parallel chord truss acted like a beam; Dario Gasparini et al., "Engineering Design of Covered Bridges," in *Covered Bridges and the Birth of American Engineering*, ed. Justin Christianson and Christopher H. Marston (HAER, 2015), 82.

31. Town, Patent (1820), 402. Town may not have had a clear idea of the neutral axis, which was not generally understood until after 1840 and was not necessary for design; David T. Yeomans, *The Trussed Roof: Its History and Development* (Ashgate, 1992), 49. Addis, *Structural Engineering*, 93–95, reports that "'neutral axis' was first popularized by Tredgold in 1822, by which time the problem of bending had been virtually solved." Peters asserts that Hodgkinson's experiments (1846–1849) show this could not have been so; see chapter 6, note 34. Dermot O'Dwyer, "Development in the Design of Beams from 1820–1860," in *Iron, Steel and Buildings: Studies in the History of Construction, Proceedings of the 7th Conference of the Construction History Society* (Construction History Society, 2020), 73–82.

32. The reactions of expert engineers (chapter 3) indicate the design was novel; an exception, accounts of British authors, indicates anti-American bias. Robert Fletcher and J. Parker Snow, "A History of the Development of Wooden Bridges," *ASCE Transactions* 99 (1934): 314–408, and J.G. James, "Evolution of Wooden Bridge Trusses to 1850," *Journal of Institute of Wood Science* 9 (1982): 116–135, 168–193; James's errors and suppositions also may indicate anti-American bias.

33. Chas. F. Zimpel, *Das Eisenbahnwesen von Nordamerika, England und anderen Ländern, practisch und populär dargestellt* (Förster, 1840), 71.

34. Town's patent (1820), 401.

35. Re observations, 1790s–1815, of arch and beam bridges supported on closely spaced piers, see Timothy Dwight, *Travels in New-England and New-York* (Author, 1821–1822; reprint, Belknap, 1969), 1: 358–359; 2:48, 69, 77; 4:26–27.

36. Addis, *Structural Engineering*, 134–162.

37. Yeomans, "Early Carpenters' Manuals," 18.

38. *Mémoire sur une nouvelle méthode économique de construire des ponts en charpente d'une plus grande solidité, inventée tout récemment et mise en pratique par Monsieur Wiebeking* [1812], 1, in

Archives Nationales, F14 3185, dossier: Ponts et chaussées, Proposition d'inventeurs, Ports, 1802–1829.

39. Carl Friedrich Wiebeking, *Theoretisch-Practische Wasserbaukunst* 3 (Zängl, 1814), pl. 115. L. G. Booth, "The Development of Laminated Timber Arch Structures in Bavaria, France and England in the Early Nineteenth Century," *Journal of the Institute of Wood Science* 5 (1971): 3–16, 3–7.

40. For example, Leupold, *Theatrum Pontificale oder Schau-Platz der Brücken und Brücken-Baues* (Zunkel, 1726), 27 and Fulton, *Treatise*, 127. Critique by Carl Ghega, *Über nordamerikanischen Brückenbau und Berechnung des Tragungsvermögens der howeschen Brücken* (Fuss Witwe, Prandel, 1845), 2.

41. Town brochures. Typescript provided by Richard Sanders Allen of letters ("originals in the possession of R. H. Goodell, Lancaster, PA") of April 6, 10, 11, 1832, from Ithiel Town to President and Managers of the Columbia Bridge Company, as part of a bid for the Columbia-Wrightsville Bridge over the Susquehanna. The lattice was rejected and the low bid selected; Robert H. Goodell, "The Second Columbia Bridge," *Journal of Lancaster County Historical Society* 57 (1953): 6–8.

42. Town (1839), 7.

43. Bruce Robbins, "Commodity Histories," *PMLA* 120 (2005): 454–463.

44. During construction (1833–1835) of the Washington-Baltimore line, an early railroad, workers cut trees in nearby woods for ties; Franz Anton Gerstner, *Die innern Communicationen der vereinigten Staaten von Nordamerika*, ed. L. Klein (Förster,1842), 2: 218. Re white pine, "exported to Britain in great quantities" and "very extensively employed in the erection of bridges, particularly *frame* and *lattice* bridges, a construction peculiar to the United States," see Stevenson, "On the Building Materials of the United States of North America" (read before Society of Arts for Scotland), *Civil Engineers and Architects Journal* 4 (1841): 267.

45. Mario Rinke, *Vom Konstrukt zum Typus: Der Wandel der strukturellen Form von Tragwerken im 18. and 19. Jahrhundert* (Deutscher Kunstverlag, 2016), 130, 134.

46. Joachim Radkau and Ingrid Schäfer, *Holz* (Rowohlt, 1987), 150, cites Werner Sombart, *Der moderne Kapitalismus* (1928), 2: 1137ff. Per J. R. H. Moore, *Industrial History of the American People* (Macmillan, 1914), 34–35, a tree shortage was felt by the seventeenth century.

47. *America's Wooden Age*, ed. Brooke Hindle (Sleepy Hollow Restorations, 1975), 3. Radkau, *Holz*, 150–151. *Material Culture of the Wooden Age*, ed. Brooke Hindle (Sleepy Hollow, 1981), 4; Brooke Hindle, "Artisan During America's Wooden Age," in *Technology in America: A History of Individuals and Ideas,* ed. Carrol W. Pursell Jr. (MIT Press, 1986), 16.

48. Iñaki Iriarte-Goñi and María Isabel Ayuda, "Wood and Industrialization: Evidence and Hypotheses from the Case of Spain, 1860–1935," *Ecological Economics* 65 (2008): 177–186.

49. Michael Williams, *Americans and Their Forests: A Historical Geography* (Cambridge University Press, 1989), 3–4.

50. David Stevenson, "On the Building Materials of the United States," 266

51. Williams, *Americans and Their Forests*, 17.

52. William Strickland, *Journal of a Tour in the United States of America, 1794–1795* (New York Historical Society, 1971), 75.

53. William Priest, *Travels in the United States of America; Commencing in the Year 1793, and Ending in 1797* (Johnson, 1802), 87. Review of Adam Seybert's 1818 *Statistical Annals of the United States of America*; Sidney Smith, "Statistical Annals of United States of America," *Edinburgh Review* 33 (1820): 75.

54. "I will give direction to the boats not to molest the timber under the circumstances you have stated"; letter, Thomas Hardy to Ithiel Town, London, August 14, 1813, New York Public Library, Thomas Addis Emmet Collection, EM. 5311; also *Leading Business Men of New Haven County* (Mercantile, 1887), 30.

55. *Report of the Superintendent of Public Works, to the Legislature of South Carolina for the Year 1823* (Faust), reprinted in David Kohn and Bess Glenn, eds., *Internal Improvement in South Carolina 1817–1828* (privately printed, 1938), 234.

56. Robert McAlister, *Lumber Boom of Coastal North Carolina* (History Press, 2013), chap. 2. Robert S. Starobin, "The Economics of Industrial Slavery in the Old South," *Business History Review* 44 (1970): 152; Marcus W. Jernegan, "Slavery and the Beginnings of Industrialism in the American Colonies," *American Historical Review* 25 (1920): 223–234, 229–230, 239.

57. Cathrine W. Bishir, "Location Book of the Raleigh and Gaston Railroad," *Journal of Society of Industrial Archeology* 5 (1979): 53; Carl R. Lounsbury, "The Wild Melody of Steam: The Mechanization of the Manufacture of Building Materials, 1850–1890," in *Architects and Builders in North Carolina: A History of the Practice of Building*, ed. Catherine W. Bishir et al. (University of North Carolina Press, 1990), 195. Nathan Rosenberg, "America's Rise to Woodworking Leadership," in Hindle, *America's Wooden Age*, 46–48.

58. Franz Anton Gerstner, *Berichte aus den Vereinigten Staaten von Nordamerica, über Eisenbahnen, Dampfschiffahrten, Banken und andere öffentliche Unternehmungen* (Melzer, 1839), 15–16; Charles F. Hummel, "The Business of Woodworking: 1700 to 1840," in *Tools and Technologies: America's Wooden Age*, ed. Paul B. Kebabian and William C. Lipke (Fleming Museum, University of Vermont, 1979).

59. Zimpel, *Eisenbahnwesen*, 8; Lounsbury, "Wild Melody," 201. The timber for railway bridge-building in Britain was imported from the Baltic states, Scandinavia, and North America; Nicholas A. Bill, "Timber Bridge Construction on British and Irish Railways, 1840–1870: Scale of Construction and Factors Influencing Material Selection," *Construction History* 31 (2016): 80.

60. D. K. Brown, "Structural Improvements to Wooden Ships Instigated by Robert Seppings," *Naval Architect* (1979): 103–104.

61. Lounsbury, "Wild Melody," 196–201.

62. Charles W. Peale, *Essay on Building Wooden Bridges* (Bailey, 1797), 5. F. B. Brock, "Truss-Bridges," *Engineering News* 9 (1882): 371. Charles Coleman Sellers, *Mr. Peale's Museum: Charles Willson Peale and the First Popular Museum of Natural Science and Art* (Norton, 1980), 94–95. "Board" is shorter and thinner than "plank."

63. Williams, *Americans and Their Forests*, 5. Lounsbury, "Wild Melody," 196–201, 207–208. John Anthony Eisterhold, "Lumber and Trade in Seaboard Cities of the Old South: 1607–1860" (PhD diss., University of Mississippi, 1970), 74; Harry Roy Merrens, *Colonial North Carolina in the Eighteenth Century* (University of North Carolina Press, 1964), 96–101.

64. Eugene S. Ferguson, "The American-ness of American Technology," *Technology and Culture* 20 (1979): 7; Cochran, *Frontiers of Change*, 66. Thomas C. Cochran, "Cotton Textiles and Industrialization," in *Science and Society in Early America*, ed. Randolph Shipley Klein (American Philosophical Society, 1986), 259. Williams, *Americans and Their Forests*, 94–95, 167–169. The "first" sawmill in North America dates to 1611, according to Hindle, *Emulation and Invention* (Norton, 1981), 8. Charles E. Peterson, "Sawdust Trail: Annals of Sawmilling and the Lumber Trade from Virginia to Hawaii . . . to the Year 1860," *APT Bulletin* 5/2 (1973): 107–108. John O. Curtis, "Introduction of the Circular Saw in Early 19th Century," *APT Bulletin* 5/2 (1973): 162. During early use, lumber cut on circular saws was of small dimensions; Jensen, "Board and Batten," 45. Williams, *Americans and Their Forests*, 5. Lumber list for 1838 lattice bridge; Bishir, "Location Book," 55: Lounsbury, "Wild Melody," 198–199. Conwill et al., "Bath-Haverhill Bridge," 14.

65. "Suitable timber can be easily procured and sawed at common mills, as it requires no large or long timber"; Town (1821), 7. Zimpel, *Eisenbahnwesen*, 102, wrote of Town and Long bridges: "They require only timber of ordinary, easy-to-get dimensions, and all the wood required can be obtained without much fuss." German translator of British article about Long's bridge: "If the Americans, whose forests still cover most of their land, so delicately split the wood which they use in their bridges, what should the Bridge-Building Commissions think in those states where wood is a luxury, yet is squandered on miserable little bridges"; "Oberst S. H. Long's hölzerne Brüke aus kleinem holze," *Dingler's Polytechnisches Journal* 38 (1830): 176 (from *Mechanics' Magazine*, August 28, 1830, 433–434). Also "Beiträge zu Holzkonstruktionen bei Brücken, Viadukten, u.s.w. in Nordamerika," *Allgemeine Bauzeitung* 10 (1845): 7, 8, 11.

66. Tom F. Peters, *Time is Money: Die Entwicklung des modernen Bauwesens* (Hofmann, 1981), 176.

67. Rinke, *Vom Konstrukt*, 120.

68. Gasparini and Simmons, "American Truss Bridge Connections I," 119.

69. Rob Tarule, "Mortise and Tenon Timber Frame: Tradition and Technology" in Kebabian and Lipke, *Tools and Technologies*, 28–42. Also Lee H. Nelson, "Early Wooden Truss Connections vs. Wood Shrinkage: From Mortise-and-Tenon Joints to Bolted Connections," *Association for Preservation Technology Bulletin* 27 (1996): 11–23.

70. Rinke, *Vom Konstrukt*, 128–29. Massimo Laffranchi and Paolo De Giorgi, "Some Remarks on the Grubenmanns' Wooden Bridge Structures," in *John Soane and the Wooden Bridges of Switzerland*, ed. Angelo Maggi and Nicola Navone (Archivio del Moderno, Accademia di Architettura, Universit. della Svizzera italiana; John Soane's Museum, 2003), 118, 126.

71. Poussin, *Travaux*, 63.

72. Philibert de l'Orme, *Nouvelles inventions pour bien bastir et à petits fraiz* (1561), in *Traités d'architecture*, ed. Jean-Marie Pérouse de Montclos (Léonce Laget, 1988); Gregory K. Dreicer, "Nouvelles inventions: L'interchangeabilité et le génie national," *Culture Technique* 26 (1992): 213–220.

73. Douglas Harnsberger, "In Delorme's Manner: A Study of the Applications of Philibert Delorme's Dome Construction Method in Early 19th Century American Architecture" (master's thesis, University of Virginia, 1981) and B. Hays, "The Transfer of Thin Wood Vaulting from France to America," in *History of Construction Cultures: Proceedings of the 7th International Congress on Construction History*, ed. João Mascarenhas-Mateus and Ana Paula Pires (CRC, 2021), 57–62.

74. Amand-Rose Émy, *Traité de l'art de la charpenterie* (Carilian-Goeury et Anselin, 1837; 1841), 2:161. Dreicer, "Nouvelles inventions," 218.

75. US Patent and Trademark Office, B. F. Sturtevant, "Blank for Shoe Pegging Machine" (August 16, 1859); Ross Thomson, *The Path to Mechanized Shoe Production in the United States* (University of North Carolina Press, 1989), 34–35, 125–129; "How Shoe-Pegs Are Made," *Scientific American* 21 (1869): 362, quoted in Thomson, 34; Ross Thomson, "Crossover Inventors and Technological Linkages: American Shoemaking and the Broader Economy, 1848–1901," *Technology and Culture* 32 (1991): 1018–1046.

76. Nicholas A. Bill, "Design and Construction Deficiencies of Timber Railway Bridges," *Proceedings of the Institution of Civil Engineers, Engineering History and Heritage* 167, no. 4 (2014): 184, 187, 190. As iron manufacture became more reliable and costs came down, innovators were able to test, use, and observe the material in parts that could be produced in quantity. Eventually, forces in metal structures could be continuously carried through rigid connections; Rinke, *Vom Konstrukt*, 128–129, 163.

77. Nails were used to attach the cladding. On treenails in Burr bridges, see Karl Culmann, "Der Bau der hölzernen Brücken in den Vereinigten Staaten von Nordamerika . . . in den Jahren 1849 und 1850," *Allgemeine Bauzeitung* 16 (1851): 75 (hereafter Culmann, "Holz"). Concerning bridges by Burr, Long, and Town, Chevalier states, "Ils sont tous remarquables en ce qui'ils n'exigent presque pas de fer"; Chevalier, *Lettres* 2:100. Also Lee H. Nelson, *The Colossus of 1812: An American Engineering Superlative* (ASCE, 1990), 16. William Falconer, *An Universal Dictionary of the Marine* (Cadell, 1780), 298, asserts, "tree-nails are justly esteemed superior to spike-nails or bolts, which are liable to rust, and loosen as well as to rot the timber."

78. Tom F. Peters, "Thoughts about the Culture of the Balloon-Frame and the Steel High-Rise Skeleton," *Proceedings of the 83rd Association of Collegiate Schools of Architecture Annual Meeting* (ACSA, 1995), 479.

79. Jamie Garnock, *Trellis: The Creative Way to Transform Your Garden* (Rizzoli, 1991), 15. Re the use of iron, see Riedel der Ältere, "Nachricht von eisernen Brücken," *Sammlung nützlicher Aufsätze und Nachrichten, die Baukunst betreffend* 1 (1797; 2nd ed.: Maurer, 1811): 114. Elizabeth Wilkinson and Marjorie Henderson, *The House of Boughs: A Sourcebook of Garden Designs, Structures, and Suppliers* (Viking Penguin, 1985), 53, 96. Anne L. Helmreich, "Trellis," *History of Early American Landscape Design*, National Gallery of Art, https://heald.nga.gov/mediawiki/index.php/Trellis#Jefferson.

80. "Description of the Building Lately Erected for Public Worship, by Trinity Church, in the City of New-Haven, By Mr. Ithiel Town, Architect," in *Sermon, Preached at the Consecration of Trinity Church in the City of New-Haven*, ed. John Henry Hobart (Trinity Church Vestry, Oliver Steele, 1816), 25–29; http://anglicanhistory.org/usa/jhhobart/new_haven1816.html. "Each series of truss-braces may be termed a lattice; hence it is, by some, termed the double lattice bridge. The diamond-truss bridge, double diamond-truss, or double diamond bridge, would be more appropriate and better understood"; Town (1839), 13.

81. During the late eighteenth century, Ivan Petrovich Kulibin (1735–1818) designed and built large-scale models of heavy, wooden arch bridges of latticework, which required thousands of pieces of iron; Marie Joseph Gaston-Rosnay patented an iron extensible latticework bridge (Paris, 1799, unknown if built); James, "Evolution of Wooden Bridge Trusses," 133–134, 181. A British reviewer of Stevenson's *Sketch* remarked that Town's bridge "reminds us of the model of a bridge made by a Russian peasant, intended to cross the Neva at Petersburg, by one arch" (though it was nothing like Town's); review of David Stevenson, *Sketch of the Civil Engineering of North America*, *Architectural Magazine and Journal* 5 (1838): 473–474. Re Gaston-Rosnay, see James, "Evolution of Iron Bridge Trusses," 78–79, and James, "Some Steps in the Evolution of Early Iron Arched Bridge Design," *Transactions of Newcomen Society* 59 (1987–88): 162; James says it "was not taken seriously." Fletcher and Snow, "History," 403, mention a lattice bridge built in 1813 (which still existed in 1933) over Otter Creek in Pittsford, Vermont, perhaps the only claim for a prior lattice in the United States; it is mentioned nowhere else, and James, "Evolution of Wooden Bridge Trusses," 191n121, states that Richard Sanders Allen disproved the claim. Other lattice bridges in Pittsford might be source of error; Town of Pittsford, Vermont, "Covered Bridges," https://pittsfordvermont.com/about-pittsford-vt/covered-bridges, accessed 4 January 2023. Three horizontal latticeworks, atop each other, nailed onto a frame between the gables of the church at Baar, Switzerland (1645) are an example of a structural use; Josef Killer, *Die Werke der Baumeister Grubenmann* (Birkenhäuser, 1985), 65, 67.

82. Leupold, *Theatrum Pontificiale*, 79–80, table 13, fig. 2. Appeared in Grollier de Servières (1719); see James, "Evolution of Wooden Bridge Trusses," 127.

83. *Plan, Durchschnitt und Aufriß der drey merkwerdigsten hoelzernen Bruecken in der Schweiz, nämlich der beruehmten Schaffhauser-Bruecke, ueber den Rhein, jener zu Wettingen, ueber die Limmat, und der zuletzt erbaueten zu Mellingen, ueber die Reuß* (Mechel, 1803); "diamond-shaped," 13; *Plans, coupes et élévations des trois ponts de bois les plus remarquables de la Suisse, savoir: Celui de Schaffhouse, sur le Rhin, celui de Wettingen, sur la Limmat et cellui de Mellingen, sur la Reuss; accompagnées chacun d'une description détaillée* (Mechel, 1803). The French and German editions are not identical. Friedrich Gerhard Wahl, *Aus Erfahrungen gesammleter theoretisch-praktischer Unterricht in dem Strassen und Brückenbau* (Hahn, 1786), pl. 6. Émy, *Traité*, 2:400. Thomas Pope, *A Treatise on Bridge Architecture* (Niven, 1811), states, "A specimen [model] of this Bridge may be seen at Trenton" (194, 60–61, pl. 70); Michel Chevalier, *Lettres sur l'Amerique du Nord* (Gosselin, 1836), 2:100; Chevalier, *Histoire*, 2:36; Herman Haupt, *General Theory of Bridge Construction* (Appleton, 1851), 45–46. Most general bridge treatises cite Grubenmann bridges; Killer, cited in Pierre Dubas, "Beitrag der Baumeister Grubenmann zur Entwicklung des Holzbrückenbaus," offprint, *Schweizer Ingenieur und Architekt* 101 (1983), 4. The bridge over the Limmat at Wettingen (supposedly built 1765–1766) also had a latticed floor

structure, but it may never have been built—this text is so poorly translated that it is unclear; C. Ceraldi and E. Russo Ermolli, "Swiss Covered Bridges of the Eighteenth Century. Special Case: Bridge at Schaffhausen," S. Huerta, ed., *Proceedings of First Construction History Society Conference* (Juan de Herrera, 2003), 569–578. Also, Nicola Navone, "Eighteeneeth-Century Reputation of the Grubenmann Brothers," in Maggi and Navone, *John Soane and the Wooden Bridges*, 31–55; and Massimo Laffranchi and Paolo De Giorgi, "Some Remarks on the Grubenmanns' Wooden Bridge Structures," in Maggi and Navone, *John Soane and the Wooden Bridges*, 115–135.

84. *Weekly Raleigh Register*, February 22, 1803, 3.

85. David Yeomans, "Soane and Swiss Bridges," *Construction History* 19 (2003): 49–51, 60–61. Fulton, *Treatise*, 117–119.

86. Burr patents 662X (lost) and 2,796X. See publications of Richard Sanders Allen; Nelson, *Colossus of 1812*; Fletcher and Snow, "History," 329–332. Also, Hubertis M. Cummings, "Theodore Burr and His Bridges across the Susquehanna," *Pennsylvania History* 23 (1956): 476–486; Emory L. Kemp and John Hall, "Case Study of Burr Truss Covered Bridge," *Engineering Issues* 101 (1975): 391–412; Dennis M. Zambala, "Barrackville Covered Bridge," Washington, DC: Historic American Building Record, 1973, with addendum, Lola Bennett, 2002.

87. Culmann, "Holz," 75–77, pls. 391–393; Chevalier, *Histoire* 2:558, 565. Also Dario Gasparini et al., "The Development of the American Truss," in Christianson and Marston, *Covered Bridges*, 121–124; Lawrence E. Jones and Brian S. Wiens, "Structural Rehabilitation of a Historic Covered Bridge: Bridgeport Covered Bridge," in *Proceedings of the 4th International Conference on Timber Bridges*, ed. Bettina and Steffen Franke (Bern University of Applied Sciences/Empa, 2022), 85–91.

88. Culmann, "Holz," 75 ("iron" meant "steadfast"). Culmann showed that in some later bridges, roadways were hung from the arch by iron rods rather than pinned to it.

89. Allen, *Covered Bridges of the Northeast*, 15. Town (1839), 6.

90. Eda Kranakis, *Constructing a Bridge: An Exploration of Engineerging Culture, Design, and Research in Nineteenth-Century France and America* (MIT, 1997), 34–37, 57–58, 72–85. James Finley, "Description of the Patent Chain Bridge," *Port Folio* 3 (1810): 441–453.

91. Robert Seppings, "On a New Principle of Constructing His Majesty's Ships of War" *Philosophical Transactions of the Royal Society of London* 104 (1814): 289–291 (offprint, Bulmer, 1814). Thomas Wright, "Thomas Young and Robert Seppings: Science and Ship Construction in the Early Nineteenth Century," *Transactions of the Newcomen Society* 53 (1981–1982): 55–72.

92. Robert Seppings, "On the Great Strength Given to Ships of War by the Application of Diagonal Braces," *Philosophical Transactions of the Royal Society of London* 108 (1818): 2. Seppings was reacting to Thomas Young and Dupin; on Seppings's sources, see Wright, "Thomas Young," 57, 62–63, 65. Calhoun, *Intelligence*, discusses Seppings and Town separately.

93. Seppings, "On a New Principle" (1814): 286–287.

94. Town (1821), 7. One could use "even the *softest pine*"; Town (1839), 5.

95. Conversation with Prof. Jonathan Ochshorn, January 22, 1993. Quote from Tom F. Peters, "American Culture of Construction," *Perspecta* 25 (1989): 155.

96. Town (1839), 12, and Town (1831), 7. Similarly, on redundancy, see Philibert de l'Orme, *Nouvelles Inventions*, bk. 1, chap. 6, fol. 8.

97. John E. Greiner (1895), in Gasparini and Simmons, "American Truss Bridge Connections I," 199.

98. Richard Guy Wilson, "Thomas Jefferson's 'Bibliomanie' and Architecture," in *American Architects and their Books to 1848*, ed. Kenneth Hafertepe and James F. O'Gorman (University of Massachusetts Press, 2001), 59–89. "To Jefferson's mind, architecture was largely a matter of formulae and mathematics"; see I. T. Frary, *Thomas Jefferson: Architect and Builder* (Garrett and Massie, 1931), 4; Travis McDonald, "Jefferson Builder," in *Jefferson and Palladio: Constructing a New World*, ed. Guido Beltramini and Fulvio Lenzo (Officina Libraria, 2015), 137–148. Re "Grecian" and Town, see W. Barksdale Maynard, *Architecture in the United States, 1800–1850* (Yale University Press, 2002), 254–255, 219–279. Jefferson's library was much smaller than Town's; Quinn, "Library of Ithiel Town," 18. In an advertisement, Town says, "While the Grecian architecture . . . possesses this important advantage over the trifling, unmeaning innumerable little parts, in the prevailing modern taste in building, it is, *very fortunately* at the same time, a much more permanent and economical style"; Ithiel Town, "Architecture," *Evening Post* (New York), January 3, 1826, 4; see chap. 7.

99. Re Town's familiarity with Durand, see Ellwood C. Parry III, "Thomas Cole's Imagination at Work in *The Architect's Dream*," *American Art Journal* 12 (1980): 44–45. Town's "façades and floor plans reveal rigid standards of organization and an unmistakable central focus"; David P. Handlin, "New England Architects in New York, 1820–40," *American Quarterly* 19 (1967): 689. Town in 1840 gave a copy of Durand's *Recueil* to Thomas Cole as partial payment for *Architect's Dream*; Rebecca Ayres Schwartz, "Thomas Cole on Architecture: Picturing the Gothic" (PhD diss., University of Delaware, 2016), 30. Re Davis and Durand, see Susanne Brendel-Pandich, "From Cottages to Castles: Country House Designs of Alexander Jackson Davis," in Peck, *Alexander Jackson Davis*, 59. Re geometry, mechanization, and the mode of thinking reflected in the Greek Revival, see Jonathan Hale, *The Old Way of Seeing: How Architecture Lost Its Magic* (Houghton Mifflin, 1994), 26–44. Re Durand, see Ulrich Pfammatter, *The Making of the Modern Architect and Engineer*, trans. Madeline Ferretti-Theiligl (Birkhäuser, 2000), 53–67, and David Park Curry, "Replication, Pattern, and Symbolic Form: Connecticut State Capitol in the Context of Nineteenth-Century Public Design" (PhD diss., Yale University, 1981), 32–41. Re trade catalogs, see Alex Sowa, "Mimesis Recast: Artistic Production in the Aftermath of Neoclassicism," in Dobraszczyk and Sealy, *Function and Fantasy*, 87. Re botany, see Philip Steadman, *The Evolution of Designs: Biological Analogy in Architecture and the Applied Arts*, rev. ed. (Routledge, 2008), 27–30, which I discovered just prior to submission of this manuscript. Re West Point, see Marvin J. Andersen, "The Architectural Education of Nineteenth-Century American Engineers: Dennis Hart Mahan at West Point," *Journal of Society of Architectural Historians* 67 (2008): 222–226, 235–236.

100. Addis, *Structural Engineering*, 14–17, 22–23.

101. This applies to structural and technological systems, such as railroads; Tom F. Peters, *Transitions in Engineering: Guillaume Henri Dufour and the Early 19th Century Cable Suspension Bridges* (Birkhäuser, 1987), 9–11, 183. Re overlay, see Laffranchi and De Giorgi, "Some Remarks,"

116–122, and the discussion of Douglas in chap. 3. Building on Hauri and Peters, see Mario Rinke, "Formal and Structural Multiplicity in Early Truss Design," in *Structures and Architecture: Beyond Their Limits*, ed. Paulo J. S. Cruz (CRC, 2016), 1334.

102. Hans H. Hauri, "Thoughts on the Historical Development of Methods for Dimensioning Bridges," in *The Development of Long-Span Bridge Building*, ed. Tom F. Peters, 3rd ed. (ETH, 1981), 156. Also, Mario Rinke and Toni Kotnik, "The Changing Concept of Truss Design Caused by the Influence of Science," *Structures and Architecture*, ed. Paulo J. S. Cruz (CRC, 2010), 1959–1967. "Strut" and "brace" in this discussion are identical in meaning.

103. Hauri, "Thoughts," 156.

104. Laffranchi and De Giorgi, "Some Remarks," 128, 119. Also, Rinke and Kotnik, "Changing Concept," 1960–1961. Reportedly, the designer of Upper Bridge over Patapsco River based this structure, which features an accretion of struts, on the Schaffhausen Bridge; Richard Sanders Allen, *Covered Bridges of the Middle Atlantic States* (Bonanza Books/Greene, 1959), 21–22.

105. Laffranchi and De Giorgi, "Some Remarks," 120.

106. Rinke, "Formal and Structural Multiplicity," 1335.

107. Town indicated that lattice diagonals should be "incline[d] so as to make an angle . . . of about 45° or any angle that may be necessary for a brace (as they do the office of a brace)"; Town, Patent (1820).

108. Mario Rinke and Toni Kotnik, "From Construct to Type: The Transformation of Constituents in the Development of Trusses," in *Structures and Architecture: Concepts, Applications, and Challenges*, ed. Paulo J. S. Cruz (CRC, 2013), 1947, 1952.

109. E.g., R. Moebius, *Sammlung von Zeichnungen zu eisernen Gittern aller Art* (Moebius, 1837), https://katalog.slub-dresden.de/id/0-1123661855; C. F. von Ehrenberg, *Baulexikon* (Sauerländer, 1840), 276; Paulin Desormeaux, *Manuel complet du treillageur et du menuisier des jardins* (Roret, 1836). Sometimes French and German writers used the English word "lattice"; the French terms for "truss" are *ferme* or *poutre droite*.

110. Schönhöfer, "Hinweg mit der französischen Standesbezeichnung: 'Ingenieur'!" *Zeitschrift für Architecktur und Ingenieurwesen* (1915): 22, and Georg Christoph Mehrtens, "Zur Verdeutschung des Fremdwortes 'Ingenieur,'" *Eisenbau* 6 (1915): 78–79.

CHAPTER 3

1. Michel Chevalier, *Les voies de communication aux États-Unis [excerpted from Journal de l'Industriel et du Capitaliste* and Chevalier's *Lettres sur l'Amérique du Nord* (Paris: F. G. Levrault, 1837), 48.

2. "Each minute teems a new one." Review of David Stevenson, *Sketch of the Civil Engineering of North America, Civil Engineers and Architects Journal* 1 (1837–1838): 308.

3. C. Sumner, review of Francis J. Grund, *The Americans, in their Moral, Social, and Political Relations, North American Review* 46 (1838): 106–107. Between 1810 and 1860, two or three US travelers' accounts were published each year; Renata R. Wasserman, ed., *Exotic Nations: Literature*

and Cultural Identity in the United States and Brazil, 1830–1930 (Cornell University Press 1994), 156n6.

4. Rondo Cameron, "The Diffusion of Technology as a Problem in Economic History," *Economic Geography* 51 (1975): 224.

5. Thomas C. Cochran, *Frontiers of Change: Early Industrialism in America* (Oxford University Press, 1981), 10.

6. Michael Drolet, "Industry, Class and Society: A Historiographic Reinterpretation of Michel Chevalier," *English Historical Review* 123 (2008), 1233.

7. William J. Gilmore, *Reading Becomes a Necessity of Life: Material and Cultural Life in Rural New England, 1780–1835* (University of Tennessee Press, 1989), 17–27, 351–352; re publications in rural New England, 204–209; re Webster, 19. Daniel Webster, "Bunker Hill Monument," in *Great Speeches and Orations of Daniel Webster*, ed. Edwin P. Whipple (Little, Brown, 1879), 131.

8. *Western Carolinian* (Salisbury), February 8, 1825, 1. Roads, reprinted from *Bucks County Patriot* (Pennsylvania); China, from *Grieg's World Displayed* (possibly London).

9. David R. Meyer, *Networked Machinists: High-Technology Industries in Antebellum America* (Johns Hopkins University Press, 2006), 20.

10. William Charvat, *Literary Publishing in America, 1790–1850* (University of Massachusetts Press, 1959), 17–21. In the United States, authors, publishers, and readers were distributed rather than concentrated in a capital (such as London).

11. Thomas Smits, *European Illustrated Press and Emergence of a Transnational Visual Culture of the News, 1842–1870* (Routledge, 2020), 2–3, 9–11.

12. Benedict Anderson, *Imagined Communities*, rev. ed. (Verso, 2006), 67, 86–87.

13. Daniel R. Headrick, *Tools of Empire* (Oxford University Press, 1981), 131. David N. Livingston, *Putting Science in Its Place* (University of Chicago Press, 2003), 145–147. On direction communication, see e.g., Andreas Kahlow, "The Roeblings: Migration, Knowledge Transfer, and Tacit Knowledge," in *Construction Matters: Proceedings of the 8th International Congress on Construction History*, ed. Stefan Holzer et al. (VDF, 2024), 532–539.

14. Fritz Machlup, *Knowledge: Its Creation, Distribution, and Economic Significance* (Princeton University Press, 1980), I; 170–173.

15. Francis E. Griggs Jr., "Forerunner in Iron Bridge Construction: Interview with Squire Whipple," *Civil Engineering Practice* 3 (1988): 24–26; Daniel Hovey Calhoun, *American Civil Engineer: Origins and Conflict* (Technology Press/MIT, 1960), 123–140; Barrie M. Ratcliffe, "Origins of Paris-Saint-Germain Railway," *Journal of Transport History* 1/4 (1972): 197–219; Ratcliffe, "Building of Paris-Saint-Germain Railway," *Journal of Transport History* 2/1 (1973): 20–40.

16. Earl Heydinger, "Suppression of 'Bad News' About Early Railroads," *Railway and Locomotive Historical Society Bulletin* 119 (October 1968): 63–66.

17. [Karl] Bernhard, *Reise Sr. Hoheit des Herzogs Bernhard zu Sachsen-Wiemar-Eisenach durch Nord-Amerika in den Jahren 1825 und 1826* (Hoffmann, 1828), 120. US ed.: Bernhard, Duke of

Saxe-Weimar Eisenach, *Travels through North America During the Years 1825 and 1826* (Carey Lea & Carey, 1828). Bernhard, *Reise*, 309, 267, and *Travels*, 206. "Lattice truss" might have been a closer translation. This apparently was Town's second lattice, built 1819–20 over the Cape Fear; see chap. 5. Charles Joseph Latrobe, in *Rambler in North America: 1832–1833*, 2d ed. (Selley and Burnside, 1836), 2: 79 mentions without comment "a bridge constructed upon the lattice principle"; it was built around 1828 at Clarke's Ferry near Duncan's Island, south of Harrisburg, PA.

18. Bernhard, *Reise*, 267; 309. A model of each patent was required by the Patent Office, which burned in 1836. Bernhard provides a detailed account of models he saw there. A reviewer of *List of Patents Granted by the United States . . . Alphabetically Arranged from 1790 to 1820 . . . to 1826* in *North American Review* 23 (1826): 299, noted that "twenty or thirty models [were] crowded on a shelf not more than three feet long and half as deep, so that it is almost impossible distinctly to examine them, and observe their forms and proportions."

19. Town (1821), 8, wrote of "models which are (or soon will be) placed in most of the principal cities of the United States." [Ithiel Town,] "Some Account of Ithiel Town's Improvement," *American Journal of Science* 38 (1840): 277, mentions models in "the Patent Office, Washington City, and at the American Institute, back of the City Hall, City of New York."

20. Guillaume Lambert, *Voyage dans l'Amérique du nord, en 1853 et 1854, avec notes sur les expositions universelles de Dublin et de New-York* (Hayez, 1855), 150.

21. Diffusion of ideas "will ordinarily leave a minimum of historical evidence"; A. L. Kroeber, "Stimulus Diffusion," *American Anthropologist* 42 (1940): 1–2, 20.

22. *Moniteur Universel*, December 3, 1803, 283 (from *l'Observateur)*. Frank Griggs Jr., "Permanent Bridge," *Structure* (October 2013): 39–41; https://www.structuremag.org/?p=817; without evidence, this article deems the structure "America's First Covered Bridge."

23. [Editorial,] *Journal des chemins de fer* 4 (1845): 571.

24. "Sciences et lettres," *Moniteur Universel* (1846): 2394. Ralph R. Shaw, "Engineering Books Available in America Prior to 1830," *New York Public Library Bulletin* 37 (1933): 38–61, 157–60, 209–22, 301–10, 539–55 (offprint, New York Public Library, 1933). Michigan University Transportation Library, "Engineering Books Available in America Prior to 1830: Material in the Transportation Library Not Contained in the Bibliography Prepared by Ralph W. Shaw" (MS, 56 pp., Michigan University, [1933]) contains thousands of titles Shaw missed.

25. [Editorial], *American Railroad Journal* 1 (1832): 33.

26. Gabriel Tarde, *Laws of Imitation*, 2d ed. (1895) Elsie Clews Parsons, trans. (Holt, 1903), 344.

27. See Newton, *Town and Davis*, 46, Town advertised in the *National Intelligencer* and the New York *Weekly Express*; I was unable to locate these advertisements. "Bridges, improvement on—Ithiel Town, Fayetteville, North-Carolina, Jan. 28," in "Patents Issued in 1820: Useful Inventions, &c.," *National Calendar* 2 (1821): 143. Announcements of individual bridges also appeared, such as "Northampton Bridge" [reprinted from *Hampton Gazette*], *Christian Journal and Literary Register* 11 (February 1827): 61; Wm. Polke, "Report of the Michigan Road Commissioner," *Journal of House*

of Representatives of the State of Indiana (December 17, 1832): appendix A, 4; "Harlem Rail Road" [from *New York Gazette*], *Niles' Weekly Register*, October 24, 1835, 121.

28. Town made a pitch to members of Congress. Town (1839) was issued with an illustrated cover (February 1841) containing a letter that began: "The Hon. [space left blank for name] Member of Congress is respectfully requested to forward to some ingenious and influential person in his district, this description of the well-tested manner of construction." Re Town's travels, see R.W. Liscombe, "A 'New Era in My Life:' Ithiel Town Abroad," *Journal of Society of Architectural Historians* 50 (1991): 5–17.

29. "America" [letter to John Weale], 52; Letter, Ithiel Town, New York, to John Weale, London, [Sept?] 17, 1836, British Architectural Library, Royal Institute of British Architects, Manuscripts & Archives Collection, Weale Papers.

30. Re *Civil Architecture*, see John Fitchen, *The New World Dutch Barn: The Evolution, Forms, and Structure of a Disappearing Icon* (Syracuse University Press, 1968), 88; Talbot Hamlin, *Greek Revival Architecture in America* (Oxford University Press, 1944), 164; James F. O'Gorman, *Edward Shaw of Boston: Antebellum Architect* (American Philosophical Society, . In the 1831 edition, Shaw presented the lattice and Waterloo bridges. (Town had visited Waterloo Bridge in 1829; letter, Town to A. J. Davis, London, December 21, 1829, Metropolitan Museum.) Shaw, even in later editions, showed Town's original patent (1820).

31. Ithiel Town, "A Description of Ithiel Town's Improvement in the Construction of Wood and Iron Bridges," *American Journal of Science* 3 (1821): 158–166, 3 pls.; Ithiel Town, "Some Account of Ithiel Town's Improvement in the Construction, and Practical Execution of Bridges for Roads, Railroads, and Aqueducts," *American Journal of Science* 38 (1840): 276–297; also, "A Description of Ithiel Town's Improvement in the Principle, Construction and Practical Execution of Bridges," *Journal of American Institute* 4 (1840): 409–424. These were reprints of Town (1821) and (1839), respectively. The *American Journal of Science* did not customarily include articles on bridges; besides two by Town, only the Long bridge appeared during this period. The firm of Nathanial Jocelyn, friend of Town and painter of portraits of Town and his daughter Etha, provided engraved illustrations of the bridge; Toby Maria Chieffo-Reidway, "Nathaniel Jocelyn: In the Service of Art and Abolition" (PhD diss., College of William and Mary, 2005), 20. Also Simon Baatz, "'Squinting at Silliman': Scientific Periodicals in the Early American Republic, 1810–1833," *Isis* 82 (1991): 233. Portions of Town (1839) were integrated into an article about the lattice bridge in "Architects and Architecture of New York," *Brother Jonathan*, June 17, 1843, 181–182; July 1, 1843, 241–244; July 8, 1843, 271–274.

32. "Benjamin Silliman, Sr." in *The History of Science in the United States: An Encyclopedia*, ed. Marc Rothenberg (Garland, 2001), 516–517; William Martin Smallwood, *Natural History and the American Mind* (Columbia University Press, 1941; reprint, AMS, 1967), 79. Also, Gilmore, *Reading Becomes a Necessity of Life*, 205, 207. Silliman probably sent the journal to universities and scholars worldwide.

33. "Notices Industrielles," *Bulletin de la société d'encouragement pour l'industrie nationale* 34 (1835): 134; "Neue Art hölzerne Brüken zu bauen," *Allgemeine Polytechnische Zeitung* (1836): 231; "New Bridges," *London Journal of Arts and Sciences* 7 (1836): 104; "Lattice Bridges," *Mississippi Free Trader* (Natchez), September 8, 1836, 2.

34. My interpretation of the image is indebted to Angela Miller, *The Empire of the Eye: Landscape Representation and American Cultural Politics, 1825–1875* (Cornell University Press, 1993), 190–195; also, 13, 82–87. Jocelyn used techniques comparable to ones Monet would later use in bridge portraits; Robert L. Herbert, "Industry in the Changing Landscape from Daubigny to Monet," in *From Millet to Léger: Essays in Social Art History* (Yale University Press, 2002), especially 150–153.

35. Town (1821), 4.

36. Re "snail-like," see English navy captain Basil Hall, *Forty Etchings: From Sketches Made with the Camera Lucida, in North America, in 1827 and 1828* (Cadell, Simpkin and Marshall, 1829), "Bridge across the Congaree in [Columbia,] South Carolina," pl. 17. Re "minimum," see W. R. C., "Notes on Practical Engineering: Bridges," *American Railroad Journal* 117 (1844): 9. Town (1839), 15–16, referred to his bridge as "a very plain form" and showed an illustration of a lattice bridge clad "to appear like a light and elegant iron [arch] bridge."

37. Ben Marsden and Crosbie Smith, *Engineering Empires* (Palgrave Macmillan, 2005), 244–245.

38. Anderson, *Imagined Communities*, 175, 185. Re national interpretations of geometry, see Amir Alexander, *Proof! How the World Became Geometrical* (Farrar, Straus and Giroux, 2019).

39. Many sources cited here used the term. In England, too, Town bridges were called "American timber bridges"; Carl Ghega, *Über nordamerikanischen Brückenbau und Berechnung des Tragungsvermögens der howeschen Brücken* (Kaul Fuss Witwe, Prandel, 1845), 10. Even in the middle of the century, iron lattices bridges were labeled "American bridges"; e.g., Georges Boulangé, "Notes recueillies pendant une visite rapide de quelques chemins de fer d'Allemagne," *Annales des ponts et chaussées* 3/1 (1854): 65, 71, 76.

40. [Friedrich List], "Neuestes," *Eisenbahn-Journal und National-Magazin* 27 (1836): 97–99.

41. W. O. Henderson, "Friedrich List: Railway Pioneer," in *Trade and Transport: Essays in Economic History in Honour of T. S. Willan* (Manchester University Press, 1977): 136–156. Also, William Notz, "Frederick List in America," *American Economic Review* 16 (1926): 249–265; John F. Bell, "Frederick List, Champion of Industrial Capitalism," *Pennsylvania Magazine of History and Biography* 66 (1942): 56–83. List knew Lafayette. Joseph Lehmann, "Friedrich List's Eisenbahn-Journal," *Zeitung des Vereins Deutscher Eisenbahn-Verwaltungen* 1 (1861): 137. See also Eugen Wendler, *Friedrich List as a Railway Pioneer in the USA* (Springer, 2020), 47–55.

42. Erik Van Der Vleuten et al., "Europe's System Builders: The Contested Shaping of Transnational Road, Electricity, and Rail Networks," *Contemporary European History* 16 (2007): 321–347.

43. J. F. Normano, "Saint-Simon and America," *Social Forces* 2 (1932): 10–11; quote translated by Normano; *Oeuvres de Claude-Henri de Saint-Simon* 2 (Dentu, 1868; reprint, Anthropos, 1966), 2:133, 138–163.

44. Maurice Wallon, *Les Saint-Simoniens et les chemins de fer* (Pedone, 1908); Jean Walch "Les saint-simoniens et les voies de communication," *Culture Technique* 19 (1989): 285–294; Nicholas Papayanis, *Planning Paris before Hausmann* (Johns Hopkins University Press, 2004), 135–141; Richard Wittman, "Space, Networks, and the Saint-Simonians," *Grey Room* 40 (2010), 29.

45. Howard Douglas, *Essay on the Principles and Construction of Military Bridges, and Passage of Rivers in Military Operations*, 2nd ed. (Boone, 1832), 386–387. Pages numbers cited are from the

second edition; unless otherwise indicated, text appeared in both first (Egerton, 1816) and second editions. The third edition (1853) is not cited here. See chapter 2 on additive (accretionary) strategy.

46. Douglas, *Essay*, 400–401.

47. Douglas does identify British inventors. Some of text and lattice bridge list apparently comes from Town (1831); illustrations are recopied. Town (1839), 9–10, includes a passage from Douglas but adds "on the principle of Ithiel Town's patent." One can imagine Town's annoyance at not being clearly identified by Douglas!

48. Douglas, *Essay*, 404. He describes and illustrates Timothy Palmer's Permanent Bridge, Burr's Trenton Bridge, and Grubenmann's Schaffhausen Bridge (405–411); none appeared in the first edition.

49. Françoise Planchot, "Le Général Simon Bernard, ingénieur militaire aux États-Unis (1816–1831)," *Revue française d'études américaines* 13 (1982): 87–98. Re Bernard and Poussin, see Roger G. Kennedy, *Orders from France: The Americans and the French in a Revolutionary World, 1780–1820* (Knopf, 1989), 10, 398. Poussin worked in North Carolina after Town built early lattice bridges in the area; John Manuel, *The Natural Traveler along North Carolina's Coast* (Blair, 2003), 226. Also, Steven Rowan, "The Memoirs of Guillaume Tell Poussin: 'French Connection' in the Construction of American Roads, Canals, and Railroads," *History Faculty Works* 12 (2020), https://irl.umsl.edu/history-faculty/12.

50. G[uillau]me Tell Poussin, *Travaux d'améliorations intérieures projetés ou exécutés par le gouvernement général des États-Unis d'Amérique, de 1824 à 1831* (Anselin et Carilian-Goeury, 1834/6), 63–66, pl. 10; German translation by H. F. Lehrritter, *Öffentliche Bauwerke in den vereinigten Staaten von Amerika* (Pustet, 1836–1837), 2:219–223, pl. B.

51. Poussin, *Travaux*, 49, 66. Publications discussed in this chapter were widely known; authors sometimes cited each other; e.g., Ghega, *Über nordamerikanischen Brückenbau*, vi, cites Poussin and Gerstner.

52. G[uillau]me-Tell Poussin, *Chemins de fer américains: Historique de leur construction*, 2nd ed. (Carilian-Goeury, 1836), xiv.

53. "Nordamerikanische Brücken," *Allgemeine Bauzeitung* 4 (1839): 87–91, pl. 291.

54. Review of David Stevenson, *Sketch of the Civil Engineering of North America*, *Civil Engineers and Architects Journal* 1 (1837–1838): 308. Re Stevenson's expertise, see Finch, "Early American Engineering," in *Engineering Classics of James Kip Finch* (Cedar, 1978), 111, 115–16. Stevenson does not criticize use of wedges in the Longs bridge; David Stevenson, "Notice Relative to Long's American Frame-Bridge," *Edinburgh Philosophical Journal* (April 1841): 299–300; read before Royal Society, May 15, 1839: *Transactions of the Royal Scottish Society of Arts* 2 (1844): 81–84.

55. David Stevenson, *Sketch of the Civil Engineering of North America* (John Weale, 1838), 224.

56. Stevenson, "Notice," 297

57. Stevenson, *Sketch*, x, 231. He showed what appears to be the Permanent Bridge, Philadelphia and Burr bridge, Trenton.

58. Town (1839), 9.

59. Stevenson was impressed by the moving of brick houses in New York City, as was Basil Hall, *Travels in North America in the Years 1827 and 1828* (Cadell; Simpkin and Marsh), 1:38–45. Stevenson had read Hall; Stevenson, *Sketch*, 229.

60. Michel Chevalier, *Lettres sur l'Amerique du Nord* (Gosselin, 1836), 2: 4. Drolet, "Industry, Class and Society," 1238–1240, and Michael Drolet, "Michel Chevalier: Vision of Modern Europe?" *Books and Ideas*, January 29, 2018, https://booksandideas.net/Michel-Chevalier-Visionary-of-Modern-Europe.html.

61. Félix Tourneux, review of Chevalier, *Histoire*, in *Moniteur Universel* (1840): 2119. Also, Tresch, *Romantic Machine*, 206–207. Drolet, "Michel Chevalier: Vision of Modern Europe?"; Jeremy Jennings, "Democracy before Tocqueville: Michel Chevalier's America," *Review of Politics* 68 (2006): 398–427.

62. John William Ward, introduction to *Society, Manners, and Politics in the United States: Letters on North America*, by Michel Chevalier (Cornell University Press, 1961), ix. See also N[icolaus] H[einrich] Julius, *Nordamerikas sittliche Zustände, nach eigenen Anschauungen in den Jahren 1834, 1835 und 1836* (Brockhaus, 1839), xiv–xv; S. S., review of Guillaume-Tell Poussin, *De la puissance américaine*, *Journal des économistes* 24 (1849): 310–312. Henz, when writing of period when Gerstner's report (*Allgemeine Polytechnische Zeitung*, 1839) appeared, mentions the "wenig bekannten Werken von Tell-Poussin und Stevenson," and adds that "das Werk von Michel Chevalier . . . war damals noch nicht erschienen," implying that Chevalier's work was well known; Henz, "Urtheile amerikanischer Ingenieure über einige Eisenbahnfragen," *Verhandlungen des Vereins zur Beförderung des Gewerbfleißes* (Berlin) 21 (1842): 96–97. Re Robinson and the Philadelphia & Reading Railroad, see Emery L. Kemp and Richard K. Anderson, "Reading-Halls Station Bridge," Written Historical and Descriptive Data, HAER, National Park Service, US Department of Interior, ca. 1987, from Prints and Photographs Division, Library of Congress (HAER PA-55), 9–10.

63. Re the second volume of Chevalier's *Histoire*, a reviewer claimed that it contained the first description in Europe of the lattice bridge, "une création heureuse du génie américaine"; Henri Richelot, *Journal des économistes* 5 (1843): 191.

64. Chevalier's *Histoire* has been translated by Steven Rowan et al., *History and Description of the Routes of Communication in the United States* (John W. Barriger III National Railroad Library, University of Missouri–St. Louis, 2015). Town (1839), 8, mentions Robinson. On the lattice bridge, see Chevalier, *Histoire* 2:551–582, 564–582. Chevalier, *Lettres*, mistakenly states that Burr and Long bridges preceded Town; only Burr did. He called the lattice bridge "le plus neuf, le plus intéressant et le plus convenable pour les chemins de fer, en raison de sa fixité"; Chevalier, *Lettres*, 2:101; Long bridge is given one paragraph, 557; Burr bridge, 557–563.

65. Chevalier, *Histoire*, 2:566. Chevalier compares the superstructures of a nineteen-span lattice railroad bridge over James River, Richmond (Robinson, 1836–1838), which was 867.42 meters long, and the wooden bridge on the Paris-Saint Germain Railway at Asnières over the Seine near Paris, a five-arch bridge 162 meters long (see Introduction); Chevalier, *Histoire*, 2:571–575; see also Michel Chevalier, "Progrès à obtenir dans l'exploitation des chemins de

fer en France," *Journal des économistes* 12 (1880): 174–175. Chevalier, *Histoire*, 2:577, unfavorably compares the Asnières bridge to the double-lattice bridge over Schuylkill near Reading, on Mount Carbon-Philadelphia Railroad. Richmond bridge and Chevalier's report are described in Richard B. Osborne, "Professional Biography of Moncure Robinson," *William and Mary Quarterly*, 2nd ser., 1 (1921): 237–260; [reprint from 1889], 248, 256.

66. See "Letters of Moncure Robinson to his Father, John Robinson, of Richmond, Va., Clerk of Henrico Court," *William and Mary College Quarterly Historical Magazine* 8 (1928): 71–95, 143–156; 9 (1929): 13–33. On Robinson's career, see Osborne, "Professional Biography," 237–260. (It seems Robinson knew James Fenimore Cooper; 256.) Engineers learned their profession in Robinson's service; Chevalier, *Histoire*, 2:433.

67. Moncure Robinson, "Obituary Notice of Michel Chevalier," *American Philosophical Society, Proceedings* 19 (1880–1881), 36. Chevalier cited in Archives Nationales, F14 9285, in a plan "No. 17. Chemin de fer de Commentry à Monluçon."

68. Drolet, "Industry, Class and Society," 1260–1261.

69. Franz Anton Gerstner, *Berichte aus den Vereinigten Staaten von Nordamerica, über Eisenbahnen, Dampfschiffahrten, Banken und andere öffentliche Unternehmungen* (Melzer, 1839). Front page announcement in *American Railroad Journal*, November 15, 1938, 297. There was a serialization in *Journal of the Franklin Institute* 31 (1841). See also, Frederick C. Gamst, "Franz Anton Ritter von Gerstner," *Railroad History* 163 (1990): 13–27.

70. Gerstner, *Berichte*, 7; re British railways, 56, 60–61.

71. Franz Anton Gerstner, *Die innern Communicationen der vereinigten Staaten von Nordamerika*, ed. L. Klein (Förster, 1842); a recent translation is Franz Anton Ritter von Gerstner, *Early American Railroads: Franz Anton Ritter von Gerstner's 'Die innern Communicationen' 1842–1843*, ed. Frederick C. Gamst and David J. Diephouse, trans. John C. Decker (Stanford University Press, 1997). Gerstner's wife Clara's published account does not report on technology: Clara Gerstner, *Beschreibung einer Reise durch die Vereinigten Staaten von Nordamerika in den Jahren 1838–40* (Leipzig, 1842).

72. Gerstner, *Die innern Communicationen*, 1:107–108, 202–203, 235; 2:61, 126, 166–167, 170, 187, 245–246, 256, 261, 264, 277, 291.

73. Gerstner, *Die innern Communicationen*, 1:202–203. Harlem Bridge is mentioned in "Harlem Railroad Company" [from *New York American*], *New York Farmer, and American Gardner's Magazine* 8 (November 1835): 346; "Harlem Rail Road" [from *New York Gazette*], *Niles' Weekly Register*, October 24, 1835, 121; *Annual Report of the Railroad Commission of the State of New-York for Fiscal Year Ending Sept. 26, 1856* (C. van Benthuysen, 1857), 474–475.

74. Chevalier, *Histoire*, 2:333, cites Poussin's *Travaux* and mentions the US survey work of Bernard and Poussin (e.g., Chevalier, *Histoire*, 2:44). Reviews made these books known: a review of Ghega's *Über amerikanischen Brückenbau* stated, regarding Town's bridge: "Beschreibungen dieser Brücken enthalten die Werke von Poussin, Gerstner, Michel Chevalier, David Stevenson"; K, "Brückenbau," *Eisenbahnzeitung* (1845): 330. Town's brochures cited works that described his bridge. Ghega, *Über nordamerikanischen Brückenbau*, vi, had read Poussin, as had Chas. F. Zimpel, *Das Eisenbahnwesen von Nordamerika, England und anderen Ländern, practisch und populär dargestellt*

(Förster, 1840), 99. Gerstner's reports attracted attention in Germany because little was known "außer den wenig bekannten Werken von Tell-Poussin und Stevenson," per Henz, "Urtheile amerikanischer Ingenieure über einige Eisenbahnfragen," *Verhandlungen des Vereins zur Beförderung des Gewerbfleißes* (Berlin) 21 (1842): 96. Julius Pollack, "Metallbrücke zur Uebersetzung des Royalkanals bei Dublin für die Dublin-Drogheda-Eisenbahn nach Town's nordamerikanischem Konstrukzionssystem gebaut im Jahre 1845," *Allgemeine Bauzeitung* 13 (1848): 18, cited Chevalier, Gerstner, and Ghega, along with *Journal of the Franklin Institute* and *The Builder*.

75. See the commentary on Chevalier and Poussin in Henri-Charles Emmery, "Analyse et extrait: Des *Lettres sur l'Amérique du nord*, du M. Michel Chevalier; des deux ouvrages de M. Poussin, intitulés, le premier, *Travaux d'améliorations intérieures, projetés ou exécutés aux Etats-Unis, de 1824 à 1831*, et le second, *Chemins de fer américains*," *Annales des ponts et chaussées* 1/2 (1837): 101. Other *Annales* articles: David Stevenson, trans. Henri-Charles Emmery, "Travaux publics de l'Amérique du Nord," *Annales des ponts et chaussées* 1/2 (1839): 141–226; Michel Chevalier, with commentary by Léon Lalanne, "Histoire et description des voies de communication aux États-Unis, et des travaux d'art qui en dépendent par M. Michel Chevalier: Analyse et Extraits." *Annales des ponts et chaussées* 2/2 (1843): 245–321, pls. 56–57.

76. "Travaux publics: sur quelques nouveaux systèmes de ponts," *L'Illustration* 4 (1844): 22–23, 74–75.

77. G. H. Dufour, review of Poussin, *Travaux*, in *Bibliothèque Universelle* (Geneva) 57 (1834): 160. Dufour writes that the book "peut être également profitable à l'homme d'état, à l'ingénieur et à l'industriel" (166).

78. Chevalier, (*Histoire*, 2: 568), claimed to have seen bridges over the Nottoway, Opequon at Winchester, and the Schuylkill near Reading the construction periods he gives for the latter two (December 1836 to September 1838 and 1839 to 1840, respectively) occurred after his departure from the United States. Chevalier, *Histoire* 2: 425, mentions the construction of the Raleigh & Gaston Railroad; had he seen the lattice bridges, it is likely he would have called attention to them.

79. Zimpel, *Eisenbahnwesen*, iv; "Americanische Systeme," 102; He deemed the Town bridge "simpler . . . yet more effective than Long's system," although he described the latter in detail, 99. Zimpel refers to Poussin's book; he saw the lattice bridge as a *Sprengwerk* rather than a *Hängewerk*, as Bernhard (above) had. This is comparable to Chevalier's view of the bridge as an arch. Re Berlin-Frankfurt Railway, see [Eugène] Bontoux, "Mission d'Allemagne," MS, École nationale des ponts et chaussées.

80. "Beiträge zu den Holzkonstrukzionen bei Brücken, Viadukten, u.s.w. in Nordamerika," *Allgemeine Bauzeitung* 10 (1845): 7–11, pl. DCXLVI (see chapter 8).

81. His description appears drawn from Town's brochure. Town had by this time visited England and was corresponding with publisher John Weale. Hosking, *Essay and Treatise*, lxxix. Others in England were more aware; for example, a lattice bridge at Vogelenzang, Netherlands, was described and illustrated in a translated article, Frederick Willem Conrad "Account of the Railway from Amsterdam to Rotterdam, and of the Principal Works

upon It," *Institution of Civil Engineers: Minutes of Proceedings* 3 (1844): 175, 188. On January 20, 1846, John Rennie mentioned lattice bridges ("of American origin") of Moorsom and McNeil (see chap. 6): "Address to Annual General Meeting," *Institution of Civil Engineers: Minutes of Proceedings* 5 (1846): 34.

82. Joseph Hyacinthe Garella, "Note sur deux ponts provisoires en charpente construits dans le système américain de M. Town," *Annales des ponts et chaussées* 2/2 (1842): 371–383, pl. 34. In a letter dated August 19, 1851, to the Ministère de Travaux Publics, Garella, who wanted to visit the tubular bridges in Great Britain, complained that in the *Annales des ponts et chaussées*, one couldn't find "en tems [*sic*] convenable" information on important engineering works constructed in Europe: "Ce n'est que dans les journaux ordinaires, et dans des feuilletons très incomplets, qu['on] apprennent seulement le nom, sans qu'aucun détail scientifique puisse les renseigner exactement et leur permettre de proposer pour les ouvrages qu'ils ont à projeter en France, des combinaisons analogues à celles qu [illegible]issent si bien au dela de nos frontières"; Archives Nationales F14, 2231.

83. Letter, Town to A. J. Davis, January 12, 1829, Metropolitan Museum. It is not known if Town could read French; he could not speak it, according to Morse, who recounted a carriage ride from Paris to Rome with Town and Nathaniel Jocelyn, in *Samuel F. B. Morse: His Letters and Journals*, ed. Edward Lind Morse (Houghton Mifflin, 1914) 1:319. It is unknown if Town promoted his invention while abroad. R. W. Liscombe ("New Era," 5–6) asserted that Town "seized the opportunity to publicize his 1820 patent truss: He indicated in one letter, dated London, December 21, that he carried copies of the *Description*." As such evidence does not appear to exist, I contacted Liscombe, who replied: "I think that you will see from my text that I did not present specific evidence regarding Town's carriage of his brochure"; letter, R. W. Liscombe to Gregory Dreicer, February 25, 1992.

84. [Joseph Alexandre] de Saint-Claire, "Sur une travée en charpente construite au Vaudreuil, route royale de Mantes à Rouen, dans le système des ponts américains de M. Town," *Annales des ponts et chaussées* 2/1 (1841): 303–309, pl. 9. This bridge served as a test; the conclusion was that, "ainsi qui l'avait annoncé d'ailleurs l'ouvrage de M. Poussin, que le système des ponts de bois américains est certainement susceptible d'un grand nombre d'applications utiles." De Saint-Claire's bridge was initially a series of "St. Andrew's crosses"; when the bridge sagged, five overlapping crosses were added to the middle (thus creating a lattice). Couche, "Travaux d'art," 335, believed that instead of this the chords should have been strengthened to resist bending; De Saint-Clair reports the measure was effective.

85. Samuel Charles Brees, *Railway Practice* (Williams, 1840), pl. 9.; Moorsom, in J. R. Mosse, "American Timber Bridges," *Institution of Civil Engineers: Minutes of Proceedings* 22 (1862–1863): 324–325; list of early lattice bridges in Nicholas Bill, "Timber-Lattice Bridges on Railways in the UK: 1840–1870," *Engineering History and Heritage* 166 (2013): 137.

86. Moorsom "visited and studied every railway and canal working or in progress of construction in England." His plan for a steeper grade railway won over Stephenson and Brunel. Moorsom seems to have favored "American" technology, which he employed on the Birmingham and Gloucester line. Ghega, *Über nordamerikanischen Brückenbau*, 9–10, mentions lattice bridges

on this railway; Carl Ghega, *Die Baltimore-Ohio Eisenbahn über das Allegheny-Gebirg mit Besonderer Berücksichtigung der Steigungs- und Krümmungsverhältnisse* (Kaulfuss Witwe, Prandel, 1844), 131–132. "Surrey Zoological Garden," *Morning Post* (London), July 19, 1848, 5, reported that "an American bridge, built upon principles which have commanded the respect of the scientific," was on display in Surrey Zoological Garden, London.

87. "Geschiedkundige Herinnerlingen betrekkelijk den Ijzeren Spoorweg van Amsterdam naar Haarlem, Leiden . . . en Rotterdam: . . . Kunstwerken," *De Nederlandsche Stoom Post*, August 1, 1847, 131–132; *Tijdschrift van het Koniklijk Institut van Ingenieurs, Notulen* (1869–1870): 167. This 54-meter bridge functioned for sixteen years and required reinforcing. Letter, Geert Verbong to Gregory Dreicer, July 10, 1991.

88. Engineer James Thomson worked with Macneill; see his presentation, *Proceedings of the Philosophical Society of Glasgow* 1 (1843): 182; James Thomson, "Observations upon Iron Lattice Bridges," *Journal of the Franklin Institute* 3rd ser., 7 (1844): 229–230, from *Transactions of the Royal Scottish Society of Arts* 2 (1844): 78–79 (communicated December 11, 1843).

CHAPTER 4

1. As was happening in other fields; for example, Tiago Saraiva, *Fascist Pigs: Technoscientific Organisms and the History of Fascism* (MIT Press, 2016), 11.

2. Joseph-Mathieu Sganzin, *Programme ou résumé des leçons d'un cours de constructions*, ed. Émile Reibell, 4th ed. (Carilian-Goeury & Dalmont, 1839–1841), 334–336, pl. 43. Reibell cites Poussin; drawings are labeled "Ponts américains de Mr. Town."

3. [Polytechnischen Schule, München], *Sammlung ausgeführter Construction aus dem Gebiete des Strassen- Brücken- und Wasserbaus, von den Eleven des Ingenieurcurses der k. polytechnischen Schule in München* (Wolf, [1843–]1852), pl. 9.

4. De Sappel, in "Mission de Belgique, Holland, Prusse, etc.," manuscript, École nationale des ponts et chaussées, described and drew Conrad's lattice bridge in Holland.

5. Louis Charles Mary, "Ecole centrale des arts et manufactures: Cours de construction des ponts, manuscript" (Paris, 1861), 273–277, pl. 34. Plate volume (1843) does not include Town bridge; text volume unavailable. The 1861 edition includes bridges by Town, Long, and Howe; very likely these were included in Mary's course much earlier. Ulrich Pfammatter, *Making of the Modern Architect and Engineer* (Birkhäuser, 2000), 132–150, and Valérie Nègre, "Architecture et construction dans les cours de l'École centrale des arts et manufactures (1833–1864) et du Conservatoire national des arts et métiers (1854–1894)," in *Bibliothèques d'atelier: Édition et enseignement de l'architecture, Paris 1785–1871* (Institut National d'Histoire de l'Art, 2011), 3.

6. Georg Mehrtens, *Notizen über die Fabrikation des Eisens und der Eisernen Brücken* (Author, 1882), 38 and Richard Shelton Kirby et al., eds., *Engineering in History* (McGraw-Hill, 1956), 227. The lenticular truss of, for example, Laves (from Hannover) was said to be similar to Town's bridge; "Ueber Town's Gitterbrücken (*lattice bridges*) and Long's Rahmenbrücken (*frame bridges*)," *Polytechnisches Centralblatt* (Leipzig) 6 (1840): 100.

7. Example of inventors'continued focus on lattice structures: Katia Bertoldi, "Structural Design Principles for Diagonal Bracings in Truss and Beam Support Systems," US Patent 11,781,307 B2 (October 20, 2023); reference to Ithiel Town, 1.

8. Town (1825), 1. "Mr. George Smart's Patent Iron Bridge," *Register of Arts and Sciences* 2 (1825; article dated October 23, 1824): 49–51; James, "Evolution of Iron Bridge Trusses," 80–81. One reviewer of Stevenson's *Sketch of the Civil Engineering of North America* suggested that the lattice idea was stolen from Smart: *Civil Engineers and Architects Journal* 37 /8: 309. Charles Manby, secretary of the Institution of Civil Engineers, mentioned that Smart's bridge was analyzed in Olinthus Gregory, *Mathematics for Practical Men* (Baldwin, Craddock, and Joy, 1825), 231–32; George Willoughby Hemans, "Description of a Wrought-Iron Lattice Bridge, Lately Erected on the Line of the Dublin and Drogheda Railway," *Institution of Civil Engineers: Minutes of Proceedings* 3 (1844): 64. Nicholas A. Bill, "Timber-Lattice Bridges on Railways in the UK: 1840–1870," *Proceedings of the Institution of Civil Engineers, Engineering History and Heritage* 166 (2013): 136, resuscitated the Smart myth, citing Tomlinson's *Cyclopedia of Useful Arts*, a popular publication and font of nationalism published decades after the fact (1852). British "precedents" also include the (dysfunctional) arched and trussed bridges of John Burrows, who worked in Ottawa (1820s to 1840s) under Lieutenant-Colonel John By. E. F. Bush, "John Burrows," in *Dictionary of Canadian Biography* (University of Toronto Press, 1988), 7:121–122; John Weiler, "Colonial Connections: Royal Engineers and Building Technology Transfer in the Nineteenth Century," *Construction History* 12 (1996): 7–9; Denison, "Description of a Series of Bridges Erected across the River Ottawa," in *Papers on Subjects Connected with the Duties of the Corps of Royal Engineers* 3 (1839): 158–163.

9. "Long's Wooden Frame-Bridges," *Mechanics' Magazine* 13 (1830): 433–434; Stephen H. Long, *Beschreibung der von dem Oberstlieutenant Long erfundenen hölzernen Brücke*, trans. C. J. Gauss (Hahn, 1840).

10. Dario Gasparini and David Simmons, "American Truss Bridge Connections in the 19th Century I: 1829–1850," *Journal of Performance of Constructed Facilities* 11 (1997)," 120–121; Dario A. Gasparini and Caterina Provost, "Early Nineteenth Century Developments in Truss Design in Britain, France and the United States," *Construction History* 5 (1989): 28–31. Also, "The Myth and Reality of Colonel Long's Wedges," in Philip C. Pierce et al., *Covered Bridge Manual* (Pub. No. FHWA-HRT-04-098), prepared for U.S. Dept. of Transportation, Federal Highway Administration, Turner-Fairbank Highway Research Center, Mclean, VA, 2005, 254–255.

11. Franz Xav. Jos. Maschek, *Neue Theorie der allgemein als der zweckmäßigsten, vollkommensten und wohlfeilsten anerkannten nordamerikanischen Brücken-Konstructionen, nach Town's, Long's and Howe's System, wie auch ihrer in der neuesten Zeit erfundenen Vereinfachung und Verbesserung von Neville und Clark auf eine originelle, leicht faßliche Weise (Ohne höhere Mathematik) dargestellt* (Spurny, 1860; 1st ed., 1857). This Prague architect considered (11, 27) the lattice bridge the basis for Long, Howe, Neville bridges and for theoretical understandings. He compares the challenges of Town and Long bridges and notes (22) that in Europe the Town was believed to have greater "solidity or strength" than the Long, which had been superseded by the Howe.

12. Stephen H. Long, *Description of Col. Long's Bridges, Together with a Series of Directions to Bridge Builders* (Brown, 1836), 40.

13. Town and Long bridges were statically indeterminate; it was not possible at the time to accurately calculate stresses in each member. On Navier, Gasparini and Provost, "Early Nineteenth Century Developments," 28–30; William Addis, *Structural Engineering* (Ellis Howard, 1990), 70. Town sought information on Long's work: William Elliot, states in a letter to Town, apparently in response to Town's query, "I do not know what are Major Long's improvements on the Railroad"; letter, Washington, March 9, 1829, New York Historical Society, Misc. MSS Elliott. This was one year before Long's first bridge patent.

14. Long, *Description of Col. Long's Bridges*, 7–9, 19–20, pl. 3. Words italicized in original.

15. Town (1821), 6. Dario A. Gasparini et al., "G. W. Whistler and Howe Bridges on the Nikolaev Railway, 1842–1851," *Journal of Performance of Constructed Facilities* 30 (2016): 2.

16. Haupt, "To the Editors of the American Railroad Journal," *American Railroad Journal* 11 (1840): 196–200, provided an analysis of lattice deficiencies. Re Haupt, see Francis A. Lord, *Lincoln's Railroad Man: Herman Haupt* (Fairleigh Dickinson University Press, 1969), and Thomas Weber, *Northern Railroads in the Civil War, 1861–1865* (King's Crown, 1952), 137–168; David A. Pfeiffer, "Working Magic with Cornstalks and Beanpoles: Records Relating to the U.S. Military Railroads During the Civil War," *Prologue Magazine* 43 (2011), https://www.archives.gov/publications/prologue/2011/summer/usmrr.html.

17. [Herman Haupt], *Hints on Bridge Construction, by an Engineer* [1841], 11, footnote. This promotional publication is undated; date provided in "Herman Haupt," House of Representatives, 62d Congress, 2d Session, Report 682 (May 9, 1912): 3.

18. Herman Haupt, "Lattice Bridge," US Patent 1445 (December 27, 1839).

19. Frank Abial Flower, "General Herman Haupt," in Herman Haupt, *Reminiscences of General Herman Haupt* (Wright & Joys, 1901), xiv, states that Haupt found one exception—Benjamin H. Latrobe of B&O Railroad used the parallelogram of forces to analyze "simple triangular systems" similar to Fink and Bollman bridges. Haupt, *General Theory*, 253–254, shows Latrobe's design, which he likens to the Schaffhausen bridge because of its bracing system.

20. Haupt, *General Theory of Bridge Construction* (Appleton, 1851), 6–7; the theory was supposedly complete five years before its 1851 publication; Haupt, "Lattice Bridge"; Haupt, *Hints*; Flower, "General Herman Haupt," xv–xvi. It was less useful than Whipple's work; Stephen P. Timoshenko, *History of Strength of Materials* (McGraw-Hill, 1953; reprint, Dover, 1983), 185.

21. "Improvisation" per Mario Rinke and J. Schwartz, "Force Material Form: Tranferring Historical Construction Concepts into Contemporary Architectural Design," in *Structures and Architecture*, ed. Paulo J. da Sousa Cruz (Taylor and Francis, 2013), 1255; 1260 maintains the stereotype: "Europe" as "academic" and "North America" as "pragmatic."

22. Ithamar A. Beard, "Description of an Improvement on the Bridge Patented by Itheil Towne [sic], Esq., the term of which patent has expired," *Journal of the Franklin Institute* n.s., 25 (1840): 175–177. Town's original patent suggested the use of bolts.

23. J.H. Garella, "Note sur deux ponts provisoires en charpente construits dans le système américain de M. Town," *Annales des ponts et chaussées* 2/2 (1842): 371–383, pl. 34. Garella's use

of bolts was reported in Julius Pollack, "Metallbrücke zur Uebersetzung des Royalkanals bei Dublin für die Dublin-Drogheda-Eisenbahn nach Town's nordamerikanischem Konstrukzions-system gebaut im Jahre 1845," *Allgemeine Bauzeitung* 13 (1848): 19; 21.

24. Garella, "Note sur deux ponts provisoires," 375–376. For Navier, the cross-sectional size of top and bottom chords and the space between them were key factors; this method did not take into account diagonal member dimensions. Beam theory "was already well-developed by the end of the 18th century"; Addis, *Structural Engineering*, 70–71. Stefan M. Holzer, "Polonceau Roof and Its Analysis," *International Journal for the History of Engineering & Technology* 80 (2013): 46–47.

25. Re Howe, see Richard Sanders Allen, *Covered Bridges of the Northeast*, rev. ed. (Greene, 1983), 18–19.

26. One person argued that Howe bridges "should be denominated Long's Bridges" because they employed the same "principle"; "Origin of Railroads in the United States," *American Railroad Journal* 28 (1855): 84–85.

27. US Patent 1685 (July 10, 1840) showed a lattice with verticals; cross-bracing with verticals and arch; cross-bracing with integrated arch. Originally, Howe included iron tension rods; this feature was dropped and introduced in a subsequent patent (No. 1711, August 3, 1840). Howe states, "It has been deemed expedient to separate it . . . and to make it the subject of a distinct patent," in order to have claims for wooden and metal bridges.

28. Gasparini et al., "Whistler and Howe," 4.

29. U.S. Patent 1711 (August 3, 1840; reissued September 3, 1850). On Long and Howe bridges, see Gasparini and Provost, "Early Nineteenth Century Developments," 28–30; Gasparini and Simmons, "American Truss Bridge Connections I," 119–129; Dario A. Gasparini and Francesa da Porto, "Prestressing of 19th Century Wood and Iron Truss Bridges in the U.S.," in Huerta, *Proceedings*, 977–986.

30. Gasparini et al., "Whistler and Howe," 13–14.

31. W. R. C., "Notes on Practical Engineering: Bridges," *American Railroad Journal* 17 (1844): 9; an observation shared by others, see notes below. See also the evolutionary and scientification stories of Condit, *American Building Art*, 92. Carl Condit, *American Building* (University of Chicago Press, 1968), 61, deemed the Howe-lattice "curious." Karl Culmann, "Der Bau der hölzernen Brücken in den Vereinigten Staaten von Nordamerika . . . in den Jahren 1849 und 1850," *Allgemeine Bauzeitung* 16 (1851): 74 (hereafter Culmann, "Holz"), incorrectly states that the Howe bridge directly followed the Long.

32. Dario A. Gasparini and Melissa Fields, "Collapse of Ashtabula Bridge on December 29, 1876," *Journal of Performance of Constructed Facilities* 7, no. 2 (1993): 109–125, esp. 115.

33. Lewis M. Prevost Jr., "Description of Howe's Patent Truss Bridge, Carrying the Western Railroad over the Connecticut River at Springfield, Massachusetts," *Journal of Franklin Institute* 3rd ser., 3 (1842): 292; W. R. C., "Notes on Practical Engineering," 9, states that in the Howe-lattice at Springfield, "the braces, which are of timber, cross each other in the style of lattice work." Howe-lattice bridges built in United States: Warren, Springfield, Warehouse Point, and Holyoke, Mass.

Letter, Richard Sanders Allen to Raymond E. Wilson (Smithsonian NMAH, Raymond E. Wilson Covered Bridge Collection, 1984.0009, box 5, binder: "Trusses"). Robert Fletcher and J.P. Snow, "History of Development of Wooden Bridges," *ASCE Transactions* 99 (1934): 341–42. Re Connecticut River Bridge and Western Railroad, see Gasparini, "Whistler and Howe," 2–5.

34. Pollack, "Metallbrücke," 1–27; pls. 148–149; excerpted in *Polytechnisches Centralblatt* (Leipzig) NF2 (1848): 625. He claimed that in England, France, Belgium, and the Netherlands, the Town bridge was built "exclusively"; in Austria, the Howe was preferred. Given improvements to the lattice and the cost and difficulty of obtaining iron rods for Howe, he considered the lattice a viable alternative; Pollack, "Metallbrücke," 19. Re Russia, see Gasparini, "Whistler and Howe."

35. Friedrich August von Pauli, "Beitrag zur Kenntnis des bayerischen Eisens," *Kunst- und Gewerbe-Blatt des polytechnischen Vereins für das Königreich Bayern* 39 (1853): 4–25, pl. 3. Stefan M. Holzer, *Die König-Ludwig-Brücke Kempten* (Bundesingenieurkammer, 2012), 48–51. Charles Couche, "Travaux d'art des chemins de fer d'Allemagne," *Annales des mines* 5/5 (1854), 316–317, remarks that the Howe-lattice combined features of Town and Long and that in the United States single cross-bracing was used between tension rods, while in Europe, the number of "Saint Andrew's crosses" was doubled, creating a "real lattice," though different from Town's.

36. Builders apparently erected it on temporary scaffolding, piece by piece. Holzer, *Die König-Ludwig Brücke*, 52–60. The Royal Inspector (Baurath) of the Bavarian Building Department described the assembly of Howe frames and tightening of tension rods, around 1852: Ferdinand Jodl, *Der Strassen- und Brückenbau: Vorlegeblätter zum Selbstunterricht für Architecten und Ingenieure* (Grieben, [1860–1864]), 65.

37. Gasparini, "Whistler and Howe," 1–16. On Jouravsky in St. Petersburg, see Stephen P. Timoshenko, *History*, 141–144, 186–189. Jouravsky visited the United States in 1859–1860. (His name appears in various forms, including Jourawski).

38. Pennsylvania inventor Henry Wilton patented a lattice-arch bridge (June 24, 1839); F.B. Brock, "Truss-Bridges," *Engineering News* 9 (1882): 385–386; Fletcher and Snow, "History," 336–337. Wilton's solution would likely have been opposed by Town; Town (1839), 6–7. Culmann, "Holz," 74, pl. 390; the depiction of Howe in the United States differed from the Howe-lattice in Bavaria and Austria in the middle of the century and analyzed by Ghega.

39. On Bull Run Bridge, see Debbie Robison, "Bull Run Bridge on the Orange and Alexandria Railroad Civil War Railroad Bridge Construction and Repair," in *Northern Virginia History Notes*, September 21, 2014, http://www.novahistory.org/BullRun/BullRunBridge.htm; Ron Baumgarten, "The Civil War in Northern Virginia & Beyond," *All Not So Quiet along the Potomac*, December 11, 2012, http://dclawyeronthecivilwar.blogspot.com/2012/12/a-mystery-solved-blockhouse-at-chain.html. Victor C. Darnell, "Haupt Iron Bridge on the Pennsylvania Railroad," *Journal of the Society for Industrial Archeology* 14 (1988): 35–50; Victor Darnell, "Haupt Truss Bridge," Written Historical and Descriptive Data, HAER, National Park Service, US Department of Interior, 1986, from Prints and Photographs Division, Library of Congress (HAER PA-207), 4.

40. Carl Ghega, *Die Baltimore-Ohio Eisenbahn über das Allegheny-Gebirg mit Besonderer Berücksichtigung der Steigungs-und Krümmungsverhältnisse* (Kaulfuss Witwe, Prandel, 1844), v, ix–x;

descriptions of Long bridges, 85–88. The presence of Ghega and Löhr (see below) was announced in the *Savannah Georgian* and reprinted in *American Railroad Journal* 14 (1842): 320: "It is said that they will be accompanied by several pupils of the Imperial Polytechnic School."

41. Ghega, *Die Baltimore-Ohio Eisenbahn*, 80–81. He was apparently instrumental in Austrian State Railway's decision to use the "American system" of locomotives, cars, and track design in 1845; see review of Ghega, *Die Baltimore-Ohio Eisenbahn*, *Eisenbahnzeitung* 3 (1845): 5.

42. Carl Ghega, *Über nordamerikanischen Brückenbau und Berechnung des Tragungsvermögens der howeschen Brücken* (Kaul Fuss Witwe, Prandel, 1845).

43. Addis, *Structural Engineering*, 70.

44. Ghega, *Über nordamerikanischen Brückenbau*, vi, 11; 27, he refers to Navier, Barlow, Tredgold, Eytelwein, Long, among others, and provides English words for some concepts; 66, 57–58, on Howe. Pollack, "Metallbrücke," 18, believed, like Ghega and Löhr, that "the system of Mr. Howe is based completely on the same principle as that of Mr. Town."

45. Bill Addis, *Building: 3000 Years of Design Engineering and Construction* (Phaidon, 2007), 311, 315; Addis, *Structural Engineering*, 69–72.

46. Moritz Löhr, "Ueber den Bau hölzerner Brücken in Nordamerika," *Allgemeines Wiener Polytechynisches Journal* 2 (1843): 419–421; also *Polytechnisches Centralblatt* (Leipzig) NF 2/2 (1843): 419.

47. Löhr, "Ueber den Bau hölzerner Brücken," 420. These reasons are repeated in Hinrich Müller, *Die Brückenbaukunde in ihrem ganzen Umfange: ein Handbuch für Ingenieure und Baumeister* (Romberg, 1850–1853), 2:109. There is a similar assessment in *Report of James Laurie, Civil Engineer, of His Examination of Certain Bridges on the Railroads of the State of New-York* [1856], 25: "The lattice plan of bridging possesses less absolute strength than most others in use, for the quantity of material used. About one half the strength of the timbers being destroyed by the treenails."

48. Culmann, "Holz," 69–129, pls. 387–397; excerpted in *Polytechnisches Centralblatt* (Leipzig) NF5 (1851): 1492–1497, and *Eisenbahnzeitung* 9 (1851): 165–166; Culmann, "Bau der eisernen Brücken," 163–222, pls. 478–487. Christine Lehmann and Bertram Maurer, *Karl Culmann und die graphische Statik* (Ernst & Sohn, 2006), 50–86.

49. A simple form of statics had been used to analyze roof trusses. Rowland J. Mainstone, "Structural Analysis, Structural Insights, and Historical Interpretation," *Journal of Society of Architectural Historians* 56 (1997): 316–319.

50. "Pendent or hanging bridges, called also *Philosophical Bridges*, are those not supported either by posts or pillars, but hung at large in the air, only supported at the two ends or abutments. Instances of such bridges are given by Palladio and others"; *English Encyclopedia* (Kearsley, 1802), 2:113. David Yeomans, "Soane and Swiss Bridges," *Construction History* 19 (2003): 54, writes, "During the eighteenth century and well into the nineteenth a truss . . . did not, and could not have implied a series of braced panels, which is how engineers think of trusses today." See discussion of Navier, chapter 4.

51. Peter Barlow's *Essay on the Strength and Stress of Timber* (1817) is considered a breakthrough. George Michael Danko, "Evolution of the Simple Truss Bridge" (Ph.D. diss., University Pennsylvania,

1979), 86–104; David T. Yeomans, *The Trussed Roof: Its History and Development* (Scolar, 1992), 48. Danko, "Evolution," 136, states, "It is more than coincidence that the lattice truss appeared at the same time as a sizable body of theory on the strength of the simple beam became available in [the United States]." Town (1821) noted the importance of "science"; much later, Town (1839), 4–5, 10–11, and passim, repeatedly mentions "engineers" and "engineering," cites "Mechanics," references experience and "mathematical principles," and provides formulas for testing with scale models.

52. Addis, *Structural Engineering*, 69–72; Gasparini and Simmons, "American Truss Bridge Connections I," 127.

53. Culmann, "Holz," 80, refers readers to the Stevenson article in *Allgemeine Bauzeitung* (1839).

54. Culmann, "Holz," 86; also Culmann, "Bau der eisernen Brücken," 193. *Fachwerk* means "half timbering." When Culmann first used the new term ("Holz," 71), he did not define it. He subsequently defined it with the English term, "frame-work, as opposed to "lattice-work" bridge ("Holz," 86).

55. Culmann, "Holz," 86 and 105. Culmann's work was based on Navier and others. Holzer, "Polonceau Roof," 47. On Navier, see Rinke and Kotnik, "From Construct to Type," 1951–1952.

56. Dennis Hart Mahan, *An Elementary Treatise of Civil Engineering*, new [4th?] ed. (Wiley and Putnam, 1846), 116, noted, as Culmann would, that "strains [i.e., stresses] are borne by the tree-nails, and when used for structures subjected to variable strains and jars, it loses its stiffness and sags between the points of support." A "sinking of the center of the bridge about four inches," was reported in 1820, the lattice patent year, but Town resisted adding reinforcing; *Report of the Board of Public Works, to the Legislature of South-Carolina, for the Year, 1820*, in David Kohn and Bess Glenn, eds., *Internal Improvement in South Carolina 1817–1828: From the Reports of the Superintendent of Public Works and from Contemporary Pamphlets* (Author, 1938), 47–48. The cause was disputed in a lawsuit.

57. See chapter 3. Pollack, "Metallbrücke," 18, an *Allgemeine Bauzeitung* article that preceded Culmann's, cites Chevalier, Gerstner, and Ghega, along with the *Journal of the Franklin Institute* and *The Builder*.

58. Culmann, "Holz," 81; quotation in English in original.

59. Culmann shows the lattice (1820 patent) illustration to indicate what "everyone knows"; it appears he did not take into account the Richmond Bridge's double lattice design; Culmann, "Holz," 80. See chapter 2, note 27 and chapter 3, note 65.

60. Culmann, "Holz," 80–81, pl. 396; table, 128. Also reported in William Humber, *A Practical Treatise on Cast and Wrought Iron Bridges and Girders, as Applied to Railway Structures, and to Buildings Generally* (Spon, 1857), 92.

61. Haupt, *General Theory*, 148. Couche, "Travaux d'art," 315, also seems to favor modification rather than abandonment.

62. Chevalier, *Histoire*, 2:576; also 2:566.

63. Culmann, "Holz," 77.

64. "Jeder Bauer," in Culmann, "Holz," 81.

65. Zimpel, *Eisenbahnwesen*, 102 (see chap. 3). Alexis de Tocqueville, *Democracy in America*, ed. Francis Bowen, trans. Henry Reeve (Sever and Francis, 1862), 2:59; French and British editions, 1835; German edition, 1836; US edition, 1838. Because the settlers' work was so challenging, Culmann believed that educated immigrants needed to budget for the purchase of enslaved people; Lehmann and Maurer, *Karl Culmann*, 72–73.

66. Culmann, "Holz," 81–82, 105–109. The figure illustrating this bridge, as well as other illustrations in Culmann's reports are similar if not identical to illustrations in George Duggan, *Specimens of the Stone, Iron and Wood Bridges, Viaducts, Tunnels, Culverts, &c. of the United States Rail Roads* (Appleton, 1850). Plates in "Beiträge zu Holzkonstruktionen bei Brücken, Viadukten, u.s.w. in Nordamerika," *Allgemeine Bauzeitung* 10 (1845) are identical to several in Duggan.

67. My assessment of Culmann's and Haupt's viewpoints is based on Addis, *Structural Engineering*, 70–71. The designer initially estimates aspects of an indeterminate structure and develops design through repeated analysis and revision; letter, Jonathan Ochshorn to Gregory Dreicer, March 1, 1993.

68. Gasparini and Simmons, "American Truss Bridge Connections I," 127.

69. Culmann, "Bau der eisernen Brücken," 190. J. W. Schwedler, "Elementare Theorie der gitterförmigen Träger-Constructionen," *Civilingenieur* 5 (1859): 150–151.

70. Culmann, "Bau der eisernen Brücken," 191–194. Culmann (188–190, 193) believed that the addition of vertical members (as compressive members) meant the structures were no longer lattice bridges.

71. Town (1839), 6.

72. A. Straus, "Ueber die Berechnung der schmiedeisernen Gitterbrücken," *Eisenbahnzeitung* 10 (1852): 193 (intuition: "nach dem Gefühle"); see also Pollack, "Metallbrücke," 19–21. In response, another engineer-intern stated that the beam should be regarded as a "double T" (I-beam), with the diagonals regarded as connecting rather than loadbearing elements—thus only chords are calculated for bearing; H. Grebenau, "Entgegnung," *Eisenbahnzeitung* 10 (1852): 209.

73. Culmann, "Holz," 72, 105. Addis, *Structural Engineering*, 62–63.

74. Karl-Eugen Kurrer, *The History of the Theory of Structures: From Arch Analysis to Computational Mechanics* (Ernst & Sohn, 2008), 36; also Timoshenko, *History*, 194–97 and Addis, *Building*, 371–374.

75. "Nine-tenths" of bridge designers would not understand "complex equations"; C. Shaler Smith, *Comparative Analysis of the Fink, Murphy, Bollman & Triangular Trusses* (Woods, Steam Book and Job Printer, 1865), 4. Even Wilhelm Nordling (chapter 8) asserted the primacy of "the three equations of equilibrium of elementary statics"; Nordling, in Contamin, Nordling, et al., "Procès-verbaux" [Une nouvelle méthode pour calculer les dimensions qu'il convient de donner aux pièces des poutres en treillis], 5, January 19, February 2, 1866, *Memoires et comptes rendus des travaux de la Société des Ingénieurs civils* (1866): 35–36, 40. Whipple's contribution "consists more in the introduction of proper dimensions and proportions of parts in the truss"; Squire Whipple, "On Truss

Bridge Building," *ASCE Transactions* 1 (1872): 239–244. For capsule account of theorists' impact, see Gasparini and Fields, "Collapse," 109–110.

76. On "sensation," see Aimé Henry Résal and Joseph Vicaire, "Notice nécrologique sur M. Charles Couche," *Annales des mines* 7, no. 17 (1880): 55. Couche apparently included data from Stevenson, Gerstner, Poussin, Chevalier, and Ghega; Cullman, "Holz," was called "un intéressant travail" by Couche, "Travaux d'art," 313.

77. Couche, "Travaux d'art," 315–316.

78. E.g., *Moniteur Industriel* 13 (1844): 839/3; Pollack, "Metallbrücke"; Town's bridge was "in England, Frankreich und Belgien vervollkommnet und zugleich in Eisen nachgeahmt worden"; Maschek, 122; M. von Weber, *Die Schule des Eisenbahnwesens* (Weber, 1857), 47, describes beams constructed "nach amerikanischem System" of flat iron bars in a latticework.

79. Couche lists several wooden bridges replaced after short periods. In France, wood was regarded as suitable for temporary railway bridges. Couche cites wooden bridges placed outside planned axis of railroad tracks, so that "definitive bridges" could be built later. Couche, "Travaux d'art," 303–306, 273.

80. E. Deligny, "Notice sur la chute du pont du chemin de fer de Paris à Saint-Germain, à Asnières, incendié le 25 février 1848, et sur le pont provisoire en charpente établi par M. Eugène Flachat, ingénieur en chef," *Memoires et comptes rendus des travaux de la Société des ingénieurs civils* 1 (1848): 81–91, pls. 3–4.

81. He claimed that German engineers had rejected wood except in the case of short-span bridges of moderate importance; Couche, "Travaux d'art," 310. He stated that Prussia had almost totally rejected wood, while Austria, Württemburg, and Bavaria still used it; Couche, "Travaux d'art," 305–306, 316–326.

82. Couche, "Travaux d'art," 312.

83. Couche, "Travaux d'art," 314.

84. Re Richmond Bridge, see Couche, "Travaux d'art," 315. See chapter 2.

85. Johann Mohnié, with commentary by H. Scheffler, "Über eine verbesserte Konstruktion eiserner Gitterbrücken," *Organ für die Forschritte des Eisnbahnwesens in Technischer Beziehung* 12 (1857): 227–236, pl. 20 (also *Zeitschrift für Bauwesen* 8 [1858]: 277–284, pl. N); Mohnié, "Verbesserte Konstrucktion eiserner Gitterbrücken," *Romberg's Zeitschrift für praktische Baukunst* 17 (1857): 225–232; Prüsmann, review of "Über eine verbesserte Construction by Mohnié," *Zeitschrift des Architecten- und Ingenieur-Vereins für das Königreich Hannover* 4 (1858): 121–123.

86. Robert Crawford, "Railway System of Germany," *Institution of Civil Engineers: Minutes of Proceedings* 22 (1862–63): 16. Mallet in Crawford, "Railway System," 38.

87. Martin Trautz, *Eiserne Brücken im 19. Jahrhundert in Deutschland* (Werner-Verlag, 1991), 76, states that during the 1850s only simple shapes of iron were readily available in Germany; it was expensive to import.

88. A.E. Richard De Jonge, *Riveted Joints: A Critical Review of Literature Covering Their Development, with Bibliography and Abstracts of Most Important Articles* (American Society of Mechanical

Engineers, 1945), and the American Society of Civil Engineers and Research Council on Riveted and Bolted Structural Joints, *Bibliography on Bolted and Riveted Joints* (ASCE, 1967). Regarding riveting work, David A. Simmons, "'Continuous Clatter': Practical Field Riveting," *Journal of the Society for Industrial Archaeology* 23 (1997): 4–20.

89. "Herron's Patent Trellis Railway Structure," *American Railroad Journal* 13 (1841): 65–68, 103–05; "Herron's Patent Railway Track," *American Railroad Journal* 18 (1845): 760; "Herron's Trellis Railway Superstructure," *Journal of the Franklin Institute* 3rd ser., 7 (1844): 83–86; James Herron, "Herron's Patent American Railway Track," *Journal of the Franklin Institute* 3rd ser., 10 (1845): 366–370. The track was tested on the Philadelphia and Reading Railroad in 1844: "3 miles on lattice plank . . . took as long to lay down as the 77 miles on ties"; Richard Boyse Osborne, "Diary," manuscript, Mss. 7888-7895, 3:83; cited with permission of the Council of Trustees of the National Library of Ireland.

90. Daniel Hovey Calhoun, *Intelligence of a People* (Princeton University Press, 1973), 250, figure 9.

91. "*Lattice car* genannt"; Franz Anton Gerstner, Die innern Communicationen der vereinigten Staaten von Nordamerika, ed. L. Klein (Förster, 1842), 1:149, 2:93, 312.

92. Rainer Graefe, "Die Bogendächer von Philibert de l'Orme," in *Zur Geschichte des Konstruierens*, ed. Rainer Graefe (Deutsche-Verlags-Anstalt,1989), 177–179.

93. Polonceau, [Dispositions adoptées pour les wagons qu'il vient d'exècuter dans les ateliers de la Compagnie d'Orléans pour les voyages de S.M. l'Empereur sur cette ligne], "Procès-verbaux," August 22, 1856, *Mémoires et compte rendu des travaux de la Société des ingénieurs civils* 9 (1856): 335. Five cars of a nine-car train "sont réunis par des ponts, suivant le système américain." Perhaps it was a Howe structure. I have been unable to find the design.

CHAPTER 5

1. David N. Livingstone, *Putting Science in Its Place: Geographies of Scientific Knowledge* (University of Chicago Press, 2003), 42. Construction is omitted even where it would seem to be the subject; e.g., Bill Addis, *Building* (Phaidon, 2007) doesn't focus on construction as process or its relationship to structural design and industrialization; lumber appears as a pre-nineteenth-century artifact; timelines confirm a belief in progress. Construction doesn't appear in Antoine Picon, "First Steps of Construction in Iron: Problems Posed by the Introduction of a New Construction Material," in Rinke and Schwartz, *Before Steel*, 51–63.

2. And how is it related to what Thomas Harding calls "national amnesia" re industrialization; see Maxine Berg and Pat Hudson, *Slavery, Capitalism and the Industrial Revolution* (Polity Press, 2023), 4.

3. Informed by observations of Jaron Lanier on music and AI, "Artificial Writing, Artificial Thinking," at Townsend Center for the Humanities, UC Berkeley, April 8, 2025.

4. Karl E. Weick, "Small Wins: Redefining the Scale of Social Problems," *American Psychologist* 39 (1984): 42–43; Charles F. Sabel and Jonathan Zeitlin, "Stories, Strategies, Structures: Rethinking Historical Alternatives to Mass Production," in *World of Possibilities: Flexibility and Mass Production in Western Industrialization*, ed. Charles F. Sabel and Jonathan Zeitlin (Cambridge

University Press, 1997), 13. For a bibliographic essay on construction, see Amy E. Slaton, *Reinforced Concrete and the Modernization of American Building, 1900–1930* (Johns Hopkins University Press, 2001).

5. "Part of our Schumpeterian heritage," according to Nathan Rosenberg, "Economic Development and Transfer of Technology: Some Historical Perspectives," *Technology and Culture* 11 (1970): 567.

6. James M. Utterback, *Mastering the Dynamics of Innovation* (Harvard Business School, 1994), xviii.

7. Drew R. McCoy, *The Elusive Republic: Public Economy in Jeffersonian America* (Norton, 1980), 17.

8. Henry Clay, "In Defense of the American System," US Senate, Classic Senate Speeches, https://www.senate.gov/artandhistory/history/common/generic/Speeches_ClayAmericanSystem.htm.

9. D. F. Harrison, "Bridges and Economic Development, 1300–1800," *Economic History Review* 45 (1992): 259, and Phyllis Deane, *First Industrial Revolution*, 2d ed. (Cambridge University Press, 1963; 1969), 117. On the statistics for France, see François Caron, "Birth of a Network: First French Railway System," in *Technological Revolutions in Europe*, ed. Maxine Berg and Kristine Bruland (Elgar, 1998), 275.

10. In Europe, investment in early railway building came from the state or British investors, who often contributed expertise and hardware. Richard J. Evans, *The Pursuit of Power: Europe 1815–1914* (Viking Penguin, 2016), 152–153, and Mikulas Teich and Roy Porter, *The Industrial Revolution in National Context: Europe and the USA* (University of Cambridge Press, 1996), 102–109. Also, Arthur L. Dunham, "How the First French Railways Were Planned," *Journal of Economic History* 1, no. 1 (1941): 12–25.

11. Editor's comment following "Extract of a letter from Boston, June 19," *Belfast Evening Post*, September 11, 1786, 4.

12. *Report of the Commissioner for the Department of Public Buildings, to the Board of Public Works for the Year 1820* (Daniel Faust, 1820), reprinted in David Kohn and Bess Glenn, eds., *Internal Improvement in South Carolina 1817–1828: From the Reports of the Superintendent of Public Works and from Contemporary Pamphlets* (Privately printed, 1938), 22.

13. Guillaume Tell Poussin, *Chemins de fer américains*, 2d ed. (Carilian-Goeury, 1836): cover, title page. Carl Ghega, *Über nordamerikanischen Brückenbau und Berechnung des Tragungsvermögens der howeschen Brücken* (Kaul Fuss Witwe, Prandel, 1845), 4–5, notes that speed in construction had become the rule, especially for railroads. See also Tom F. Peters, *Time Is Money* (Hofmann,1981).

14. William Coxe, *Travels in Switzerland in a Series of Letters to William Melmoth, Esq.* (White et al., 1789), 8. Thomas Tredgold, *Elementary Principles of Carpentry*, 3rd ed. (John Weale, 1840; 2nd ed. 1828), 129.

15. "Great Railroad Bridge," *American Railroad Journal* 25 (1852): 486–487 (from *New York Farmer and Mechanic*).

16. Portage Bridge was designed so that "when any particular piece becomes defective, it can be taken out and replaced without disturbing other parts of the bridge"; "'High Bridge,' Portage,

New York," *Civil Engineer and Architects Journal*, 65–66, pl. 8. Timbers were replaced, apparently along with metal hardware; "Erie Railway," *American Railroad Journal* 40 (1867): 531.

17. "Past and Present Construction," *The Engineer* 23 (1867): 418.

18. Inventors used models to promote bridge-building proposals (see chapter 3) and wrote about issues surrounding scale models in design; Town (1831), 4; Town (1839), 10–11. William Addis, *Structural Engineering* (Ellis Howard, 1990); Bill Addis, ed., *Physical Models: Their Historical and Current Use in Civil and Building Engineering Design* (Ernst & Sohn, 2021); Denis Patrick Smith, "Role of Structural Models in the Design of British Bridges, 1800–1870" (PhD diss., Imperial College of Science and Technology, 1970).

19. "The History of the First Yadkin River Bridge," https://www.netfrontage.com/yadkinbridge/; James Brawley, "Noted Architect Designed First Yadkin Bridge," *Salisbury Post*, May 7, 1972, http://www.trading-ford.org/townbrid.html; "Beard, Lewis," NCPedia, *Dictionary of North Carolina Biography*, ed. William S. Powell, https://www.ncpedia.org/biography/beard-lewis; Jethro Rumple, *History of Rowan County, North Carolina* (Bruner, 1881; reprint, Heritage Books, 2005), 185; *Catalogue of the Graduates of Yale College, From 1702 to 1827* (Hezekiah Howe, 1827), 14. Thanks to Catherine Bishir.

20. Sources on these bridges are cited in footnotes below. The first bridge's stone piers apparently still exist; "The Piedmont's Highway: Wil-Cox Bridge," http://www.gribblenation.com/ncpics/us29/wilcox.html. Four lattice bridges were built over Pee-Dee at Cheraw, SC (1824, 1828, 1836, 1866). We know Beard's Bridge was a lattice, according to letter, Trajan to editor Philo White, September 12, 1825, in *Western Carolinian* (Salisbury), October 11, 1825, 3; Kohn and Glenn, *Internal Improvement*, 47–48. Beard's Bridge was later known as Locke's Bridge.

21. Catherine W. Bishir, "Black Builders in Antebellum North Carolina," *North Carolina Historical Review* 61 (1984): 434–438.

22. "Patents Issued in 1820," *National Calendar* 2 (1821): 143, lists: "Bridges, Improvement on—Ithiel Town, Fayetteville, North-Carolina, Jan. 28." Letter, Town to Roger S. Baldwin, Fayetteville, March 8, 1824, Yale University, Baldwin Family Papers, box 15, folder 178.

23. Kohn and Glenn, *Internal Improvement*, 47–48. The importance of durability and long span (which meant fewer piers) was also cited. This was in reference to the Congaree Creek Bridge, which sagged at its center. A "black force" who replaced "white labourers" during the "sickly season" built the road (including causeway) to and from the bridge (47); workers were "slaves" (76).

24. To enhance lumbering capabilities, sawyers were brought from forested areas such as Maine. Robert S. Starobin, "Economics of Industrial Slavery in the Old South," *Business History Review* 44 (1970): 152–155. It was so challenging to find "mechanics and laborers" for infrastructure projects in South Carolina that the options were employing enslaved workers or bringing in workers from elsewhere; "Report of the Civil and Military Engineer of the State of South-Carolina, for the Year 1818," reprinted in Kohn and Glenn, *Internal Improvement*, 21.

25. Slaton, *Reinforced Concrete and the Modernization of American Building*, 127–167.

26. US Senate, "Missouri Compromise Ushers in New Era for the Senate," https://www.senate.gov/artandhistory/history/minute/Missouri_Compromise.htm.

27. List culled from *South Carolina Gazette* in Marcus W. Jernegan, "Slavery and the Beginnings of Industrialism in the American Colonies," *American Historical Review* 25 (1920): 230; re lumber products, 223–234, 229–230, 239. Starobin "Economics of Industrial Slavery," 138, 149, 152. Also, John S. Lupold and Thomas L. French, Jr., *Bridging Deep South Rivers: Life and Legend of Horace King* (University of Georgia Press, 2004), 45, 47. Re lumber-producing labor in the region of Town's earliest bridges, see John Anthony Eisterhold, "Lumber and Trade in the Seaboard Cities of the Old South: 1607–1860" (Ph.D. diss., University of Mississippi, 1970), 75–77; re enslaved labor in milling in Georgia, see 38–40.

28. Carl R. Lounsbury, "Wild Melody of Steam: The Mechanization of the Manufacture of Building Materials, 1850–1890," in Catherine W. Bishir et al., *Architects and Builders in North Carolina* (U. of North Carolina, 1990), 101. Catherine W. Bishir, "A Proper Good Nice and Workmanlike Manner: A Century of Traditional Building Practice, 1730–1830," in Bishir et al., *Architects and Builders in North Carolina*, 99–102; on contractors' use of enslaved labor in building, see Catherine W. Bishir, "Spirit of Improvement: Changes in Building Practice, 1830–60," in Bishir et al., *Architects and Builders in North Carolina*, 157–159. There is an overview in Bishir, "Black Builders," 423–461; white builders resented free and enslaved Blacks because they provided cheaper labor, but enslavers upheld the practice (434, 449, 456–57). Re enslaved workers in canal work, see *Report of Superintendent of Public Works for the Year 1823*, reprinted in Kohn and Glenn, *Internal Improvement*, 263, 267. Re wages paid to enslavers, see *To the Legislature of South-Carolina . . . Report on Roads for 1828*, reprinted in Kohn and Glenn, *Internal Improvement*, 560.

29. Re bridges for cotton trade, see Lupold and French, *Bridging Deep South Rivers*, e.g., 9, 92, 98. Ronald Bailey, "The Other Side of Slavery: Black Labor, Cotton, and Textile Industrialization in Great Britain and the United States," *Agricultural History* 68 (1994): 35.

30. Robert S. Starobin, *Industrial Slavery and the Old South* (Oxford University Press, 1970), 35–74. "The Growth of Slavery and Southern Railroad Development," *Railroads and the Making of Modern America*, https://railroads.unl.edu/views/item/slavery_rr. Also, Keri T. Petersen, "The North Carolina Railroad, Industrial Slavery, and the Economic Development of North Carolina" (PhD diss., University of North Carolina at Greensboro, 2017); Aaron W. Marrs, *Railroads in the Old South: Pursuing Progress in a Slave Society* (Johns Hopkins University Press, 2009), 8, 55–57; Theodore Kornweibel Jr., "Railroads and Slavery," *Railroad History* 189 (Fall-Winter 2003): 34–59; Theodore Kornweibel Jr., *Railroads in the African American Experience: A Photographic Journey* (Johns Hopkins University Press, 2010), 1–37, and Steven G. Collins, "Progress and Slavery on the South's Railroads," *Railroad History* 181 (Autumn 1999): 6–25. Enslavers were compensated for use of their enslaved workers; in some cases, railroad companies owned an enslaved work force. Anthony E. Kaye, "The Second Slavery: Modernity in the Nineteenth-Century South and the Atlantic World," *Journal of Southern History* 75 (2009), 627–650; 632, re slavery as "national institution" inseparable from international capitalism, industry, and technology.

31. Petersen, "North Carolina Railroad," 69; Carl Blankenship, "Rowan Museum's Black History Exhibit Examines Slavery in County, Will Become Part of Permanent Display," *Salisbury Post*, February 20, 2022, https://www.salisburypost.com/2022/02/20/rowan-museums-black-history-exhibit-examines-slavery-in-county-will-become-part-of-permanent-display/; Historic Salisbury Foundation, "Early History of Salisbury and Rowan County" and

Reginal W. Brown, "Rowan County, North Carolina Slavery 1753–1865 Timeline," https://historicsalisbury.org/wp-content/uploads/2020/07/Week_02_The_Great_Wagon_Road_1753-1774.pdf.

32. Sven Beckert, "Cotton and the Global Origins of Capitalism," *Journal of World History* 28, no. 1 (2017): 113–115.

33. "Mr. Town's Bridges," *Western Carolinian* (Salisbury), February 8, 1825, 2, also mentions the bridge over Mill River. Re Fayetteville, Congaree, and Goose Creek bridges, "constructed on the plan of Mr. Town," see *Report of the Board of Public Works to the Legislature of South-Carolina, For the Year, 1820* (Daniel Faust, 1820), 47–48; *Plans and Progress of Internal Improvement in South Carolina* (Young and Son, 1820), 76–77; and *Report of Superintendent of Public Works for the Year 1828*, 573; all reprinted in Kohn and Glenn, *Internal Improvement*. Town (1831), 3.

34. Lupold and French, *Bridging Deep South Rivers*, 18–20, 29–34, 35–42, 58, 78, surmise that King and Godwin, familiar with the 1824 bridge, learned to build lattice bridges at the 1828 replacement bridge. Seth King, investor in the 1824 bridge and possibly the sole investor in the replacement, knew Godwin and Horace King; Lupold and French (32) believe that this may indicate their involvement in the construction of the 1828 bridge. (A testimonial to Pee-Dee bridge quality signed by Seth King and six others was reproduced in Town [1831], 5.) Re Horace King's youth around Cheraw (13–14). Fifteen workers at Cheraw were listed by first names—an indication that they were enslaved; wood was prepared by them (31). Fire destroyed the first two Cheraw bridges; the third was deliberately burned during Civil War (30–34). Re legislation that established the Cheraw Bridge Company, see *Charleston Courier*, January 22, 1822, 1. On the span at Columbus over the Chattahoochie (Dillingham Street Bridge), including interpretation of lithograph, Lupold and French, 43–63, 189; re enslaved workers, 52–53; "Dillingham Street Covered Bridge," *Encyclopedia of Alabama*, http://encyclopediaofalabama.org/article/m-3046.

35. "Construction of the Capitol," North Carolina Historic Sites, https://historicsites.nc.gov/all-sites/north-carolina-state-capitol/history/construction-capitol.

36. Re enslaved population, see Bishir, "Proper Good Nice," 50. Town posted an announcement regarding the Clarendon Bridge (he owned fifty shares of the bridge company), which included his title "Superintendent"; e.g., *Fayetteville Observer*, May 8, 15, 22, 29, 30, 1823. *The Western Carolinian* (Salisbury), August 3, 1824, 3, announced the building of South Yadkin bridge by contractor Samuel Lemley. The Clarendon Bridge was still in existence and maintained in 1844, according to the *Fayetteville Observer*, November 6, 1844, 3; it was burned during the Civil War. Re Fayetteville and the Cape Fear region, see Robert B. Outland III, "Slavery, Work, and the Geography of the North Carolina Naval Stores Industry, 1835–1860," *Journal of Southern History* 62 (1996): 32–33.

37. "The Lafayette Trail," Distinctly Fayetteville, https://www.distinctlyfayettevillenc.com/things-to-do/museums-history/cultural-heritage-trails/lafayette-trails-lafayette/; "Lafayette in Fayetteville" (2016), Lafayette Society, https://www.lafayettesociety.org/the-lafayette-trail/. Apparently, Town had met Lafayette, with whom Morse was acquainted; "Art and Travel," Library of Congress, Samuel F. B. Morse Papers, 1793–1919, https://www.loc.gov/collections/samuel-morse-papers/articles-and-essays/collection-highlights/art-and-travel/.

38. Angela Lakwete, *Inventing the Cotton Gin: Machine and Myth in Antebellum America* (Johns Hopkins University Press, 2003). For an image of Whitney's armory bridge, see John Warner Barber, *Connecticut Historical Collections* (Barber/Willard, 1836), 220.

39. Robert P. Merges, *American Patent Law: A Business and Economic History* (Cambridge University Press, 2022), 67–70. Richard Sanders Allen, *Covered Bridges of the Northeast* (Greene, 1957; rev. ed., 1983), 14–16. Builder Theodore Burr's business difficulties apparently put an end to his career; Donald Wolf, *Crossing the Hudson: Historic Bridges and Tunnels of the River* (Rutgers University Press, 2010), 13–16.

40. Cooke, "Denial of Slavery in Management Studies," 1911. Yochai Benkler, *The Penguin and the Leviathan: How Cooperation Triumphs over Self-Interest* (Crown Business, 2011), 4–5, 30–35, ties Taylorism to evolutionism. Caitlin Rosenthal, *Accounting for Slavery: Masters and Management* (Harvard University Press, 2013), 199–205.

41. Mary N. Woods, *From Craft to Profession: Practice of Architecture in Nineteenth-Century America* (University of California Press, 1999) and Julia M. Truettner, *Aspirations for Excellence* (University of Michigan, 2003), 48. See also Bishir, "Proper Good Nice," 120–129. An example of how Town earned fees: James Hayward of Boston and Maine Railroad published a bid for proposals for two lattice bridges in Durham; "To Bridge Contractors," *Boston Post*, November 2, 1840, 3.

42. Cooke, "Denial of Slavery," 1895–1899. Rosenthal, *Accounting for Slavery*. Enslaved labor in the United States produced most of the cotton processed in Britain's factories. Sven Beckert, *Empire of Cotton: A Global History* (Knopf, 2015), 101–105, 115–116; Beckert, "Cotton and the Global Origins of Capitalism," 114–116. John Bellamy Foster, Hannah Holleman, and Brett Clark, "Marx and Slavery," *Monthly Review* 72 (2020): 96–117, https://monthlyreview.org/2020/07/01/marx-and-slavery/. "Forgotten," according to Beckert, "Cotton and the Global Origins of Capitalism," 114–116. Walter Johnson, *River of Dark Dreams: Slavery and Empire in the Cotton Kingdom* (Harvard University Press, 2013), 252–254. The impact on labor would be continued in skyscraper construction; Joanna Merwood-Salisbury, "The First Chicago School and the Ideology of the Skyscraper," in *Architecture and Capitalism: 1854 to the Present*, ed. Peggy Deamer (Routledge, 2014), 35–36.

43. "Oberst S. H. Long's hölzerne Brüke aus kleinem Holze," *Dingler's Polytechnisches Journal* 38 (1830): 175 (from *Mechanics' Magazine*). The German translator of the article explains God's annoyance with the slow builders of Noah's ark.

44. Dell Upton, "Traditional Timber Framing," in *Material Culture of the Wooden Age*, ed. Brooke Hindle (Sleepy Hollow, 1981), 87–91; Hubert G. H. Wilhelm, "Germans in Ohio," 60–78, and Gerlinde Leiding, "Germans in Texas," 362–378; both in *To Build in a New Land: Ethnic Landscapes in North America*, ed. Allen G. Noble (Johns Hopkins University Press, 1992).

45. This "wooden horizontal superstructure," per Gerstner, designed by Moncure Robinson and completed approximately half a year before Gerstner's visit, is the bridge later referred to by Chevalier, Culmann, and Couche; see chapters 3 and 4.

46. The points are summarized; Franz Anton Gerstner, *Berichte aus den Vereinigten Staaten von Nordamerica, über Eisenbahnen, Dampfschiffahrten, Banken und andere öffentliche Unternehmungen* (Melzer, 1839), 15–17.

47. Gerstner's first point contrasts with Chevalier's description of US canal building; Michel Chevalier, *Les voies de communication* (Levrault, 1837), 15. Re challenges of building quickly and cheaply, see Arthur L. Dunham, "How the First French Railways were Planned," *Journal of Economic History* 1 (1941): 12–25. In a rare case of published response, the accuracy of several Gerstner claims was discussed by Detmold and Moncure Robinson: Henz, "Urtheile amerikanischer Ingenieure über einige Eisenbahnfragen," *Verhandlungen des Vereins zur Beförderung des Gewerbfleißes* (Berlin) 21 (1842): 96–123. Neither engineer expressed strong agreement with first point. David Stevenson attributed low cost to "three causes": compensation for land was not necessary; "works are not executed in so substantial and costly a style"; "wood, which is the principal material used in their construction, is got at a very small cost"; David Stevenson, *Sketch of the Civil Engineering of North America* (John Weale, 1838), 253. British observers during the early 1850s enumerated similar factors; D.L. Burn "Genesis of American Engineering Competition, 1850–1870," *Economic History* 2 (1931): 292–311; Nathan Rosenberg, ed., *The American System of Manufactures* (Edinburgh University Press, 1969). Re railroad "as an actor and as a model" for other industries, see Michael J. Piore and Charles F. Sabel, *The Second Industrial Divide: Possibilities for Prosperity* (Basic, 1984), 65–68.

48. Gerstner, *Berichte*, 17.

49. Franz Anton Gerstner, *Die innern Communicationen der vereinigten Staaten von Nordamerika*, ed. L. Klein (Förster,1842), 2:261–262. *Fourth Annual Report of the Raleigh and Gaston Rail-Road Company* (Raleigh Register, 1840), 8; re slave labor, 15; Matt Stokes, "Raleigh and Gaston Railroad," NCPedia, January 2023, https://www.ncpedia.org/raleigh-and-gaston-railroad; Petersen, "North Carolina Railroad," 81–82. Catherine Bishir, "Location Book of the Raleigh and Gaston Railroad," *IA: Journal of Society of Industrial Archeology* 5 (1979); 49–59. Several stone piers still existed in 1979. See "slaves" in index of Frederick C. Gamst, ed., David J. Diephouse and John C. Decker, trans., Franz Anton Ritter von Gerstner, *Early American Railroads: Franz Anton Ritter von Gerstner's 'Die innern Communicationen' 1842–1843* (Stanford University Press, 1997), 842.

50. Merges, *American Patent Law*, 30–31, 45–46. Brock, "Truss-Bridges," series in *Engineering News* 9–10 (1882–1883); Howard Newlon, "Bridge Patents Issued by United States Patent Office, 1790–1873," manuscript, Smithsonian Institution NMAH, Division of Engineering, file: "Bridge History, General"; David Guise, *Abstracts & Chronology of American Truss Bridge Patents, 1817-1900* (Society for Industrial Archeology, 2009); Ross Thomson, *Structures of Change in the Mechanical Age* (Johns Hopkins University Press, 2009), 178.

51. Christine MacLeod, "The Paradoxes of Patenting: Invention and Diffusion in 18th- and 19th-Century Britain, France, and North America," *Technology and Culture* 32 (1991): 885, 899, and throughout.

52. Re Blanchard's original patents, including 3,131X (January 20, 1820), see Carolyn C. Cooper, *Shaping Invention: Thomas Blanchard's Machinery and Patent Management in Nineteenth-Century America* (Columbia University Press, 1991), 40–41, 81–84. Samuel Morse and Hezekiah Augur were developing a marble carving machine, which they abandoned because of Blanchard's patent; Paul J. Staiti, *Samuel F. B. Morse* (Cambridge University Press, 1989), 102.

53. Allen, *Covered Bridges of the Northeast*, 20–32. The description of the process "came from various things I had read, been told, or which gradually dawned on me"; Allen, letter to

Gregory Dreicer, February 18, 1992. Milton S. Graton, *Last of the Covered Bridge Builders* (Clifford-Nicol, 1978), 79–108. A conversation with bridge builder Will Truax, January 15, 2024, provided an alternative account. Todd Craig Radford, "Town Lattice Truss: An Appropriate Bridge Technology for Developing Countries" (PhD diss., MIT, 2010), 295. Doug Kramer, "Historic Construction Practices & Procedures: How they *Built* Our Covered Bridges," Second National Covered Bridge Conference (Dayton, Ohio, 2013). Rolling a bridge into place: "launch" in English; *lancer* or *rouler* in French; *vorwalzen* or *vorschieben* in German.

54. Stevenson, *Sketch*, 236. Julius Pollack, "Metallbrücke zur Uebersetzung des Royalkanals bei Dublin für die Dublin-Drogheda-Eisenbahn nach Town's nordamerikanischem Konstrukzions-system gebaut im Jahre 1845," *Allgemeine Bauzeitung* 13 (1848): 17, is similar; perhaps taken from Stevenson.

55. Stephen H. Long, *Description of Col. Long's Bridges, Together with a Series of Directions to Bridge Builders* (Brown, 1836), 25; also, S. H. Long, *Description of the Jackson Bridge, Together with Directions to Builders of Wooden or Frame Bridges* (Sands & Neilson, 1830), 21. Stevenson cites Long's *Description* in *Sketches*, 234. At the Royal Scottish Society of Arts, Stevenson gave similar, more detailed description of the process (which included construction of "substantial" scaffolding in riverbed); this description refers only to the Long bridge; *Transactions of the Royal Scottish Society of Arts* 2 (1844): 84 (read before the society, May 15, 1839); also Stevenson, "Notice Relative to Long's American Frame-Bridge," *Edinburgh Philosophical Journal* (April 1841): 297.

56. Fletcher and Snow, "History," 363, 399, seem to agree.

57. Charles E. Peterson, "Early American Prefabrication," *Gazette des Beaux-Arts* 33 (1948): 38–39. Re West Indies trade, see Charles E. Peterson, "Sawdust Trail: Annals of Sawmilling and the Lumber Trade from Virginia to Hawaii . . . to the Year 1860," *APT Bulletin* 5/2 (1973): 88–91.

58. Riedel der Ältere, "Nachricht von eisernen Brücken" (dated August 18, 1796), *Sammlung nützlicher Aufsätze und Nachrichten, die Baukunst Betreffend* (1797; 2d ed., Maurer, 1811), 120.

59. John Weiler, "Colonial Connections: Royal Engineers and Building Technology Transfer in the Nineteenth Century," *Construction History* 12 (1996): 11–12. Gilbert Herbert, *Pioneers of Prefabrication: The British Contribution in the Nineteenth Century* (Johns Hopkins University Press, 1978), 4–29.

60. Lounsbury, "Wild Melody," 196.

61. John Hall, *Series of Select and Original Modern Designs for Dwelling Houses of the Use of Carpenters and Builders Adapted to the Style of Building in the United States* (1840), cited in Peterson, "Early American Prefabrication," 41–45. See also, Charles E. Peterson, "Prefabs in the California Gold Rush, 1849," *Journal of Society of Architectural Historians* 24, no. 4 (1965): 318–324.

62. John Fitchen, *The New World Dutch Barn* (Syracuse University Press, 1968), 51. Rob Tarule, "Mortise and Tenon Timber Frame," in *Tools and Technologies: America's Wooden Age*, ed. Paul B. Kebabian and William C. Lipke (R.H. Fleming Museum, University of Vermont, 1979), 41–42, states: "A search for descriptions of heavy timber tool use in seventeenth-, eighteenth-, and nineteenth-century American and English books on architecture and the trades proved largely fruitless."

63. Duncan James, "Carpenters' Assembly Marks in Timber-Framed Buildings," *Vernacular Architecture* 49, no. 1 (2018): 3, 5–6, 26. Salzman, cited in R. J. Brown, *Timber-Framed Buildings of England* (Robert Hale, 1986), 40.

64. Henry Glassie, "Barns across Southern England: A Note on Transatlantic Comparison and Architectural Meanings," *Pioneer America* 7 (1975): 16. Thomas J. Schlereth, "The New York Artisan in the Early Republic: A Portrait from Graphic Evidence, 1787–1853," *Material Culture* 20 (1989): 11.

65. Fitchen, *New World Dutch Barn*, 64, 134.

66. Fitchen, *New World Dutch Barn*, 60; 66, notes that in barns, unlike most structures, the frame remains visible on the interior. This is true also for covered bridges. Cheap labor and a lack of complex machinery sparked creativity in Ireland, a hot spot for lattice bridge design; Dermot W. O'Dwyer and Ronald Cox, "Early Irish Railway Construction," in *Studies in Construction History: Proceedings of the Second Construction History Society Conference*, ed. James Campbell et al. (Construction History Society, 2015), 121.

67. Theodore Burr described the building process for McCall's Ferry, PA, road bridge, 1814. This deck-stiffened arch bridge (see chapter 2) was built on scaffolding on floats in the Susquehanna River. Each half, built separately, was joined by sliding and rolling the floats on the river surface, which had risen and frozen during construction. Between forty and 120 men worked on any given day during erection: "Had it not been for the friendship of these people, I almost doubt whether I should ever have effected the object." "Liquor [was] handed around in great abundance"; Theodore Burr, in *Niles Weekly Register* 19 (1815–1816): 200–202, reprinted in Charles C. Sanford, *Quest for America, 1810–24* (Anchor, 1964), 427. Re Burr bridge, whose timbers were prepared in advance and assembled piece by piece on a temporary bridge, see John C. Trautwine, "Description of Viaduct Near Peters' Island[, Pennsylvania]," *Journal of the Franklin Institute* n.s., 14 (1834): 76.

68. David T. Yeomans, *The Trussed Roof: Its History and Development* (Scolar, 1992), 133. Apparently, the original white pine roof structure was replaced or reinforced with steel, 1906–1907; US Department of Interior, National Park Service, National Register of Historic Places, Inventory Nomination Form, "Center Church" (June 2, 1970), 3, and unnumbered page.

69. Town's Center Church roof truss seems to be based on a design in a builder's manual by Asher Benjamin, under whom Town may have learned to build; Yeomans, *Trussed Roof*, 128–135.

70. Reported by Henry Howe, who reputedly knew Town; George Dudley Seymour, "Ithiel Town: Architect, Engineer, Art-Collector, 1784–1844," manuscript, 1, 10–11, Yale University Library, George Dudley Seymour Papers, Group 442, Series 5, Box 105, Folder 1504. Seymour, letter to architectural historian Talbot Hamlin, January 27, 1934, states that papers found in Center Church reveal that Asher Benjamin was "preceptor" of Town and project partner, Isaac Damon, who "was paid forty dollars for a design for Center Church." Damon withdrew, leaving Town in charge. Also, letters from Seymour to Dumas Malone (May 10, 1934; December 12, 1934), Yale University Library, Seymour Papers, box 105, folder 1500. In Frank C. Deering, Saco, Maine, to Seymour, May 8, 1928, Yale University Library, Seymour Collection, group 442, series 5, box 105, folder 1501; it is reported that the spire was built within the tower

and "raised by windlass and tackle." Re this and two other cases, see James Wickham Roe, "Early New Haven Inventors," in Richard Shelton Kirby, ed., *Inventors and Engineers in Old New Haven* (Books for Libraries, 1969; original,1939), 12.

71. "New-York Theatre," *Arkansas Gazette*, December 12, 1826, 3, reports the use of lattice and cites Fayetteville and Springfield bridges. Walls were buttressed every 14 feet; Town worked with construction contractors Seth Geer and A. Riley and mason Joseph Tucker. Theodore Junior Shank, "Bowery Theatre, 1826–1836" (PhD diss., Stanford University, 1956); Robert Amell, "In and around the Bowery Theatre," Manhattan Unlocked, March 12, 2011, http://manhattanunlocked.blogspot.com/2011/03/in-and-around-bowery-theatre.html, and Daniel S. Levy, "New York City: The Life and Times of the Bowery Theater," OUP Blog, March 4, 2022, https://blog.oup.com/2022/03/new-york-city-the-life-and-times-of-the-bowery-theater.

72. Yale University Library, Baldwin Family Papers, box 15, folder 178. Roger S. Baldwin, Yale graduate and lawyer, was eventually elected Connecticut governor and US senator.

73. Thirteen days later, Town wrote to Baldwin, "We have about one third part of the Bridge at Cheraw raised and the workmen are going on rapidly with the raising of the residence." Town, Fayetteville, January 29, 1824, letter to Roger S. Baldwin; in a letter of March 8, 1824, Fayetteville, also addressed to Baldwin, Town states that the bridge is being clad and announces: "The new bridge is now over the river and is *certainly* the most perfect piece of timber work I have ever seen for a length of 400 ft.—as straight and as true as a piece of Cabinet work." Yale University Library, Baldwin Papers, box 15, folder 178. "Raising" can have a general meaning; e.g., William Oliver, *Eight Months in Illinois; With Information to Emigrants* (Bruce, Easton, and Kennedy, 1843; reprint, University Microfilms, 1966), 129, re building of a log house.

74. Town, New York, June 5, 1837, to Jon[athon] Livermore; letter, sold on ebay, 2017. The builder referred to is probably Joseph Hayward; "The Mystery of the Lost High Bridge Watch," Croton History & Mysteries, November 22, 2013, https://crotonhistory.org/2013/11/22/the-mystery-of-the-lost-high-bridge-watch/.

75. Joseph D. Conwill, "Who Built the Covered Bridges? Social Organization of Building," Best Practices, Care and Repair of Covered Bridges National Conference, University of Vermont, Burlington, Vermont, 2003.

76. Joseph D. Conwill, "Builders and Practices," in Justin Christianson and Christopher H. Marston, eds., *Covered Bridges* (HAER, 2015), 165–68.

77. Town (1825), 2; Town (1831), 5

78. MacLeod, "Paradoxes," 898.

79. Christine MacLeod, "Strategies for Innovation: The Diffusion of New Technology in Nineteenth-Century British Industry," *Economic History Review* 45 (1992): 295–296, 300–304.

80. Town (1939), back cover, in the library of the École nationale des ponts and chaussées, states, "Terms for Patent right, one dollar per foot, for the length of the bridge, if applied for and paid before commencing; otherwise, two dollars per foot will be charged, as indemnity for the risk of bad construction." Letter, William Elliot to Town, dated March 9, 1829, Washington: "I wrote

[?] a letter stating that your terms for the patent right were about 1 doll. for every running foot in length: but no right less than 100 dolls"; New York Historical Society: Misc. MSS Elliott. "My price (the lowest) is $1 per foot for the whole length of the trusses of the bridge"; Ithiel Town, letter to Jonathan Livermore, July 23, 1831, New York, Marjorie and John Geiser II Collection, Southeastern Architectural Archive, Special Collections Division, Tulane University Libraries. Gerstner, *Innern Communicationen*, 1:202, reports one dollar per foot. W[illia]m Polke, "Report of the Michigan Road Commissioner," *Journal of House of Representatives of the State of Indiana* (December 17, 1832): appendix A, 4, states the patent fee was "not to exceed 50 cents per foot." Profits were also generated by tolls; Fred Kniffen, "The American Covered Bridge," *Geographical Review* (1951): 115. Re licensing, see MacLeod, "Paradoxes," 901.

81. Town (1821), 8. Town (1839), 12, reduced the claim to one-third.

82. Robert Fletcher and J. P. Snow, "History of Development of Wooden Bridges," *ASCE Transactions* 99 (1934): 335. The lattice seems to have required less wood than competing systems: "On the Pottsville and Sunbury Railway, in Pennsylvania, the wood for small truss bridges, for crossing roads is 12 dollars per 1000 feet super [*sic*] and inch thick, is equivalent to two loads of timber in England. That for trellis work costs only 8 1/2 dollars"; "On Trellis Bridges," *Civil Engineers and Architects Journal* 3 (1840): 125–126, pl. 8 [adapted from *Revue générale de l'architecture et des travaux publics* 1 (1840): 33–42, pl. 3], per Moncure Robinson. Michel Chevalier, *Histoire et description des voies de communication aux Etats-Unis et des travaux d'art qui en dépendent* (Gosselin, 1840–41), 2: 566, reported that for ordinary "truss bridges" for roads, wood cost 27 fr. 12 c. per cubic meter, while the lattice cost 19 fr. 50 c.

83. According to the National Archives, all records of assignees were lost in 1836 Patent Office fire, along with original materials relating to Town's patents. On Town bridge builders, see Richard Sanders Allen publications; Fletcher and Snow, "History," 335–337; Thomas L. French and Edward L. French, "Horace King, Bridge Builder," *Alabama Heritage* 11 (1989): 34–47; Joseph D. Conwill et al., "Brown Bridge," Written Historical and Descriptive Data, HAER, National Park Service, US Department of Interior, 2002, from Prints and Photographs Division, Library of Congress (HAER VT-28). Town (1825), 2, crankily recounts story of a builder who tried to construct a lattice bridge without Town's instructions. Town (1831), 3, emphasizes that receiving guidance from Town was "indispensably necessary." "I wish you to be very particular to secure your bridge well with braces both to keep it from leaning sideways and to keep the top part of it straight so as that nothing can ever make it *crooked* or *to lean*, let every thing possible be done to make it secure, where it is once done, so that it may be a good bridge long after we are all *dead* who have any thing to do with it," he wrote to a builder; Town, New York, 13 November 1837, to Jonathon Livermore; letter, sold on ebay, viewable at https://www.worthpoint.com/worthopedia/ithiel-town-architect-engineer-1837-1899544022.

84. An exception is a chapter devoted to construction in Romain Morandière, *Traité de la construction des ponts* (Dunod, 1874 [text]; 1878 [plates]), 1837–1872.

85. "All the difficult and costly scaffolding and lifting machines usually required in bridgebuilding are no longer needed"; Chas. F. Zimpel, *Das Eisenbahnwesen von Nordamerika, England und anderen Ländern, practisch und populär dargestellt* (Förster, 1840), 102; he made similar claim for Long bridge (98–99). It was reported in Pollack, "Metallbrücke," 17, that "all the parts of this kind of bridge

are assembled beforehand at the work site and only then systematically put up at its destination, with the use of suspension or pile scaffolding, according to circumstances."

86. On Haupt, see chapter 4. Herman Haupt, *General Theory of Bridge Construction* (Appleton, 1851), 148–149. In *American Railroad Journal* (1840): 196, he cited "its beautiful simplicity, light appearance, and especially its economy." In his [1841] brochure, *Hints on Bridge Construction, by an Engineer*, 9, Haupt stated: "It is certain that they possess great simplicity—they are of all others the most easily framed, and the most economical."

87. Michel Chevalier, "Progrès à obtenir dans l'exploitation des chemins de fer en France," *Journal des Économistes* 12 (1880): 174–175. This was the last article written by Chevalier. In 1880, he still favorably compared the low-cost strategies of the United States to those of France.

88. Dennis Hart Mahan, *Elementary Treatise of Civil Engineering*, new [4th?] ed. (Wiley and Putnam,1846), 174. Fr. Schubert et al., *Vademecum für den praktischen Ingenieur und Baumeister* (Müller, 1850), 116.

89. Mansfield Merriman and Henry S. Jacoby, *Text-Book on Roofs and Bridges*, Part III: *Bridge Design*, 5th ed. (Wiley, 1912; 1st ed. 1888), 7. The authors were professors at Lehigh and Cornell.

90. "Machine," *Encyclopædia Britannica*, https://www.britannica.com/technology/machine.

91. James Barton, "On the Economic Distribution of Material in the Sides, or Vertical Portion, of the Wrought-Iron Beams," *Transactions of the Institution of Civil Engineers* 14 (1854–1855): 489.

92. J. H. Garella, "Note sur deux ponts provisoires en charpente construits dans le système américain de M. Town," *Annales des ponts et chaussées* 2/2 (1842): 373–74.

93. George Willoughby Hemans, "Description of a Horizontal Lattice Bridge, Formed of Wrought Iron Bars, 144 Feet Span, Over the Royal Canal, on the Dublin and Drogheda Railway," *Institution of Civil Engineers of Ireland: Transactions* 2 (1848): 101. Lentze, "Bemerkungen über die grosseren Brückenbauwerke in Frankreich, England, und Irland auf ein Reise in Winter 1844/45 gesammelt," *Verhandlungen des Vereins zur Beförderung des Gewerbfleißes* (Berlin) (1846): 107.

94. Hemans, "Description of a Horizontal Lattice Bridge," 101–103.

95. Hermann Lohse, "Notizen über einige neuere Brücken Englands," *Zeitschrift für Bauwesen* 7 (1857): 215.

96. Railroad shops for construction were at Pottstown, PA, around 1844. Osborne, "Diary," 3:80, 82.

97. Targé, "Der Saale-Uebergang der Magdeburg-Leipziger-Eisenbahn bei Grizehna unfern Calbe an der Salle," *Zeitschrift für Bauwesen* 3 (1853): 479–484, pls. 69–70; Helling, "Ueber die Aufstellung der neuen eisernen Gitterbrücke über die Saale in der Magdeburg-Leipziger Eisenbahn," *Zeitschrift für Bauwesen* 4 (1854): 168–171, pl. D. Another early launch may have been the 1852 lattice bridge over the River Suir at Aherlow on the Waterford and Limerick Railway, Ireland.

98. Herman Haupt, *Reminiscences of General Herman Haupt* (Wright & Joys, 1901), 186. "Colonel E. C. Smeed and His Work," *Engineering Magazine* 4 (1892): 115–118.

99. John H. White, "James Millholland and Early American Railroad Engineering," *US National Museum Bulletin* 252, paper 69 (1967): 8–9, 32–33. This bridge was constructed in Baltimore,

hung under a truss supported on each end by a railroad flatbed car, transported to the site, driven over the existing bridge (which was then removed), and lowered into place: "The whole interruption to the traffic not exceeding two hours."

100. E. Deligny, "Notice sur la chute du pont du chemin de fer de Paris à Saint-Germain, à Asnières, incendié le 25 février 1848, et sur le pont provisoire en charpente établi par M. Eugène Flachat, ingénieur en chef," *Memoires et comptes rendus des travaux de la Société des Ingénieurs civils* 1 (1848): 85–86; Bernard Marrey, *Les ponts modernes: 18e-19e siècles* (Picard, 1990), 141–43. Émile Clapeyron was also involved in the design; Henri Mathieu, Léon Molinos, and Henri de Dion did the calculations. Town and Long bridges were considered; the engineers decided to build a Long-type bridge with much reinforcing.

101. Lagallisserie and Vaudrey, "Passerelle américaine provisoire," *Nouvelles Annales de la Construction* 4 (1858): 78–79, pls. 29–30; "Reconstruction du pont Saint-Michel, à Paris: Passerelle américaine provisoire et chantier à étages pour la fabrication du béton," *Ingénieur* n.s., 2 (1858): 1–3, pl. 1; Morandière, *Traité*, 321, 582–583, pl. 147. Regarding another provisional Town bridge, see Aymard, "Notice sur les travaux exécutés au pont Saint-Esprit pour la construction d'une arche marinière," *Annales des ponts et chaussées* 3/2 (1859): 1–48, pls. 157–158. A provisional lattice bridge was built in Warsaw in 1860; photograph, École nationale des ponts et chaussées, Ph. 52, A-4. Sean R. Weiss, "Engineering, Photography, and the Construction of Modern Paris, 1857–1911" (PhD diss., CUNY, 2013), 29–30, 56, 108; Elizabeth Anne McCuley, *Industrial Madness: Commercial Photography in Paris, 1848–1871* (Yale University Press, 1994), 195–232.

102. Because of a typographical error in Mehrtens, some histories state 1858.

103. Mehrtens, "Zur Baugeschichte der alten Eisenbahnbrücken bei Dirschau und Marienburg," *Zeitschrift für Bauwesen* 43 (1893): 97–122.

104. Émile With, *Les accidents sur les chemins de fer, leurs causes, les règles à suivre pour les éviter* (Mallet-Bachelier, 1854), 24; quote from translation, Emile With, *Railroad Accidents: Their Causes and the Means of Preventing Them* (Little, Brown, 1856), 29.

105. [Max] Becker, "Brücke über die Kinzig bei Offenburg und ihre Zerstörung durch das Hochwasser vom 1. August 1851," *Allgemeine Bauzeitung* 17 (1852): 7. (This is not the engineer Max Becker, who emigrated to the United States in 1848.)

106. Ruppert in Becker, "Die Eiserne Brücke über die Kinzig bei Offenburg," *Allgemeine Bauzeitung* 18 (1853): 182. Ruppert comments on Hannover tests. Mohnié in his articles (see chapter 4, note 85) about his lattice bridge rebuts Culmann. Tom F. Peters suggests that Culmann, Ruppert, and Becker may have been in Karlsruhe at the same time and knew each other. In any case, they would have known of each others' work. Re debates between administrator/theorists and designers, see Peter Lundgreen, "Engineering Education in Europe and U.S.A., 1750–1930: The Rise to Dominance of School Culture and the Engineering Professions," *Annals of Science* 47, no. 1 (1989): 40–41.

107. Becker, "Eiserne Brücke," 179. Re the arch bridge, see Ulrich Boeyng, "Die Brücke über den Erlengraben bei Ettlingen—ein Denkmal aus der Frühzeit der Großherzoglich Badischen Eisenbahn," *Denkmalpflege in Baden-Württemberg* 18 (1989), 148–150.

108. Ruppert in Becker, "Eiserne Brücke," 181. Re old rails used in building construction in the United States, see Charles E. Peterson, "Inventing the I-Beam: Richard Turner, Cooper & Hewitt and Others," in *Technology of Historic American Buildings: Studies of the Materials, Craft Processes, and the Mechanization of Building Construction*, ed. H. Ward Jandl (Foundation for Preservation Technology/APT, 1983), 74–77. Rails (i.e., profiled members) could take tension and compression better than planks.

109. Ruppert in Becker, "Eiserne Brücke," 181–182, a rare account of a design process. Also, Ruppert, "Eiserne Gitterbrücke über die Kinzig bei Offenburg," manuscript, [1854] (including French translation), Bibliothèque, École nationale des ponts et chaussées, ms. 328.

110. "Gitterbrücke über die Kinzig," *Illustrirte Zeitung*, 196. Vielen Dank an Helmut Hilz!

111. Émile With, *Nouveau manuel complet de la construction de chemins de fer* (Roret, 1857), 118.

112. See Becker, above, notes 107–109; "Beobachtungen an der neuen Kinzig-Gitterbrücke bei Offenburg," *Organ für die Forschritte des Eisnbahnwesens in Technischer Beziehung* 10 (1855): 68; "Die neue Kinzigbrücke bei Offenburg," *Eisenbahnzeitung* 11 (1853): 97–99; "Die Kinzig-Gitterbrücke bei Offenburg," *Eisenbahnzeitung* 13 (1855): 13, 27; "Die neue Kinzigbrücke bie Offenburg," *Polytechnisches Centralblatt* (Leipzig) NF7 (1853): 1427–1434, pl. 23; "Gitterbrücke über die Kinzig bei Offenburg auf der großherzoglich badischen Staatseisenbahn," *Illustrirte Zeitung* (Leipzig), March 25, 1854, 196–198; Edmund Heusinger von Waldegg, "Notizen auf einer Reise durch Süddeutschland und die Schweiz über die Tannus und Hessische-Ludwigs-Eisenbahn, die Badische-Bayerische, und Würtembergische Staats-Eisenbahnen," *Organ für die Forschritte des Eisnbahnwesens in Technischer Beziehung* 9 (1854): 131–143, 191–207, pls. 13–14; Boulangé, "Notes recueillies," 65; Charles Couche, "Travaux d'art des chemins de fer d'Allemagne," *Annales des mines 5/5* (1854): 350–354; "Pont d'Offenbourg," *Annales des travaux publics de Belgique* 13 (1854–55): 455–458. See also Ernst Werner, "Die Eisenbahnbrücke von 1853 über die Kinzig bei Offenburg," *Bauverwaltung* 46 (1973): 480–483. Ruppert later developed U-shaped lattice-bars in order to increase their rigidity.

113. Angela Jursitzka and Helmut Pawelka, *Carl von Etzel* (Tyrolia-Verlag, 2017). In 1836–1837 Etzel traveled through England to study railways; he worked with Förster on *Allgemeine Bauzeitung* during the early 1840s. Later, with Ludwig Klein, he was a director of *Eisenbahnzeitung*. An early article was "Ueber die Konstrukzion von Fachwerkwänden," *Allgemeine Bauzeitung* 6 (1841): 337–342.

114. The decision-making process for the Sitter bridge (the first of three built) was reported in Couche, "Travaux d'art," 390.

115. Re the Thur, temporary spans were similar to Fink bridges (see chap. 7); they "looked like American bridges." "Uebersicht der im Bau begriffenen Eisenbahnen in der Schweiz." *Allgemeine Bauzeitung* 21 (1856): 135. Re launches, see Wilhelm Nordling, *Mémoire sur les piles en charpente métallique des grands viaducs* (Dunod, 1864; appendix, 1865), 128–132 (also published as "Mémoire sur les piles en charpente métallique des grands viaducs," *Annales des ponts et chaussées* 4/2 [1864]: 1–126; see chapter 8, note 10). Peter Röllin and Daniel Studer, "St. Gallen," in *INSA: Inventar der neueren Schweizer Architektur 1850–1920*, ed. Gesellschaft für Schweizerische Kunstgeschichte (Orell Füssli, 1996), 8:59–61.

116. "Uebersicht der im Bau begriffenen Eisenbahnen in der Schweiz," *Allgemeine Bauzeitung* 21 (1856): 2–7, 133–139, pls. 28–29; quotation, 7. The excerpt appeared as "Vorschlag zum Hinüberrollen der Gitterwände ohne Gerüste," *Polytechnisches Centralblatt* (Leipzig) 22 (1856): 1100–1107, pl. 25.

117. Merian, "Gitterbrücke über die Aare bei Bern," *Eisenbahnzeitung* 16 (1859): 197–199, pls. 6–7. In 1857, concerning a lattice bridge built under Etzel over Emme at Berthoud, Switzerland, it was reported, "25 to 30 men with levers were enough to put the bridge in place"; "Nouvelles Diverses," *Le Constitutionnel: Journal du Commerce, Politique and Littéraire*, March 24, 1857, 2. The weight was 18,000 Zentner (18,000 x 100 kilograms).

118. *Exposition universelle à Londres en 1862: Notices sur les modèles, cartes et dessins relatifs aux travaux publics réunis par les soins du ministère de l'agriculture, du commerce et des travaux publics* (Thunot, 1862), 174.

119. Nordling was likely responsible for the design. Etzel may have been involved in the choice of bridge type. Between abutments, the bridge was about 330 meters long. It rested on six piers 48.80 meters apart; the tallest was 80 meters high, from foundation to railroad track. Each pier's base was stone; the upper 44 meters of the piers consisted of towers composed of twelve cast-iron columns with wrought-iron cross bracing and latticework. Grandfey Viaduct, selected sources: *Exposition universelle à Londres en 1862*, 168–175; C. A. Oppermann, "Viaduc de Fribourg, sur la Sarine," *Nouvelles Annales de la Construction* 9 (1863): 161–163, pls. 45–48; Rühlmann, "Brücke und Viaduct über die Saane bei Freiburg in der Schweiz," *Zeitschrift des Architecten- und Ingenieur-Vereins für das Königreich Hannover* 8 (1862): 161–164, pl. 231; P. Fink, "Verhandlung des Vereins" [Eisenbahnbrücke über die Saane bei Freiburg in der Schweiz], *Zeitschrift des Oesterreichischen Ingenieur-Vereines* 14 (1862): 202; B. Dupuis, "Viaduc de la Sarine à Fribourg (Suisse)," *L'Illustration*, June 21, 1862, 392, 394; "International Exhibition: Class X, Section A : Juror's Report [Bridges], *Journal of the Franklin Institute* 3rd ser., 45 (1863): 159–165, 228–233 [from *Mechanics' Magazine*, December 1862]; "Übersicht der in der Industrieaustellung zu London im Jahr 1862 durch Zeichnungen oder Modelle ausgestellten größeren Brücken der Neuzeit," *Organ für die Forschritte des Eisnbahnwesens in Technischer Beziehung* 18 (1863): 260–69; Morandière, *Traité*, 1101–1103; Schweizerischer Ingenieur- und Architekten Verein, *Die Eisenbahnbrücke über die Saane bei Freiburg auf der Eisenbahn von Lausanne über Freiburg nach Bern* (Meyer & Zeller, 1867); Adolphe Bühler, "La reconstruction du Viaduc de Grandfey des chemins de fer fédéraux suisses," offprint from *L'entreprise: journal officiel de la société suisse des entrepreneurs* 15–20 (Zurich, 1928). The bridge was subject of a display at International Exhibition of 1862, London. Grandfey pier elevation drawings and a brief description in "Messrs. Schneider and Co.'s Works at Creusot," *The Engineer* 23 (1867): 94–95. Also, "Ferdinand Mathieu," https://fr.wikipedia.org/wiki/Ferdinand_Mathieu.

120. Louis Devance, *Gustave Eiffel: La construction d'une carière d'ingénieur* (Éditions universitaires de Dijon, 2016), 126–27. During the Sioule Viaduct construction, Eiffel patented an improvement to the launching mechanism.

121. Marco Rosignoli, *Bridge Launching* (Telford, 2002), 2, 19; Marco Rosignoli, *Bridge Launching*, 2d ed. (Telford, 2014), 7–11.

122. This was the method used in the United States, notably on the Kentucky River bridge near Dixville (C. Shaler Smith, who worked under Bouscaren, 1877).

123. The equipment used to launch the Sioule Viaduct was in use twelve years later; Théophile Seyrig, "Different Modes of Erecting Iron Bridges," *Institution of Civil Engineers: Minutes of Proceedings* 63 (1881): 199.

124. *Exposition universelle à Paris en 1867: Notices sur le modèles, cartes et dessins relatifs aux travaux réunis par les soins du ministère de l'agriculture, du commerce et des travaux publics* (Thunot, 1867), 346, and Ach. Ferary, "Viaduc de la Cère," *L'Illustration*, August 19, 1865, 121–122. Lepère, "Note sur la Construction du Viaduc de la Cère," *Portefeuille des conducteurs des ponts et chaussées et des garde-mines* 8 (1867): 17–23, pls. 9–14.

125. "Herstellung, Aufstellung und Transport der Rheinbrücke bei Kehl," *Eisenbahnzeitung* 19 (1861): 56, 13–14, pls. 1–2.

126. On Warsaw bridge, see Barbara Rymsza et al., "Kierbedź Bridge: A History of the First Permanent Bridge Across the Vistula River in Warsaw, Poland," *Icon* 23 (2017): 145–166; "Most Kierbedzia" (Kierbedzia Bridge), *Wikiwand*, https://www.wikiwand.com/pl/Most_Kierbedzia.

127. Morandière, *Traité*, 1845. For construction of the Danube bridge—Stadlau (Carl Ruppert, 1870), built by Schneider (Le Creusot) with engineer C. F. Mathieu—the first two spans "were pushed out into their position by the power of eight men working gearing that multiplied power 250 times; the spans, weighing 417 tons apiece, or together, 834 tons, were moved." Steam power was employed to move the final spans. W. Howard White, "An Account of the Erection of a Bridge over the Danube, near Vienna," *ASCE Transactions* 2 (1874): 291–293, also mentions Freiburg Viaduct. See Hermann Voigtländer's photographs in Eisenbahnarchiv, Technisches Museum Vienna and ETH-Bibliothek Zürich, Bildarchiv, Ans 07160-022-AL.

128. Seyrig, "Different Modes," 169.

129. Cézanne, "Notice sur quelques ponts métalliques des chemins de fer russes," *Annales des ponts et chaussées* 4/2 (1864): 235, 260. R. Merian, "Gitterbrücke über die Aare bei Bern," *Eisenbahnzeitung* 16 (1859): 198.

130. According to engineer Henry N. Maynard, who was involved in Crumlin construction, in Seyrig, "Different Modes," discussion, 190; disputed by Seyrig, 198–199.

131. *Monmouthshire Merlin* (December 1853), text reproduced in https://crumlinviaduct.co.uk/history%20view.htm. "Safety measures were put in place after this accident."

132. "Worker Fatalities," *Commonly Used Statistics*, US Department of Labor, Occupational Health and Safety Administration, https://www.osha.gov/data/commonstats.

133. Seyrig, "Different Modes," 200.

134. See chapter 7 and Mark Aldrich, *Death Rode the Rails: American Railroad Accidents and Safety, 1828–1965* (Johns Hopkins University Press, 2006).

135. R. A. Church, "Nineteenth-Century Clock Technology in Britain, the United States, and Switzerland," *Economic History Review* 28 (1975): 627, 629. Re traditional concept of knowledge divide, see Lissa Roberts and Simon Schaffer, "Preface," in *The Mindful Hand: Inquiry and Invention from Late Renaissance to Early Industrialization*, ed. Lissa Roberts et al. (Koninklijke Nederlandse Akademie van Wetenschappen, 2007), xiii–xxvii.

136. Nordling, *Mémoire*, 127. Re pier erection time, see "Civil Engineering in the Paris Exhibition: Iron Viaducts of La Cere, Busseau d'Ahun and du Midi," *The Engineer* 23 (1867): 386.

137. Franz Kreuter, "Wilhelm Pressel und seine Bedeutung für die Eisenbahntechnik," *Organ für die Forschritte des Eisnbahnwesens in Technischer Beziehung* 39 (1902): 291–296. Re companies based in France, see Bertrand Lemoine, *Gustave Eiffel* (Hazan, 1984), 32, 34, 36–37, 68, 70. Mathieu was on the Sitter and Grandfey teams and Moreaux worked within the Busseau d'Ahun, Cère, Bouble, and Bellon teams (see chapter 8). See also Dominique Barjot, "Innovation et travaux publics en France (1840–1939)," *Histoire, Économie et Société* 8 (1989): 403–414 ; [Geneviève Hermann,] *Fives: 200 Years of Industrial Revolutions* (Paris: Éditions de l'Étagère, 2013). On Moreaux, see https://fr.wikipedia.org/wiki/F%C3%A9lix_Moreaux; re Mathieu, see https://fr.wikipedia.org/wiki/Ferdinand_Mathieu; re Nouguier, see https://en.wikipedia.org/wiki/%C3%89mile_Nouguier. Rondo E. Cameron, *France and the Economic Development of Europe, 1800–1914*, in *Evolution of International Business, 1800–1945* (Routledge, 1961), 4:274–282.

138. The bridges were transported on 170 ships (many steam-powered); only a small number arrived with shipments intact; Cézanne, "Notice sur quelques ponts métalliques."

139. Full passage is worth reading; Cézanne, "Notice sur quelques ponts métalliques," 261.

140. "Crystal Palace," https://en.wikipedia.org/wiki/The_Crystal_Palace.

141. The public and jury seemed to have been charmed by models rather than the ideas and techniques they represented. Alfred Bommart, "Travaux publics," in *Exposition universelle de Londres de 1862: Rapports des membres de la section français du jury international sur l'ensemble de l'exposition*, ed. Michel Chevalier (Chaix, 1862), 3: 276–277. Apparently only the United Kingdom had a significant civil engineering display; Bommart found representation at 1855 and 1862 expositions improved. Jeffrey Auerbach, *The Great Exhibition of 1851: A Nation on Display* (Yale University Press, 1999), 91–118, and William J. Ashworth, "Industry and Transport," in *A Companion to Nineteenth-Century Britain*, ed. Chris Williams (Blackwell, 2004), 223.

142. John E. Findling, "America at the Great Exhibition," in *Die Weltaustellung von 1851 und ihre Folgen/The Great Exhibition and Its Legacy*, ed. John R. Davis et al. (Saur, 2002), 197–204; John R. Davis, *The Great Exhibition* (Sutton, 1999), 80–93; 157–59; Auerbach, *Great Exhibition*, 122–123; Michael Zakim, "Importing the Crystal Palace," in *American Capitalism*, ed. Sven Beckert and Christine Desan (Columbia University Press, 2018), 341. The widely known failure of a Rider bridge in 1850 may have lingered in people's minds; a firm in England would produce many for export. *A Guide to the Great Exhibition* (Routledge, 1851), 126; Hermione Hobhouse, *The Crystal Palace and the Great Exhibition: Science, Art and Productive Industry* (Athlone, 2002), 67; Dario Gasparini and David Simmons, "American Truss Bridge Connections in the 19th Century I: 1829–1850," *Journal of Performance of Constructed Facilities* 11 (1997): 125–27; Victor C. Darnell, "Pioneering Trusses of Nathaniel Rider," *Construction History* 7 (1991): 69–81. See Nathan Rosenberg, ed., *American System of Manufactures* (Edinburgh University Press, 1969), for introduction and reports of Joseph Whitworth and George Wallis.

143. Rosenberg, *American System*, 5–6.

144. The main source of my description is Peters, *Time Is Money*, 157–183, and Tom F. Peters, "Some Structural Problems Encountered in the Building of the Crystal Palace of 1851," in

Robert Thorne, ed., *Structural Iron and Steel, 1850–1900* (Ashgate, 2000), 1–15. See also Ernst Werner, *Der Krystallpalast zu London 1851* (Werner-Verlag, 1970), 50–53; John R. Davis, *Great Exhibition* (Sutton, 1999), 86–88, and Charles Fowler Jr., *Illustrated Exhibitor* (Cassell, [1851]), cited in Peters, *Time Is Money*, 258. Re thermal expansion, see Nicholas A. Bill, "Design and Construction Deficiencies of Timber Railway Bridges," *Proceedings of the Institution of Civil Engineers, Engineering History and Heritage* 167 (2014): 184, 188.

145. "'High Bridge,' Portage, New York," *Civil Engineer and Architects Journal*, 65–66, pl. 8 (*Canadian Journal*, 68–69, and plate). See chapter 8.

146. John W. Stamper, "London's Crystal Palace and Its Decorative Iron Construction," in Paul Dobraszczyk and Peter Sealy, *Function and Fantasy: Iron Architecture in the Long Nineteenth Century* (Routledge, 2016), 25–48; Wayne Charney, "Materializing through the Skylight," *Oz* 20 (1998): 48–53.

147. Re modularity, see "Haut Pont de Partage[sic] dans l'État de New-York," *Nouvelles Annales de la Construction* 1 (1855): 3, pl. 24. Re standardization, see JoAnne Yates and Craig N. Murphy, *Engineering Rules: Global Standard Setting Since 1880* (Johns Hopkins University Press, 2019), 19–51.

148. David A. Hounshell, *From American System to Mass Production 1800–1932* (Johns Hopkins University Press, 1984), 15–17, 331–336.

149. Re development of building parts, see chapter 2; Merrit Roe Smith, "Military Entrepreneurship," in *Yankee Enterprise*, ed. Otto Mayr and Robert C. Post (Smithsonian, 1981).

150. Re Portsmouth Bridge, which was criticized as too flexible, see Howard Douglas, *Essay on the Principles and Construction of Military Bridges, and Passage of Rivers in Military Operations* (Egerton, 1816; 2nd ed., Boone, 1832), 399–400. (Unless otherwise indicated, page numbers cited are from the 2nd ed.; the identical text appeared in the first edition (1816). See also Chevalier, *Histoire*, 2: 556; Fletcher and Snow, "History," 325–326; Llewellyn Nathaniel Edwards, "Evolution of Early American Bridges," *Transactions of Newcomen Society* 13 (1933): 102; George B. Pease, "Timothy Palmer, Bridge Builder," *Essex Institute Historical Collections Quarterly* 83 (1947): 102–103; Frank Griggs Jr., "Piscataqua Bridge," *Structure* 20 (August 2013): 28–30. Mahan, *Elementary Treatise of Civil Engineering*, 3rd ed. (Wiley and Putnam, 1839), 167. An article announcing opening of one of Town's earliest lattice bridges claimed, "When one piece of timber shall become decayed, it may be conveniently taken out and replaced with new"; "Cheraw Bridge," *Niles Weekly Register*, July 17, 1824, 329, reprint in *Cheraw Intelligencer*. See also William Scarth Moorsom, in J.R. Mosse, "American Timber Bridges," *Institution of Civil Engineers: Transactions* 22 (1862–63): 325.

151. Chevalier, *Histoire*, 2:566.

152. Conwill et al., "Brown Bridge," 14, cites Joseph C. Nelson's report that a 100-foot lattice bridge required 1,500 holes and 1,000 pegs (treenails).

153. Gianine Figliozzi, "Robot Team Builds High-Performance Digital Structure for NASA," https://www.nasa.gov/general/robot-team-builds-high-performance-digital-structure-for-nasa/; Kenneth C. Cheung and Neil Gershenfeld, "Reversibly Assembled Cellular Composite Materials," *Science* 341 (2013): 1219–1221; and Christine E. Greg2111g et al., "Ultra-Light and Scalable Composite Lattice Materials," *Advanced Engineering Materials* 20 (2018): 1800213.

154. Ben Marsden and Crosbie Smith, *Engineering Empires* (Palgrave Macmillan, 2005), 9–11.

155. Seyrig, "Different Modes," 169–172. Photos in the Albertina collection indicate that Howe Bridges (containing wood) were also built, such as the Gluhavnaier and Balina Bridges.

156. Alfred D. Chandler Jr., *The Visible Hand: Managerial Revolution in American Business* (Harvard University Press, 1977), 242.

157. Piore and Sabel, *Second Industrial Divide*, 48. Re industrialization myth, see Charles Sabel and Jonathan Zeitlin, "Historical Alternatives to Mass Production: Politics, Markets and Technology in Nineteenth-Century Industrialization," *Past and Present* 108 (1985): 133–176, and Sabel and Zeitlin, "Stories, Strategies, Structures."

158. Helmut Hilz, "Heinrich Gerber (1832–1912): Eisenbrückenbau und Unternehmertätigkeit in Süddeutschland im 19. Jahrhundert" (Ph.D. diss., Ludwig-Maximilians-Universität, München, 1992), 271.

159. Chandler, *Visible Hand*, 241. Chandler asserts that "in modern mass production . . . economies resulted more from speed than from size [of manufacturing establishments]" (244); true perhaps also for construction.

160. A. De Toni and S. Tonchia, "Manufacturing Flexibility: A Literature Review," *International Journal of Production Research* 36 (1998): 1587–1617, and Fernando F. Suarez, Michael A. Cusumano, and Charles H. Fine, "An Empirical Study of Flexibility in Manufacturing," *MIT Sloan Management Review* 37 (1995): 25–32. Regarding the focus of historians of early US industrialization, Thomas C. Cochran, "Cotton Textiles and Industrialization," in Randolph Shipley Klein, ed., *Science and Society in Early America* (American Philosophical Society, 1986), 264–66.

161. Abdulaqder M. Tah et al., "Comparison of Various Bracing System [*sic*] for Self-Supporting Steel Lattice Structure Towers," *American Journal of Civil Engineering* 5 (2017): 60–68.

162. Dolores Greenberg, "Energy, Power, and Perceptions of Social Change in the Early Nineteenth Century," *American Historical Review* 95 (1990): 693–714. Joachim Radkau, *Technik in Deutschland* (Suhrkamp, 1989), 61, indicates that historian David S. Landes, *The Unbound Prometheus: Technological Change and Industrial Development in Western Europe from 1750 to the Present* (Cambridge, 1969), 54, considers coal instead of water power a sign of "rationality."

163. On builders' post-1850 adoption of steam-powered machinery, see Robert M. Vogel, "Building in the Age of Steam," in *Building Early America*, ed. Charles E. Peterson (Chilton, 1976), 119–134; Alf O. Johansson, "Why Change to Steam Power? Institutional Constraints, Risk-Aversion and Path-Dependence in the Swedish Sawmill Industry, 1850–1900," in Berg and Bruland, *Technological Revolutions in Europe*, 207–229.

164. Amy E. Slaton, "Mere Maintenance: Stratified Industrial Labor and the Reproduction of Human Difference," Maintainers conference paper, Stevens Institute, April 8–10, 2016.

165. John Gregory, *Industrial Resources of Wisconsin* (Langdon & Rounds,1853), 251–252. Re Gregory, see O'Dwyer and Cox, "Early Irish Railway Construction," 122.

166. Ken Alder, *Engineering the Revolution: Arms and Enlightenment in France, 1763–1815* (University of Chicago Press, 1997), 128. Berg and Hudson, "Rehabilitating the Industrial Revolution," 31–32.

CHAPTER 6

1. Geoffrey C. Bowker and Susan Leigh Star, *Sorting Things Out* (MIT Press, 1999), 5, 33–39.

2. For opposing views of national engineering, see Robert A. Jewett, "Structural Antecedents of the I-Beam, 1800–1850," *Technology and Culture* 8 (1967): 362, and Eda Kranakis, *Constructing a Bridge* (MIT Press, 1997), 211.

3. Amy Slaton, correspondence with author, January 6, 2022.

4. Per Google search, February 11, 2023, this category doesn't yet exist.

5. Stine Grodal et al., "The Coevolution of Technologies and Categories during Industry Emergence," *Academy of Management Review* 40 (2015): 423–455; especially 438–440, re "producers, users, and commentators." (A version of this chapter was published five years before Grodal.)

6. William Fairbairn, *Three Lectures on the Rise and Progress of Civil and Mechanical Engineering* (Pike, 1859), 30–35.

7. Nathan Rosenberg and Walter Vincenti, *The Britannia Bridge: The Generation and Diffusion of Technological Knowledge* (MIT Press, 1978), and Stanley Smith, "The Development and Use of the Tubular Beam, 1830–1860," *History of Technology* 14 (1992): 108. Smith states that structural designs for which Stephenson is given credit were developed by employees and contractors. Conway had two 400-foot spans, each box 18 feet 6 inches high and 14 feet 8 inches wide, with nine plate girders on top and seven under forming cells. Britannia's center spans were 460 feet and the two side spans were 230 feet; over 1,500 tons each. Four years, per [Latimer Clark, with permission of Robert Stephenson], *General Description of the Britannia and Conway Bridges on the Chester & Holyhead Railway* (Chapman and Hall, 1849), 34.

8. Stephenson did not mention an iron tubular (circular) lattice pedestrian bridge installed prior to 1843 in a Boston Hotel, Tremont House; it was cited as a predecessor of Stephenson's tube; Julius Weisbach, *Principles of the Mechanics of Machinery and Engineering*, ed. Walter R. Johnson, English translation (Lea and Blanchard, 1849) 2:88; reprinted in "Weisbach's 'Mechanics of Machinery and Engineering,'" *Mechanics' Magazine* 50 (1849): 282. In 1841, tens of thousands of people had visited the bridge (patented by architect Isaiah Rogers), when it temporarily linked Faneuil Hall and Quincy Market in Boston; David Guise, "Isaiah Rogers," *Structure* (November 2014), https://www.structuremag.org/article/isaiah-rogers. "A structure of this kind, though on a rude and miniature scale, appears to have existed for years on the Cambridge line of the Eastern Counties Railway," designed by Stephenson; "Wales," *Gentleman's Magazine* (London) 32 (1849): 196.

9. Robert Stephenson, "Introductory Observations on the History of the Design," in *The Britannia and Conway Tubular Bridges*, ed. Edwin Clark (Day and Son / John Weale, 1850), 1:22–24. Re Montrose, Hammersmith, and Menai suspension bridges (a section of the latter resembled a section of a tube that Stephenson would build), see C. W. Pasley, "Description of the State of the Suspension Bridge at Montrose," *Transactions of the Institution of Civil Engineers* 3 (1840): 219–227, fig. 3. Inspired by Montrose Bridge, Stephenson "conceived that sufficient strength might be obtained by the combination of the suspension chains with deep trellis trussing." Because of the flammability and maintenance of wood, Stephenson "was led to consider the application

of [wrought iron], but substituting for the vertical wooden trellis trussing . . . wrought-iron plates riveted together with angle-iron"; William Pole, in *The Life of Robert Stephenson*, ed. J. C. Jeaffreson (Longman, Green, 1864), 81–82. The suspension idea was abandoned so late that towers were constructed to accept suspension chains. Re Roebling, see Robert M. Vogel, "Roebling's Delaware & Hudson Canal Aqueducts," *Smithsonian Studies in History and Technology* 10 (1971), 5–11. L. T. C. Rolt, *George and Robert Stephenson: The Railway Revolution* (St. Martin's, 1960, 1962), 126–127 reports, without a source, that in 1827 Stephenson walked from New York City to Montreal; if he visited New Haven, CT, or Springfield, MA, he would have seen early lattice bridges.

10. Before Stephenson's tubes were complete, a British journal noted regarding a Howe bridge in Canada: "It will thus be seen that the bridge is a piece of construction in timber, which might call to mind Mr. Stephenson's project for the Britannia Bridge, at least in its appearance, when passing over it"; "The Construction and Principles of Timber Bridges," *Builder* 5 (1847): 435. On Grubenmann bridges, see chapter 2, and Ceraldi, below.

11. Stephenson in Clark, *Britannia,* 25–26. Montrose Bridge was the suspension structure foremost in his mind. Not mentioning Fairbairn's work was undoubtedly deliberate.

12. Clark, *Britannia*; William Fairbairn, *An Account of the Construction of the Britannia and Conway Tubular Bridges* (John Weale, 1849); G. Dreysdale Dempsey, *Tubular and Other Iron Girder Bridges* (John Weale, 1850); Rosenberg and Vincenti, *Britannia Bridge*, 57–59. John Rapley labels Clark's book a "work of propaganda" written under Stephenson; John Rapley, *Britannia and Other Tubular Bridges and the Men Who Built Them* (Tempus, 2003), 8. Fairbairn broke with Stephenson around May 1848 and was no longer connected with project implementation.

13. Fairbairn, *Account of the Construction*, 176–179. Richard Byrom, "William Fairbairn—Experimental Engineer and Mill-Builder" (PhD diss., University of Huddersfield, 2015), 196–198. See also, "Inventor of the Tubular Bridge," *Scientific American* 6 (1851): 129 [from *Glasgow Daily Mail*].

14. Professor Cowper's son, Edward Alfred Cowper, was on the staff of Henderson Fox, the firm involved in the design and building of the Crystal Palace. Cowper's lecture noted the power of standard parts in construction, and how construction was carried out with minimal scaffolding by using the structure's columns. The illustration seems to show a large representation of a tubular bridge behind the lecturer; "Professor Cowper's Illustrations of the Palace of International Industry," *Illustrated London News* 18, January 4, 1851, 9–10. Robert Stephenson was a commissioner of the exposition for which the Crystal Palace was built.

15. [C. W. Pasley and Robert Stephenson], "The Britannia Tunnel Bridge over the Menai Straits: General Pasley's Letter, and Mr. Stephenson's Reply," *Journal of the Franklin Institute* 3rd ser., 14 (1847): 301 [from *Mechanics' Magazine*]; Clark, *Britannia,* 65–66. On Macneill's bridges, see below.

16. Re Osborne, see chapter 5. Richard Boyse Osborne, "Diary," manuscript, 4: 21 and 4:28–29; cited with permission of the Council of Trustees of the National Library of Ireland. From 1845, in Ireland Osborne purportedly introduced "American" ideas to the railroad. He built an

iron Howe bridge in 1846 at Ballysimmon on the Waterford and Limerick Railroad. He later returned to the United States, knew William Howe, and built what is called the first all-iron Howe bridge. Thomas R. Winpenny, "The Engineer as Promoter: Richard B. Osborne, the Camden and Atlantic Railroad, and the Creation of Atlantic City," *Essays in Economic and Business History* 22 (2004): 302.

17. Paul Sibly, "The Prediction of Structural Failures" (PhD diss., University of London, 1977), discusses Stephenson's mendacities, supported by, for example, engineer Joseph Locke, 33–38; beam length, 9, 26, 48. After the collapse, this type of beam was not used again. Rolt, *George and Robert Stephenson*, 302–303, claims, without citing a source, that the railroad company compelled Stephenson to lie.

18. Peter R. Lewis and Colin Gagg, "Aesthetics versus Function: The Fall of the Dee Bridge," *Interdisciplinary Science Reviews* 29 (2004), 177–191; R. A. Buchanan, *The Engineers: A History of the Engineering Profession in Britain, 1750–1914* (Kingsley, 1989), 184–185. Lewis and Gagg, "Aesthetics," 178, note that Pasley was dismissed after the collapse of the Dee bridge, which he had approved.

19. UK Parliamentary Papers, *Report of the Commissioners Appointed to Inquire into the Application of Iron to Railway Structures* (Her Majesty's Stationery Office, 1849), 340. Re issues such as "wabbly-ness," see Gregory K. Dreicer, "The Long Span. Intercultural Exchange in Building Technology: Development and Industrialization of the Framed Beam in Western Europe and the United States, 1820–1870" (Ph.D. diss., Cornell University, 1993); 231, 250–257, and Sibly, "Prediction of Structural Failures," 22.

20. P. J. Geraghy, "Dublin and Drogheda Railway: The First Great Irish Speculation," *Dublin Historical Record* 66 (2013), 108. Lewis and Gagg, "Aesthetics,"182–184; UK Parliamentary Papers, *Report of the Commissioners*, 359. Stephenson's observation that the tube was a "closed-up" lattice showed he had considered both types and indicated what he was to confirm: the tube, like the lattice, was "wabbly" and that a solution included vertical members. According to Michael R. Bailey, ed., *Robert Stephenson: The Eminent Engineer* (Ashgate, 2003), 91, 158, Stephenson was skilled at giving evidence and negotiating; historians, such as Bailey and Lewis and Gagg, have begun to analyze the dark side of this. Re metal and the collapse, see William M. Taylor, "Iron, Engineering and Architectural History in Crisis: Following the Case of the River Dee Bridge Disaster, 1847," *Architectural Histories* 1 (2013): 6; also Sibly, "Prediction of Structural Failures," esp. 43–51. Stephenson had had ballast added to the bridge shortly before it failed.

21. The dissemination of test results was the prime legacy, according to Rosenberg and Vincenti, *Britannia Bridge*, 29; beyond data about the use and performance of wrought iron, the "tubular bridge . . . merits not much more than a footnote as far as the history of bridge building is concerned" (67–68).

22. By February 1846, tests confirmed the rectangular iron-plate tubular design. Between July 1846 and May 1847, tests were conducted on a one-sixth scale model; Pole, *Robert Stephenson*, 186. The tests were criticized; Smith, "Development," 123; "Railway Bridges," *The Engineer* 9 (1860): 157; Jourawski, "Remarques sur les poutres en treillis et les poutres pleines en tôle," *Annales des ponts et chaussées* 3/2 (1860): 128; [Auguste] Dallot in "Procès-verbaux"

(January 19, 1866), *Memoires et comptes rendus des travaux de la Société des Ingénieurs Civils* (1866): 50.

23. P. J. Geraghty, "The Battle of the Boyne: A Victorian Engineering Controversy," *Journal of County Louth Archaeological and Historical Society* 25 (2004): 402, 404; Canice O'Mahony, "Iron Rails and Harbour Walls: James Barton of Farndreg," *Journal of County Louth Archaeological and Historical Society* 22 (1990): 135; James Barton, "On the Economic Distribution of Material in the Sides, or Vertical Portion, of the Wrought-Iron Beams," *Transactions of the Institution of Civil Engineers* 14 (1854–1855): 443–452.

24. James Thomson, "Observations Upon Iron Lattice Bridges," *Journal of Franklin Institute* 37(1844): 229–230, reprinted from James Thomson, *Proceedings of the Royal Scottish Society of Arts* 2 (1844): 78–79. Reference to Thomson's presentation, *Proceedings of the Philosophical Society of Glasgow* 1 (1843): 182.

25. Smith, "Development," 101, 123–24. Smith claims criticism came mainly from younger engineers at ICE, Irish engineers, and Zerah Colburn (see note below), but criticism came from many sources. Britannia Bridge's "economic unsoundness" "was repeatedly shown by almost every continental writer on design over the next 20 years"; J. G. James, "Some Steps in the Evolution of Early Iron Arched Bridge Design," *Newcomen Transactions* 59 (1987–88): 153.

26. P. J. Geraghty, "Battle of the Boyne: A Victorian Engineering Controversy," *Journal of County Louth Archaeological and Historical Society* 25 (2004): 400–401, 404; P. J. Geraghty, "Sir John Macneill (1793–1880): King of the Irish Railways," *Transactions of the Newcomen Society* 78 (2008): 207–234.

27. P. J. Geraghty, "The Dublin and Belfast Junction Railway: The Great Link Uniting Dublin and Belfast," *Journal of County Louth Archaeological and Historical Society* 29, no. 1 (2017): 116–118; and O'Mahony, "Iron Rails," 135.

28. Sibly, "Prediction of Structural Failures," 20; Re "favourite," see "Memoirs: William Fairbairn," *Minutes of the Proceedings of the Institution of Civil Engineers* 39 (1875), 259, and Byrom, "William Fairbairn."

29. Eugène Flachat, "Introduction," in *Notice sur les ponts avec poutres tubulaires en tôle*, ed. L. Yvert (Carilian-Goeury et Victor Dalmont, 1851), xix–xxiv; Flachat was involved in the design and construction of early box girder bridges in France at Clichy (1851) and Asniéres (1852); Bernard Marrey, *Les ponts modernes: 18e-19e siècles* (Picard, 1990), 165–168. Karl Culmann, "Der Bau der eisernen Brücken in England und Amerika," *Allgemeine Bauzeitung* 17 (1852):179–180. Herman Haupt, "On the Resistance of the Vertical Plates of Tubular Bridges," *Journal of the Franklin Institute* 3rd ser., 26 (1853): 217–221. Dmitrii Ivanonich Jouravsky, *Remarques sur la résistance d'un corps prismatique et d'une pièce composé en bois ou en tôle de fer à une force perpendiculaire à leur longueur* [extract and translation from a Russian-language publication] (St. Petersburg, 1855), ii, 38–48. Émile With, *Nouveau manuel complet de la construction de chemins de fer* (Librarie Encyclopédique de Roret, 1857), 115–116; With did state that "American bridges" (i.e., lattice bridges) had similar rivet issues and that further study was required. Bommart, "Travaux publics," in Michel Chevalier, ed., *Exposition universelle de Londres de 1862: Rapports des membres de la section français du jury international sur l'ensemble de l'exposition* (Chaix, 1862) 3: 406–407. Re Culmann and Haupt, see chapter 4.

30. Smith, "Development," 122, 125–26. Dermot W. O'Dwyer and Ronald Cox, "Early Irish Railway Construction,"in James Campbell, et al., eds., *Studies in Construction History: Proceedings of the Second Construction History Society Conference* (Construction History Society, 2015), 127–128, note competition between English and Irish contractors and cite an 1849 report about testing in Ireland, probably by William Fraser, "An Account of Some Experiments on the Strength of Models of Timber Lattice Beams," *Transactions of the Institution of Civil Engineers Ireland* 3 (1849): 127–142; pl. X. Engineering School of Trinity College Dublin was the training ground for engineers in debates about to be discussed; Dermot W. O'Dwyer and Ronald Cox, "The Boyne Viaduct: Early Indeterminate Lattice Girder Analysis and Design," in *Building Knowledge, Constructing Histories*, ed. Ine Wouters et al. (CRC, 2018), 2:1006–1007.

31. William Thomas Doyne, "Description of a Wrought-Iron Lattice Bridge, Constructed Over the Line of the Rugby and Leamington Railway," *Transactions of Institution of Civil Engineers* 9 (1849–50): 353–359, fig. 15. Doyne states that the lattice, in comparison to tubular and Warren bridges, allowed easier manufacture and assembly, with concomitant savings; Doyne and William Bindon Blood, "An Investigation of the Strains upon the Diagonals of Lattice Beams, with the Resulting Formulae," *Transactions of Institution of Civil Engineers* 11 (1851): 110–112. Another case of Bidder speaking out against the lattice and for the tube: William Scarth Moorsom, "Description of the Viaduct Erected over the River Nore, near Thomastown, in the County of Kilkenny Railway," *Minutes of the Proceedings of the ICE* 11 (April 13, 1852): 434. "Doyne, William Thomas (1823–1877)," *Australian Dictionary of Biography*, National Centre of Biography, Australian National University (1972), http://adb.anu.edu.au/biography/doyne-william-thomas-3441/text5245.

32. James Barton, "On the Economic Distribution of Material in the Sides, or Vertical Portion, of the Wrought-Iron Beams," *Transactions of Institution of Civil Engineers* 14 (1854–55): 443–452. Barton worked under John Macneill, the railroad's chief engineer. The main span of the Boyne Viaduct was 264 feet, with two side spans of 138 feet (Barton, 450–451). He asserted that bars were cheaper than plates. Re debate around analysis of lattice, tube, and Warren, see D.A. Gasparini and Caterina Provost, "Early Nineteenth Century Developments in Truss Design in Britain, France and the United States," *Construction History* 5 (1989): 25–28, and Smith, "Development," 120–21.

33. O'Mahony, "Iron Rails," 135, 137, 139; Geraghty, "Battle," "Sir John Macneill," and "Dublin and Belfast Junction Railway." The Britannia tube was 30-feet deep where it passed through the central tower and 22 feet at abutments; moreover, it included camber (rise in the middle) of 9 inches to account for settlement under its own weight; Clark, *Britannia and Conway Tubular Bridges*, 2:558–559. The height of Victoria spans varied; James Hodges, *Construction of the Great Victoria Bridge in Canada* (John Weale, 1860), 78–79. The article from *The Engineer* is reprinted as "American Lattice Bridge," *Scientific American* 16 (1867): 374. Re calculations and design, see O'Dwyer and Cox, "Boyne Viaduct," 1006.

34. O'Mahony, "Iron Rails," 138–139; Stephen P. Timoshenko, *History of Strength of Materials* (McGraw-Hill, 1953; reprint, Dover, 1983), 127. Hodgkinson's paper appeared in *Memoirs and Proceedings of the Manchester Literary and Philosophical Society* 4 (1824). Isaac Todhunter and Karl Pearson, *A History of The Theory of Elasticity*, vol. 2, pt. 1 (1893; reprint, 1960), 679–680. In addition to Hodgkinson, Pole and Tate apparently used the neutral axis concept

in the design phase; S. Tyson, "Notes on the History, Development and Use of Tubes in the Construction of Bridges," *Industrial Archaeology Review* 2 (1978): 152–53. In another context, Lewis and Gagg, "Aesthetics," 183, 184, 185, note that Stephenson made "remarkable," "strange," and "curious" statements that indicate gaps in knowledge or honesty. Also, Tom F. Peters, *Building the Nineteenth Century* (MIT Press, 1996), 163–164. O'Dwyer and Cox, "Boyne Viaduct," 1009–1010, note Stephenson's awareness of unknowns and (mis) understandings See chapter 2, note 31.

35. On Fairbairn's mathematical ability, see Byrom, "William Fairbairn," 209–211; Sibly, "Prediction of Structural Failures," 44; Smith, "Development," 108–110, 116–117, 127–128.

36. Dermot O'Dwyer, "Development in the Design of Beams from 1820–1860," in *Iron, Steel and Buildings: Studies in History of Construction*, ed. James W. P. Campbell et al. (Construction History Society, 2020), 80–81. On shear, see O'Dwyer and Cox, "Boyne Viaduct," 1009–1010.

37. Geraghty, "Battle," 415–25; also Geraghty, "Dublin and Belfast Junction Railway: The Great Link," 96–136; and O'Mahony, "Iron Rails," 135. Barton had a large role in the Boyne Viaduct; his claim regarding design and construction seems justified: also O'Mahony, "Iron Rails," 140. I was unable to obtain John Macneill, *Boyne Bridge: Its History, as Originally Designed and Constructed by John Macneill* (Hodges, Smith, 1860).

38. Geraghty, "Battle," 402, 404, 407.

39. Geraghty, "Dublin and Belfast Junction Railway," 116–118; Geraghty, "Battle," 405–408; O'Mahony, "Iron Rails," 135, re the decision date.

40. Geraghty, "Battle," 417.

41. Robert Stephenson and Alexander Ross, "The Victoria Bridge at Montreal," *Artizan* 14 (1856): 59–61; Stephenson repeated this argument from the previous year; Barton, "Economic Distribution," 472–473. Stephenson: "I must ask you to bear in mind that I am not addressing you as an advocate for a tubular-bridge."

42. James Barton, "On the Comparative Strength of Lattice and Tube Bridges," *Engineering* 9 (1860): 189 [reprint of 1856 article]. There is a detailed rebuttal of Stephenson's arguments in Grand Trunk Railway Company of Canada, Victoria Bridge, *Letter to the Shareholders of the Grand Trunk Railway of Canada, by Robert McCalmont, Esq., Ex-Director, with a Report to Him by Mr. Liddell . . . Together with Counter-Reports by Messrs. Stephenson, I. K. Brunel, Edwin Clark, and A. M. Ross; and Replies to Each by Mr. Liddell* (Clowes, 1856); analysis of Stephenson's claims re sides of lattice, 21. Dermot O'Dwyer, "Development in the Design of Beams from 1820–1860," in *Iron, Steel and Buildings*, 80–81, notes a factor was "lack of understanding" regarding forces in web of beam. Re confusing speech, see Sibly, "Prediction of Structural Failures," 33–38.

43. John A. Roebling, *Final Report of John A. Roebling, Civil Engineer, to the Presidents and Directors of the Niagara Falls Suspension and Niagara Falls International Bridge Companies* (Lee, Mann, 1855), 11–12.

44. Rosenberg and Vincenti, *Britannia Bridge*, 30–43. Nicholas Cadiat and Alphonse Oudry, *Note relative à l'emploi du fer dans les ponts: Comparaison des ponts en tôle supportés par des arcs avec les*

ponts en forme de poutre [from *Annuaire de la soc. des anciens élèves des écoles nationales des arts et métiers* (1851)] (De Soye et Bouchet, 1853), 13, emphasized that the heaviness of plate construction was important because "everyone knows that, where it is subject to vibrations, the most ductile and fibrous iron in the end will become granular and break": see also, Timoshenko, *History*, 193.

45. A.E. Richard De Jonge, *Riveted Joints: Critical Review of Literature Covering Their Development, with Bibliography and Abstracts of Most Important Articles* (American Society of Mechanical Engineers, 1945), 12–13, 51–54.

46. Millholland claimed that at the time, Stephenson was experimenting with cylindrical tubes: "The cylindrical tubes failing, they adopted this plan of bridge"; letter, from James Millholland, May 1, 1849, in Herman Haupt, *General Theory of Bridge Construction* (Appleton, 1851), 255–256; John White, "James Millholland and Early American Railroad Engineering," *U.S. National Museum Bulletin* 252, Paper 69 (1967): 8–9, 32–33. Haupt also built plate bridges on Pennsylvania Central Railroad; Zerah Colburn, "American Iron Bridges," *Transactions of Institution of Civil Engineers* 22 (1862–1863): 542–543. From August 1845, Fairbairn experimented with rectangular beams; Jewett, "Structural Antecedents, 361." The form of Stephenson's tubes was determined in December 1846; Eaton Hodgkinson, "Experimental Inquiries to Determine Strength of Wrought-Iron Tubes," in UK Parliamentary Papers, *Report of the Commissioners Appointed to Inquire into the Application of Iron to Railway Structures* (Her Majesty's Stationery Office, 1849), 118. Re box structures by Fairbairn and others during the 1830s and 1840s, see Smith, "Development," 110. Stephenson completed a box beam bridge on the North-Western Railway in March 1847; Sutherland, "Introduction of Structural Wrought Iron," 80–81.

47. Project planning began in 1840s. The six Dirschau spans were each 397 feet long; a portion still exists. Lentze, "Bemerkungen über die grosseren Brückenbauwerke in Frankreich, England, und Irland auf ein Reise in Winter 1844/45 gesammelt," *Verhandlungen des Vereins zur Beförderung des Gewerbeleißes* (1846): 88–114, figs. 12–17; Lentze, "Die im Bau begriffenen Brücken über die Weichsel bei Dirschau und über die Nogat bei Marienburg," *Zeitschrift für Bauwesen* 5 (1855): 445–458, figs. 42–45, 49, 50a, 58–65, O, P, Q, R; Königliches Ministerium für Handel, Gewerbe und öffentliche Bauten, *Die im Bau befindlichen Brücken über die Weichsel bei Dirschau und die Nogat bei Marienburg* (Ernst und Korn, 1855). Martin Trautz, *Eiserne Brücken im 19. Jahrhundert in Deutschland* (Düsseldorf: Werner-Verlag, 1991), 63–67. On the comparison of testing and analysis of Britannia and Vistula bridges, see Kurrer, *History of the Theory of Structures* (Ernst & Sohn, 2008), 92–93; J. W. Schwedler, "Elementare Theorie der gitterförmigen Träger-Constructionen," *Civilingenieur* 5 (1859): 146–147. See also Wieland Ramm, *Zeugin der Geschichte: Die Alte Weichselbrücke in Dirschau* (Technische Universität Kaiserlautern/Krüger, 2004).

48. Jan Musekamp, *Shifting Lines, Entangled Borderlands* (Indiana University Press, 2024), 90-92.

49. M. Becker [includes portions of Carl Ruppert report], "Die Eiserne Brücke über die Kinzig bei Offenburg," *Allgemeine Bauzeitung* 18 (1853): 179–180. The same advantages, in addition to "the certainty of the [stresses] proceeding in the various channels allotted to them," were still cited two decades later; William G. Strype, "Description of the Iron Lattice Girder Road Bridge, Recently Erected over the River Boyne, at the Obelisk," *Transactions of Institution of Civil Engineers of Ireland* 9 (1871): 67–78, fig. 10. There is a response to Ruppert in Prüsmann, "Zur

Frage, ob die eisernen Gitterbrücken oder die Blechbrücken den vorzug verdienen," *Notizblatt, Zeitschrift des Architecten- und Ingenieur-Vereins für das Königreich Hannover* 3 (1854): 207–210; see also, G. Boulangé, "Notes recueillies pendant une visite rapide de quelques chemins de fer d'Allemagne," *Annales des ponts et chaussées* 3/1 (1854): 65. Wind resistance was a concern: engineer Julius Naeher traveled from Baden to the Britannia, climbed to the top with Edwin Clark and stood in the middle; he reported that the tube did not shake or sag when a train passed through, but "very frightening rocking" resulted from wind; Julius Naeher, "Ueber die Verbindung der Einzelnen Röhrentheile der Brittaniabrücke und die Tragfähigkeit derselben," *Allgemeine Bauzeitung* 17 (1852): 76.

50. Trautz, *Eiserne Brücken*, 69. William Fairbairn, *On the Application of Cast and Wrought Iron to Building Purposes* (John Weale, 1854; 2nd ed. 1857–58); his biographer, Pole, told the story. *On the Application*, 4th ed. (1870), 319–330, 344–347, figs. 4–6. and William Pole, *The Life of Sir William Fairbairn, Bart* (Longmans, Green, 1877): 217–37. Also "Bridge Over the Rhine at Cologne," *Artizan* 8 (1850): 104. Quote from letter, [Christian Charles Josias] Bunsen to Fairbairn, March 20, 1850, in Pole, *Fairbairn*, 230. Re rejection of tubular bridge, see C. Couche, "Travaux d'art des chemins de fer d'Allemagne," *Annales des mines* 5/5 (1854): 391. Re Friedrich Wilehelm IV, see Erik Grawert-May, "Der Garten des Deus ex machina," *Tagesspiegel*, October 15, 1995.

51. An 1850–1851 international design competition drew sixty-one entries. First prize went to Wilhelm Schwedler's suspension bridge, second to W. S. Moorsom; Pole, *Fairbairn*, 233, and Moorsom in Barton, "Economic Distribution," 487. Submissions were considered inappropriate; the government again examined bridges in England. Fairbairn submitted tubular and box girder designs with spans greater than Britannia's (570 feet). Preliminary lattice design prepared by Regierungs- und Baurath Wallbaum was approved in December 1854 and reworked by Hermann Lohse, who took over the project. Hermann Lohse, "Notizen über einige neuere Brücken Englands," *Zeitschrift für Bauwesen* 7 (1857): 216. Re testing of lattice girders, see Lohse, "Versuche über das Zerknicken der Eisenstäbe in Gitterträgern," *Zeitschrift für Bauwesen* 7 (1857): 573–582, pl. T. Re Moorsom's model, see "Exposition Universelle: Machines," *Morning Chronicle*, English ed., May 9, 1851, 2; French ed., May 10, 1851, 3; German ed., May 10, 1851, 4.

52. "Memoirs: William Fairbairn," 259; Pole, *Fairbairn*, abridged ed. (1878): 105. On the cathedral, see Mitchell Schwarzer, "The Sources of Architectural Nationalism," in *Nationalism and Architecture,* ed. Raymond Quek et al. (Ashgate, 2012), 29.

53. Bommart, *Rapport sur les travaux publics représentés à l'exposition universelle de Londres de 1862* (Imprimerie Centrale des Chemins de Fer, 1863), 288–295; Société Nationale des Chemins de fer Français, "Le pont de chemin de fer sur le Rhin, additif au catalogue de l'exposition 'Les debuts du chemin de fer à Strasbourg' à l'occasion du 125e anniversaire du pont de Kehl," manuscript, [SNCF, 1986]. Benckiser, in Pforzheim, built the bridge superstructure; the Compagnie des chemins de fer de l'Est was responsible for its foundations. Also, Émile Clapeyron, "Superstructure des ponts en fer," in Chevalier, *Exposition universelle de Londres*, 3: 417. Gustave Eiffel visited the Kehl construction site, November 1859, about ten months before the bridge launch, possibly regarding the use of compressed air in foundation construction; Bertrand Lemoine, *Gustave Eiffel* (Hazan,1984), 26.

54. Wilhelm Nordling, "Mémoire sur les piles en charpente métallique des grands viaducs," *Annales des ponts et chaussées* 4/2 (1864): 6–7.

55. Stephenson in "The Grand Trunk Railway of Canada: Victoria Bridge, Montreal," *Civil Engineers and Architects Journal* 17 (1854): 211. Barton, "Economic Distribution," 489. "[Moorsom] has commenced the construction of trellis-bridges as early as 1837, and could corroborate Mr. Barton's statement of the great facility of raising the trellis-girder entire, into their positions, in which respect, no advantage could be claimed for the tubular over the trellis-beam"; Barton, "On the Economic Distribution," 489.

56. [Latimer Clark, with permission of Robert Stephenson], *General Description*, 39; 24–40 for a detailed description of tube raising, including the participants Brunel and Locke. "The accidents that have occurred on the work have, fortunately, not been numerous, seven persons only have lost their lives"; masons erected a monument nearby (30).

57. Temporary scaffolding on one side was "100 feet high, 230 feet long, and 73 feet broad . . . and contained . . . 70,000 cubic feet of timber [and] not less than 20 tons of iron bolts"; tubes were raised between October 1849 and June 1850. A US engineer witnessed a tube raising: John B. Jervis, "A Memoir of American Engineering," *ASCE Transactions* 6 (1877): 65. Thomas Jackson, *The Tourist's Guide to Britannia Bridge*, 7th ed. (Smith, 1851), 16, 20–21, 26–39; quotation, 35.

58. For an eyewitness description of the Dirschau and Marienburg construction sites, see Winterstein, "Bericht über die Reise des Architekten-Vereins und des Vereins für Eisenbahnkunde nach Königsberg," *Zeitschrift für Bauwesen* 4 (1854): 561–568; Mehrtens, "Zur Baugeschichte der alten Eisenbahnbrücken," 105. Re the savings that lattices allowed, see Doyne, "Description," 355.

59. "'High Bridge,' Portage, New York," *Civil Engineer and Architects Journal*, 65–66, pl. 8. See Introduction and chapter 8.

60. Henry N. Maynard, *Handbook to the Crumlin Viaduct* (Wilson, 1862); reconstruction , 21–22; recost, 23; Fitzmaurice quotation, 30. J. G. James, "The Origins and Worldwide Spread of Warren-Truss Bridges in the Mid-Nineteenth Century, Part 1: Origins and Early Examples in the UK," *History of Technology* 11 (1986): 102–106 and 122n8. Re construction, see the *Monmouthshire Merlin* (December 1853) report, reproduced in John Croeso, *Crumlin Viaduct*, "History Page," https://crumlinviaduct.co.uk/history%20view.htm.

61. Smith, "Development," 115, 121. Stephenson was responsible for two bridges in Egypt near Benha over the Nile's Damietta branch and the Karrineen Canal at Birket-el-Saba (contracted 1850–1851, built 1853–1855). Trains rode atop, not through, 6.5-foot deep box girders manufactured in England and assembled on site; Bau-Inspektor Wild, "Bridge at Benha, on the Alexandria and Cairo Railway," *Civil Engineers and Architects Journal* 16 (1853): 425, fig. 36. Pole, *Robert Stephenson*, 2:68–69, 173–179. Re alteration of the Brotherton Bridge, see R. J. M. Sutherland, "The Introduction of Structural Wrought Iron," *Transactions of the Newcomen Society* 36 (1963–1964): 82–83; *Sessional Papers Printed by Order of the House of Lords* 66 (1852–53), *Appendix to Report of Lords of Committee of Privy Council for Trade*, appendix 3, "York and North

Midland Railway," 63–74. In the Netherlands, around 1855, Clark was behind change of lattice to plate design in a bridge, but it was not a tube; Augustus J. Veenendaal Jr., *Railways in the Netherlands: A Brief History, 1834–1994* (Stanford University Press, 2001), 15.

62. Stanley Triggs et al., *The Victoria Bridge: The Vital Link* (McCord Museum of Canadian History, 1992), 41. Re St. Anne's Tubular Bridge (1858) over the Ottawa River, Montreal Island, whose construction must have been linked to Victoria, see Robert W. Passfield, "Construction of the Victoria Tubular Bridge," *Canal History and Technology Proceedings* 20 (2001): 40; re Ross, 13.

63. Smith, "Development," 101–108; re original plan, see Passfield, "Construction," 8–9, 12–13, 15–16; F. N. Boxer, *Hunter's Hand Book of the Victoria Bridge* (Hunter and Pickup, 1860), 21–29, provides hyperbolic support for Stephenson, the bridge, Anglo Saxons, industry, etc.

64. Charles Legge, *Glance at the Victoria Bridge and the Men Who Built It* (Lovell, 1860), 21, 30.

65. Hodges, *Construction*, 55, 62. The 1.75-mile overall length contained over a mile of side-by-side tubes. The Britannia design was altered, for example by the substitution of built-up I-beams for stiffening cells in bridge roof; Smith, "Development," 101–108, and Triggs, *Victoria Bridge*, 55–68; Stephenson (95) refers to Roebling's Niagara Bridge (opened mid-March, 1855) being completed "within the last few months."

66. Hodges, *Construction*, "Appendix," 85–104, includes a defense by Stephenson and Ross provided to Grand Trunk Railway Board of Directors; Triggs, *Victoria Bridge*, 43. "Embarrassment" per G. P. de T. Glazebrook (1938), cited in James A. Vance, *North American Railroad* (Johns Hopkins University Press, 1995), 247. Re "mess," see "Railway History in Canada," in *Canadian Encyclopedia*, http://www.thecanadianencyclopedia.ca/en/article/railway-history/; "Grand Trunk Railway," in https://en.wikipedia.org/wiki/Grand_Trunk_Railway. Also the amazing "Victoria Bridge," *American Railroad Journal* 33 (1860): 197–199, 213, 221–223, 241–244, 261–263, 281–283, 301–303, 321–323, 345–347, 369–371, 389–390, 413–414, 443, 789–790. Re shipping to Canada, see Triggs, *Victoria Bridge*, 61. Re cost, see Smith, "Development," 123, and review of Peter W. Barlow, *Observations on the Niagara Railway Suspension Bridge*, *Civil Engineer and Architect's Journal* 24 (1861): 53–54.

67. "The Rebuilding of Victoria Bridge," *Canadian Rail* 470 (1999): 59–87.

68. Grand Trunk Railway Company of Canada, Victoria Bridge, *Letter to Shareholders of Grand Trunk Railway of Canada, by Robert McCalmont, Esq., Ex-Director, with a Report to Him by Mr. Liddell* (Clowes, 1856).

69. Grand Trunk Railway, *Letter*, ii.

70. Grand Trunk Railway, *Letter*, 52. Re Stephenson and Brunel, see Christine MacLeod, "The Nineteenth-Century Engineer as Cultural Hero," in *Brunel: In Love with the Impossible*, ed. Andrew Kelly and Melanie Kelly (Institute of Civil Engineers, 2006), 60–79. In 1852, Brunel built a railway bridge over the River Wye at Chepstow; it was called "tubular" but bore no resemblance to the Britannia Bridge.

71. Grand Trunk Railway, *Letter*, 56.

72. "In less than ten years from its completion, the Crumlin Viaduct has had practically to be *re-rivetted*, in great part, if not all over"; "Civil Engineering in the Paris Exhibition—Iron Viaducts of La Cere, Busseau d'Ahun and du Midi" *Engineer* 23 (1867): 513; James, "Origins and Worldwide Spread," 65–66, 106; John Griffen and Thomas C. Clarke, "Loads and Strains of Bridges," *ASCE Transactions* 2, no. 1 (1874): 98–99.

73. Grand Trunk Railway, *Letter*, vii. Note that Liddell had worked with both George and Robert Stephenson.

74. In Grand Trunk Railway, *Letter*, cost analyses appear repeatedly. Ben Marsden and Crosbie Smith, *Engineering Empires* (Palgrave Macmillan, 2005), 151, refer to an "oligarchy of railway engineers, including Robert Stephenson, Locke, and Brunel."

75. William Irwin, *New Niagara: Tourism, Technology, and the Landscape of Niagara Falls, 1776–1917* (Pennsylvania State University Press, 1996), 37, 44–49.

76. Hodges, *Construction*, 95–96; final plate: Stephenson subsequently proposed a second railroad bridge across Niagara gorge: box beam rather than tubular!

77. Charles Sabel and Jonathan Zeitlin, "Historical Alternatives to Mass Production: Politics, Markets and Technology in Nineteenth-Century Industrialization," *Past & Present* 108 (1985): 161–164.

78. "Quebec Bridge and Forth Bridge," *Engineering News* 58 (1907): 391.

79. The box configuration also stiffens against buckling, which is a risk to the compression chord of a conventional beam; conversation with John Ochsendorf.

80. Herman Haupt, "Lattice Bridges," *American Railroad Journal* 11 (1840): 196–200; see also Haupt, *General Theory*, 151, 153. Re early tubular invention, see Tyson, "Notes," 143.

81. Re "efficiency," see Stephenson in Clark, *Britannia*, 1:25–26; re "hollow," see Clark, *Britannia*, 1:42. Re quills, etc., see [Latimer Clark, with permission of Robert Stephenson], *General Description*, 8–9. The bridge was not designed as a continuous span, which can reduce deflection, although spans were linked; "Continuous Girders," *The Engineer* 23 (1867): 273. "Wales," *Gentleman's Magazine* (London) 32 (1849): 196. Tom F. Peters, *Building the Nineteenth Century* (MIT Press, 1996), 168. Re other tubular structures, whether or not understood as such, see Tom F. Peters, "Relative Value of Invention and History of Tall Buildings," in *Second Century of the Skyscraper*, ed. Lynn S. Beedle (Van Nostrand Reinhold, 1988), 26–27. Bigelow reported that "a tube or hollow beam" had greater rigidity than a solid shape; this was based on the works of Robison, Barlow, Young, Rennie, Tredgold, and Gregory; Jacob Bigelow, *Elements of Technology* (Hilliard, Gray, Little, and Wilkins, 1829), 46, 52. Roebling wrote of his Niagara span that the "tubular or box plan of the Bridge has added much to its stiffness, vertically as well as horizontally"; John A. Roebling, *Final Report of John A. Roebling, Civil Engineer, to the Presidents and Directors of the Niagara Falls,* 14.

82. Exterior cladding may have had damping (vibration-reducing) and rigidity-enhancing functions. An engineer reported: "The cladding of the load-bearing exterior wall [i.e., lattice structure] with 3/4″ strong boards, apart from better preservation of the wood, will simultaneously contribute to greater stability of all the load-bearing parts, by connecting them closely;

it will thereby completely offset the disadvantage that results from increasing the weight of the bridge"; Chas. F. Zimpel, *Das Eisenbahnwesen von Nordamerika, England und anderen Ländern, practisch und populär dargestellt* (Förster, 1840), 101–102. Stephenson would have been familiar with the Grubenmann and "box bridges" in Switzerland; Massimo Laffranchi and Paolo De Giorgi, "Some Remarks on the Grubenmanns' Wooden Bridge Structures," in *John Soane and the Wooden Bridges of Switzerland* (Archivio del Moderno, Accademia di architettura, Università della Svizzera Italiana/Sir John Soane's Museum, 2003), 117, 127, 130; and C. Ceraldi and E. Russo Ermolli, "Swiss Covered Bridges of the Eighteenth Century. A Special Case: Bridge at Schaffhausen," S. Huerta, ed., *Proceedings of First Construction History Society Conference* (Juan de Herrera, 2003)," 575, 577. A rectangular form could be more readily analyzed and built than a circular form; Rosenberg and Vincenti, *Britannia Bridge,* 19.

83. Couche, "Travaux d'art," 356. Re Couche, see chapter 4.

84. O'Mahony, "Iron Rails," 137.

85. Bindon B. Stoney, "Description of Some Experiments Made at the Boyne Viaduct in 1854," *Transactions of Institution of Civil Engineers of Ireland* 5 (1860): 136. Ronald Cox and Dermot O'Dwyer, *Called to Serve: Presidents of the Institution of Civil Engineers of Ireland 1835–1968* (Engineers Ireland, 2014), 93.

86. Vertical members of the bridge at Cologne stiffened and were directly connected to cross beams at the top and bottom of the structure; they were intended to spread the load to the entire upper and lower chords and the rest of the lattice, and they resisted horizontal forces. Lohse, "Die Rhein-Brücke bei Cöln," *Zeitschrift für Bauwesen* (1863), 339. Pole states, "the girders are formed of open lattice-work, instead of plates, as Mr. Fairbairn had proposed; but in other respects there has not been much material departure from Mr. Fairbairn's designs"; i.e., a tubular bridge with lattice sides; Pole, *Fairbairn*, 237. Trautz, *Eiserne Brücken*, 71, 74. Couche refers to the lattice frame as a "poutre" ("beam"); he reported that lattice diagonal intersections at center of the frame were often not connected with pegs because this was the area of greatest bending; Couche, "Travaux d'art," 336–337.

87. O'Dwyer and Cox, "Boyne Viaduct," 1008.

88. Rosenberg and Vincenti, *Britannia Bridge*, 28; Smith, "Development," 114. Fairbairn said that others claimed that plates without the verticals "would crumple like a piece of leather." Fairbairn, *On the Application*, cited in Rosenberg and Vincenti, *Britannia Bridge*, 98n169. Karl Culmann, "Der Bau der eisernen Brücken," 193–194.

89. Chevalier, *Exposition*, 2:497–498, 576–579, fig. 19. Also, Ithamar A. Beard, "Description of an Improvement on the Bridge Patented by Itheil Towne [*sic*], Esq., the term of which patent has expired," *Journal of the Franklin Institute* n.s., 25 (1840): 175–177, and J. H. Garella, "Note sur deux ponts provisoires en charpente construits dans le système américain de M. Town," *Annales des ponts et chaussées* 2/2 (1842): 371–383, pl. 34.

90. Haupt, *General Theory* (1851), 151, cited in Dario Gasparini, Rachel H. Sangree, and Mathew Reckard, "Development of the American Truss," in Justin Christianson and Christopher H. Marston, eds., *Covered Bridges* (HAER, 2015), 127.

91. John A. Roebling, *Long and Short Span Railway Bridges* (Van Nostrand, 1869), 39, 50.

92. Sean T. James, "Town Lattice Covered Bridges: Regional Variations," Second National Covered Bridge Conference (Dayton, Ohio, 2013), https://www.intrans.iastate.edu/wp-content/uploads/sites/12/2019/03/James_Town-Lattice-Covered-Bridges-Regional-Variations.pdf.

93. Wilhelm Schwedler, "Resultate über die Construction der eisernen Brücken," *Zeitschrift für Bauwesen* 15 (1865): 331–340, fig. O1. Various configurations depicted in Philippe Croizette Desnoyers, *Cours de construction des ponts* (Dunod, 1885), figures 33–36 (see fig. 4.13).

94. Jennifer Tann, "Space, Time and Innovation Characteristics: The Contribution of Diffusion Process Theory to the History of Technology," *History of Technology* 17 (1995): 152.

95. Todd Craig Radford, "Town Lattice Truss: An Appropriate Bridge Technology for Developing Countries" (PhD diss., MIT, 2010).

96. Beam theory "was already well-developed by the end of the 18th century"; William Addis, *Structural Engineering* (Ellis Howard, 1990), 70. According to Addis, Ghega's theoretical work was based on the beam analogy. Carl Ghega, *Über nordamerikanischen Brückenbau und Berechnung des Tragungsvermögens der howeschen Brücken* (Kaul Fuss Witwe, Prandel, 1845). Ghega mostly focused on the Howe-lattice bridge, Springfield.

97. M. Rinke and T. Kotnik, "From Construct to Type—The Transformation of Constituents in the Development of Trusses," in *2nd International Conference on Structures and Architecture* (CRC, 2013), 1951–1952; Gasparini and Provost, "Early Nineteenth Century Developments in Truss Design," 22; Gasparini et al., "Engineering Design of Covered Bridges," 80–83; Holzer, "Polonceau Roof," 47.

98. See chapter 2 re "hollow solid." Garella, "Note sur deux ponts provisoires," 375–376. Pollack reported Garella's use of bolts; Julius Pollack, "Metallbrücke zur Uebersetzung des Royalkanals bei Dublin für die Dublin-Drogheda-Eisenbahn nach Town's nordamerikanischem Konstrukzionssystem gebaut im Jahre 1845," *Allgemeine Bauzeitung* 13 (1848): 19, 21.

99. Whipple, *Work on Bridge Building*, 119–20. Francis E. Griggs Jr., "Analytical Modeling: Its Beginning to 1850," in *American Civil Engineering History: Pioneering Years*, ed. Bernard G. Dennis Jr., *ASCE Conference Proceedings* 124 (2002): 97–103.

100. "Wände oder resp. Balken"; Julius Pollack, "Metallbrücke," 1, 22.

101. [Robert Stephenson], "Iron Bridges," in *Encylopaedia Britannica*, 8th ed. (Little, Brown, 1856), 12:597. Engineer Dallot felt that, whether lattice or plate, the rigidity was so great that they could be considered the same way; Dallot, in "Séance de 19 Janvier 1866," *Mémoires et compte rendu des travaux de la Société des ingénieurs civils* 33 (1866): 50.

102. Thomas Tancogne-Dejean et al., "Additively-Manufactured Metallic Micro-Lattice Materials for High Specific Energy Absorption Under Static and Dynamic Loading," *Acta Materialia* 116 (2016): 14–28, https://www.sciencedirect.com/science/article/pii/S1359645416304153 (see fig. 6.16); Tess Boissonneault, "New Plate-Lattice Architectures Could Enable Stiffest 3D Printed Structures," VoxelMatters, December 17, 2018, https://www.voxelmatters.com/mit-shrink-3d-nanoscale. Also, Thomas Leslie, *Chicago Skyscrapers, 1934–86: How Technology, Politics, Finance,*

and Race Reshaped the City (University of Illinois Press, 2023), 126–127, 210, 216–218. See also chapter 5, note 161.

103. Couche, "Travaux d'art," 349; 359–383, re fundamental bridge design questions of the time; "English Abuse of America," in Henry T. Tuckerman, *America and Her Commentators* (Scribner, 1864), 252–292.

104. [Charles Clinker], *Three Months in the United States and Canada: By the Editor of the "Western Gazette"* (Western Gazette and Pulman's Weekly News, 1886), 37 (appeared in serial form, November 1883 to September 1885). A Royal Institute of British Architects member stated, "It will be perceived that the principle of the lattice girder does not materially differ from that of the large tube . . . but the lightness and elegance of appearance . . . are greatly in favour of the lattice bridge"; John Clayton, "On the Bridges and Viaducts of the Present Day," *Artizan* 14 (1856): 278.

105. See chapter 3.

106. Clark, *Britannia*, 1: 42.

107. Karl Culmann, *Die graphische Statik* (Meyer & Zeller, 1866), 527–528, cited and translated (though slightly edited here) in Kurrer, *History*, 35–36.

108. Accompanied by contractor Charles Mare, Stephenson drove the last rivet; "exactly the 200,000,000th that have been used"; it "was followed by the waving of hats and deafening acclamations of the work-people." "Opening of the Britannia Bridge for Traffic," *Illustrated London News*, March 23, 1850, 192, 194. Preserved in the collection of the Science Museum (London) is a burnt segment of tube, with the painted words, "LAST RIVET / MARCH 5 1850 / R. STEPHENSON," https://collection.sciencemuseumgroup.org.uk/objects/co413782/railway-bridge-section

109. See Introduction; see also, Nicholas A. Bill, "Timber Bridge Construction on British and Irish Railways, 1840–1870: Scale of Construction and Factors Influencing Material Selection," *Construction History* 31 (2016): 75–98. As Britannia was being built, a railway company whose chief engineer was Robert Stephenson built Royal Border Bridge, a stone viaduct (still in use) that consists of twenty-eight arches over the River Tweed near the Scottish-English border. Construction (1847–1850) took so long that a temporary wooden bridge was erected so rail traffic could proceed while it was being completed; "Royal Border Bridge," https://en.wikipedia.org/wiki/Royal_Border_Bridge.

110. Sven-Erik Åström, "Technology and Timber Exports from the Gulf of Finland, 1661–1740," *Scandinavian Economic History Review* 23 (1975): 1–14, and Åström, "English Timber Imports." Berg and Hudson, *Slavery, Capitalism*, 131.

111. Gregory K. Dreicer, "Building Myths: The 'Evolution' from Wood to Iron in the Construction of Bridges and Nations," *Perspecta* 31 (2000), 130–141. Re material choices and their meanings, see Robert Fletcher and J. P. Snow, "History of Development of Wooden Bridges," ASCE *Transactions* 99 (1934): 314–408; Eric Schatzberg, "Decline of the Wooden Airplane in the United States," *Technology and Culture* 35 (1994): 34–69; Steven Lubar, "New, Useful, and Nonobvious," *American Heritage of Invention and Technology* (1990): 14.

112. Paul N. Wilson, "Britannia Tubular Bridge," *Industrial Archaeology* 9 (1972): 238–240, and Björn Åkesson, *Understanding Bridge Collapses* (Taylor and Francis, 2008), 165–169. Victoria Bridge had roofing made of "wood covered with tin," Hodges, *Construction*, 72.

113. P. S. A. Berridge, *Girder Bridge after Brunel and Others* (Maxwell, 1969), 157, mentions lattice and timber near its end, in reference to the Crumlin Viaduct, whose design the author states was inadequate; he notes a high rate of bridge failure in the United States. "Standard histories do little more than codify the [professional] community's own mythology and thus cannot reflect critically on the basic assumptions of that community"; Schatzberg, "Decline," 35.

114. Edwin T. Layton Jr., "Mirror-Image Twins: Communities of Science and Technology in 19th-Century America," *Technology and Culture* 12 (1971): 568; Henry Petroski, *Invention by Design: How Engineers Get from Thought to Thing* (Harvard University Press, 1996), 1.

115. Bertrand Lemoine, *L'architecture du fer, France: XIXe siècle* (Champ Vallon, 1986), 9.

116. Jewett, "Structural Antecedents," throughout; 352–353, on Fairbairn; Charles E. Peterson, "Inventing the I-Beam: Richard Turner, Cooper & Hewitt and Others," *APT Journal* 12, no. 4 (1980): 3–28; Sara E. Wermiel, "Introduction of the Rolled I-beam in the U.S.A. in the 1850s, Revisited," James Campbell et al., eds., *Iron, Steel and Buildings. Proceedings of the Seventh Conference of the Construction History Society* (CHS, 2020), 93–104. Fairbairn and Hodgkinson developed an asymmetric I-section for a cast-iron beam, because of its high compressive and low tensile strength; Bill Addis, "Iron Revolution: How Iron Replaced Traditional Structural Materials Between 1770 and 1870," in Mario Rinke and Joseph Schwartz, eds., *Before Steel* (Niggli, 2010), 42, 45.

117. S. Whipple, "Extracts from a Paper Read by S. Whipple, Civil Engineer, Member of the Society, On Truss Bridge Building," *ASCE Transactions* 1 (1872): 239–245, without citing source or date, states that an early (1840s?) discussion of lattice design was "mentioned as the result of 'an ingenious mode of constructing a *Deep Beam*,' as it was called."

118. Theodor Weishaupt, "Die Gitterbrücken der Preussischen Eisenbahn," *Zeitschrift für Bauwesen* 3 (1853): 475.

119. It was recognized by Herman Haupt, who wrote that it "act[ed] on the principle of a hollow cylinder." Haupt disapproved of it, however, because doubling the lattices only increased the section of the top and bottom chords by half; Haupt. *General Theory*, 151. The lattice double frame was "like a box girder"; Snow, "Wooden Bridge Construction," 36. Also, Lola Bennett and Dorottya Makay, "Contoocook Railroad Bridge," Written Historical and Descriptive Data, Historic American Buildings Survey, National Park Service, US Department of Interior, 2003, from Prints and Photographs Division, Library of Congress (HAER NH-38), 8, 15; Bennet and MacKay determined that paired frames "interacted less than expected" (43–44).

120. Schwedler, "Resultate," 335–336, pl. 1.

121. Alexandre Berthault-Ducreux, "Exposé des faits et des principes sur lesquels repose la solution des principales questions que soulèvent les chemins de fer et les autres voies de communication," *Annales des ponts et chaussées* 2/1 (1845), 49. Quote from Georges-Louis Leclerc, Comte de Buffon, *Discours sur le style* (1753).

122. Stephenson, "Iron Bridges," 12: 607 (probably ghostwritten by Stephenson associate G. D. Phipps, according to Rapley, *Britannia*, 116). Thomas Telford had succeeded similarly; Charles Hadfield, *Thomas Telford's Temptation* (Baldwin, 1993), 11.

123. For example, C. E. Fowler, "Bridge Evolution as Relating to Southern California," *Bridges and Framed Structures 1* (May 1899): 117. Henry Petroski, "Britannia Tubular Bridge," *American Scientist* 80 (1992): 220, offers a different variety of erasure: "Today the design of the [tubular] bridge looks utterly anomalous, without obvious structural antecedents or descendants."

124. R. J. Johnston, Peter J. Taylor, and Michael J. Watts, "Global Change at the End of the Twentieth Century," 9–10; Peter J. Taylor et al., "Remapping the World: What Sort of Map? What Sort of World?," in Johnston et al., *Geographies of Global Change*, 378–379 (Blackwell, 1995); Robert A. Beauregard, "Theorizing the Global-Local Connection," in *World Cities in a World System*, ed. Paul L. Knox and Peter J. Taylor (Cambridge University Press, 1995), 232–248.

125. Linda Colley, *Britons: Forging the Nation 1707–1837* (Yale University Press, 1992), 1, 5–7, chapters 4–6; 373, cites Peter Scott, *Knowledge and Nation* (Edinburgh University Press, 1990), 168. Colley notes that little research has been done on technology during the period of creation of British identity, despite an "unceasing torrent of books and articles on the impact of industrialisation on Britain," Colley, *Britons*, 3, 385n4.

126. George Buchanan, "On the Strength of Materials for Railway Bridges," *Journal of the Franklin Institute* 3rd ser., 17 (1849): 229–230 (from *Proceedings of the Royal Scottish Society of Arts* [1848] via *Civil Engineers and Architects Journal* [January 1849]).

127. Perdonnet cited in With, *Nouveau manuel*, 231. Hero worship in British engineering biographies, such as Buchanan, *Engineers*, 19, is beginning to weaken: "The tubular form . . . was a brilliant concept to meet a seemingly impossible challenge, but today it is hard not to wonder whether this level of innovation was essential," delicately wonders James Sutherland, "Iron Railway Bridges," in Bailey, *Robert Stephenson* (2003), 334. George Stephenson, like his son, was aware of his public image; Marsden and Smith, *Engineering Empires*, 243–244.

128. [Latimer Clark, with permission of Robert Stephenson], *General Description*, 10–11. The lions are preserved below the current roadway level.

129. Michele Speitz, "The Infrastructural Sublime and Imperial Landscape Aesthetics: Robert Southey, Poet Laureate, and Thomas Telford, Father of Civil Engineering," *European Romantic Review* 32 (2021): 50–53. Telford was first president of the Institution of Civil Engineers, 1820–1834; Stephenson was president, 1855–1857.

130. Ludmilla Jordanova, "Science and Nationhood: Cultures of Imagined Communities," in *Imagining Nations*, ed. Geoffrey Cubitt (Manchester University Press, 1998), 192–211; re innovation, 202. Re technological categories, see Grodal et al., "Coevolution of Technologies and Categories," 439–440. Re the painting, see Arthur Lucas, *John Lucas, Portrait Painter, 1828–1874* (Methuen, 1910), 64–67, pl. 47. Lucas made initial sketches during a meeting preceding the floating of the last tube; hence the presence Moorson, Claxton, sailor, and laborer.

131. [Latimer Clark, with permission of Robert Stephenson], *General Description*, 20.

132. Christine MacLeod, *Heroes of Invention: Technology, Liberalism and British identity, 1750–1914* (Cambridge University Press, 2007), 199–201, 210–211, 217, 233, 318–19; 2, 47–48, on Christianity. Re design and God's benevolence, see Stephen F. Eisenman, "Design in the Age

of Darwin," in *Design in the Age of Darwin* (Northwestern University, 2008), 4–5; also Christine MacLeod, "Nineteenth-Century Engineer as Cultural Hero," in Kelly and Kelly, eds., *Brunel*, 60–79. "Robert Stephenson," https://www.westminster-abbey.org/abbey-commemorations/commemorations/robert-stephenson#i14350, accessed 1 August 2021.

133. Christine MacLeod, *Heroes of Invention,* 1–6 and passim.

134. June Hargrove, "Shaping the National Image: Cult of Statues to Great Men in the Third Republic" in *Nationalism in the Visual Arts*, ed. Richard A. Etlin (University of New England Press, 1991), 49–63.

135. At a meeting of ICE (Ireland), the Boyne Viaduct (lattice) was called "that magnificent monument to Irish engineering skill, enterprise, and industry"; *Transactions of the Institution of Civil Engineers of Ireland* 5 (1860): 134. Michael Bailey, "Wider Horizons," in Bailey, *Robert Stephenson*, 144–149.

136. A "digression" within a discussion of Nordling's viaducts; "Great Iron Viaducts," 409.

137. For a typical list, see Robert Friedel et al., *Edison's Electric Light: Biography of an Invention* (Rutgers University Press, 1986), x. Re "forgotten," see George Dudley Seymour, "Ithiel Town: Architect," *Art and Progress* 3 (1912), 715.

138. On the fate of Town's home, see "Town-Sheffield Mansion//c. 1836-1957," *Buildings of New England*, January 19, 2024, https://buildingsofnewengland.com/tag/ithiel-town. Re brick patent, see chapter 2. Re Town's daughter, Roger Hale Newton, *Town and Davis, Architects* (Columbia University Press, 1942), 30, 49, 128. An 1826 Nathaniel Jocelyn painting of Etha Town (1807–1871), who married recent Yale graduate William Thompson Peters, indicates the wealth she enjoyed. Town's name does not appear in the official history, William H. Wisely, *American Civil Engineer, 1852–1974* (ASCE, 1974), 6–13; he is cited in Daniel Hovey Calhoun, *American Civil Engineer* (Technology Press/MIT Press, 1960), 69.

139. The First Presbyterian Church was rebuilt after Fayetteville's Great Fire of 1831. Two lattice diagonals have been cut to install a duct. Thanks to Catherine Bishir and Andrew Edmonds. Catherine W. Bishir, "Town and Davis" (2009), North Carolina Architects & Builders: A Biographical Dictionary, https://ncarchitects.lib.ncsu.edu/people/P000031; John L. Sanders, "Ithiel Town" (2009), https://ncarchitects.lib.ncsu.edu/people/P000032. The oldest, much-altered, lattice bridge is Haverhill-Bath Bridge (1829) over the Ammonoosuc, New Hampshire.

140. Byrom, "William Fairbairn," 202. Gregory Dreicer, "Building Bridges and Boundaries: The Lattice and the Tube, 1820–1860," *Technology and Culture* 51, no. 1 (2010): 126–163.

141. Byrom, "William Fairbairn," 11.

142. Clark even noted that, had it been possible, the plates would have been installed diagonally; Edwin Clark, *Britannia and Conway Tubular Bridges* (Day and Son/Weale, 1850), 2:570. L. Gordon in Julius Weisbach, *Principles of the Mechanics of Machinery and Engineering* [translation of *Lehrbuch der Ingenieur-und Maschinen-Mechanik*], trans. L. Gordon (London, 1847–1848), 2: 404. Gordon noted (398) that the Grubenmann bridge at Wettingen included "the introduction of a *roof* as an integral part of *the constructive strength of the* bridge" (emphasis in original). Auguste

Perdonnet, "Histoire des chemins de fer," *L'Ami des sciences* 6 (1860): 746. Byrom, "William Fairbairn," 202, proposes that "decoupling" of bridge pier and superstructure construction began with the Britannia; this "first recorded example" tale erases decades of bridge building.

143. Byrom, "William Fairbairn," 192, 203.

144. Smith, "Development," 108–115; Byrom, "William Fairbairn," 203–204, 206, 211; the market for these bridges took off in 1847 as a result of the Dee Bridge failure.

145. David Thelen, "Making History and Making the United States," *Journal of American Studies* 32 (1998): 373–397.

146. Robert Thorne, "Introduction," in Robert Thorne, ed., *Structural Iron and Steel, 1850–1900* (Ashgate, 2000), xxii.

147. Fairbairn recognized intercultural exchange even as he expressed frustration at his unsuccessful attempts to sell bridges abroad; letter, Fairbairn to Alexander von Humboldt, 15 April 1850, in Fairbairn, *On the Application*, 4th ed., 345. Also, Robert Stephenson cited in Derrick Beckett, *Stephensons' Britain* (David & Charles, 1984), 175.

148. Ludmilla Jordanova, "Science and National Identity," in Roger Chartier and Pietro Corsi, *Sciences et langues en Europe* (Office for Official Publications of the European Communities, 1996), 230.

149. David E. Sanger et al., "Profits and Pride at Stake, Race to Vaccine Intensifies: Government, Academic and Business Labs Vie in Rush that May Imperil Safety," *New York Times*, May 3 2020, 1, 14–15.

150. Storytellers often blur the distinction between technology and industry; for example, technologies associated with industries based in Britain are sometimes labeled "British." D.L. Burn, "Genesis of American Engineering Competition, 1850–1870," *Economic History* 2 (1931): 292–311.

151. J.R. Mosse, "American Timber Bridges," *Transactions of Institution of Civil Engineers* 22 (1862–63): 305, 317.

152. Stephenson's railway consulting work in European countries including Austria, Denmark, and Sweden foundered in 1850s, for reasons perhaps related to the tube. Bailey, "Wider Horizons," 144–149, 161.

153. Émile With, *Les accidents sur les chemins de fer, leurs causes, les règles à suivre pour les éviter* (Mallet-Bachelier, 1854), 23, 139; Émile With, *Railroad Accidents* (Little, Brown, 1856), 28. In With's translation of Stephenson and Brunel's *Report of Commissioners* testimony, "lattice bridge" is repeatedly translated as "pont américain."

154. John Anderson, in his 1857 report on machinery in the United States, cited in David A. Hounshell, *From the American System to Mass Production, 1800–1932* (Johns Hopkins University Press, 1984), 64. Re iron, see Guillaume Lambert, *Voyage dans l'Amérique du nord, en 1853 et 1854, avec notes sur les expositions universelles de Dublin et de New-York* (Hayez, 1855), reference to England, 149–150.

155. James, "Some Steps in the Evolution," 153. It was not "Britain" but entrepreneurs and engineers who designed and built bridges.

156. Neil Smith and Ward Dennis, "Restructuring of Geographical Scale: Coalescence and Fragmentation of the Northern Core Region," *Economic Geography* 63 (1987): 162–165, 180.

157. Nathan Rosenberg, "Why in America?," *Exploring the Black Box: Technology, Economics, History* (Cambridge University Press, 1994), 109.

158. E.g., Kranakis, *Constructing a Bridge*, re James Finley; D.L. Burn, "Genesis;" Margaret T. Hogden, *Change and History: A Study of the Dated Distributions of Technological Innovations in England*, Viking Fund Publications in Anthropology 18 (Wenner-Gren, 1952), throughout and 189–190, 200. Also, I.K. Steele, *The English Atlantic 1675–1740: An Exploration of Communication and Community* (Oxford University Press, 1986), and J. M. Blaut, "Diffusionism: A Uniformitarian Critique," *Annals of Association of American Geographers* 77 (1987): 30–47.

159. Jonson Miller, *Engineering Manhood: Race and the Antebellum Virginia Military Institute* (Lever, 2020), 8.

160. David Armitage, "Is There a Pre-History to Globalization?" in *Comparison and History: Europe in Cross-National Perspective*, ed. Deborah Cohen and Maura O'Connor (Routledge, 2004), 165–176. Re recent (1983) origin of "globalization," see Barnaby J. Feder, "Theodore Levitt, 81, Coined the Term 'Globalization,'" *New York Times*, July 6, 2006, B7; also O'Rourke and Williamson, *Globalization and History* (MIT Press, 1999). Re globalization debates and histories, see Warwick E. Murray and John Overton, *Geographies of Globalization*, 2nd ed. (Routledge, 2015).

161. "'Made in China' is mostly made elsewhere—by multinational companies in Japan, South Korea, Taiwan and the United States that are using China as the final assembly station"; David Barboza, "Some Assembly Needed: China as Asia's Factory," *New York Times*, February 9, 2006, C1, C7. See also, Peter S. Goodman, "Chinese Firms Spend Billions Building Factories in Mexico," *New York Times*, February 11, 2023, B1, B4, and Andrew Carnegie, *The Gospel of Wealth and Other Timely Essays* (Century, 1901), 252.

162. Adam Smith, *The Wealth of Nations*, ed. Edwin Cannan (1776; reprint, Modern Library, 1937), 11–12; Sanam Yar, "Who Made Your Clothes?," *New York Times*, December 19, 2019, https://www.nytimes.com/2019/12/19/style/garment-workers-around-the-world.html; E. Benjamin Skinner, "The True Cost of a $12 T-Shirt," *New York Times*, April 24, 2023, https://www.nytimes.com/2023/04/24/opinion/fast-fashion-apparel-worker-conditions-rana-plaza.html.

163. Chris Buckley et al., "One Nation Under Xi: How China's Leader Is Remaking Its Identity," *New York Times*, October 12, 2022, A16.

164. On Braudel's narrative, see Peter J. Taylor, *World City Network: A Global Urban Analysis* (Routledge, 2004), 14.

165. See early nineteenth-century example in V. S. Virginskii, *Robert Fulton, 1765–1815* (Amerind, 1976), 70–93. Re "Americanness" of "modern industrial technology," see the case of Gribeauval, *Dictionnaire de biographie française* 13 (1985), 1209–1210; Smith, "Military Entrepreneurship," 67–68; Hounshell, *From American System*, 25–28.

CHAPTER 7

1. Initial quote: Timothy Dwight, *Travels in New-England and New-York* (Author, 1822), 4: 231; (reprint, Belknap, 1969), 4: 161. James Fenimore Cooper noted that the American continent was sometimes confused with a small nation; [James Fenimore Cooper], *Notions of the Americans: Picked Up by a Travelling Bachelor* (Carey, Lea and Carey, 1828), 1:17–18. Re railroads as magical, see William Cronon, *Nature's Metropolis: Chicago and the Great West* (Norton, 1991), 72–73; Leo Marx, "Railroad-in-the-Landscape," in *The Railroad in American Art: Representation of Technological Change*, ed. Susan Danly and Leo Marx (MIT Press, 1988), 185–189. See also, Marina Warner, *Once upon a Time: A Short History of Fairy Tale* (Oxford University Press, 2014), xiv, and Kristian Moen, *Film and Fairy Tales: The Birth of Modern Fantasy* (Bloomsbury, 2013), xiv–xvi.

2. Marina Warner, *Fantastic Metamorphoses, Other Worlds* (Oxford University Press, 2002), 17–18; also, Warner, *Once upon a Time*, 49, and Peter J. Bowler, *The Invention of Progress: The Victorians and the Past* (Basil Blackwell, 1989).

3. [Cooper], *Notions of the Americans*, 2:152, cited in Charles C. Sanford, *Quest for America 1810–24* (Anchor, 1964), 393.

4. Caroline Winterer, *Culture of Classicism* (Johns Hopkins University Press, 2002), 128; Michael Zakim, "Importing the Crystal Palace," in Beckert and Desan, *American Capitalism* (Columbia University Press, 2018), 341. Read Bain, "Technology and State Government," *American Sociological Review* 2 (1937), 860; David E. Nye, *America as Second Creation* (MIT Press, 2003) and Franklin Kelly, *Frederic Edwin Church and the National Landscape* (Smithsonian, 1988), 55. Re Cooper's renown and unpopular *Notions of the Americans* (1828), see "James Fenimore Cooper and the Image of America," in Wasserman, *Exotic Nations* (Cornell University Press, 1994), 154–185. The elitist attitudes of Cooper and Morse eventually caused them to "turn sour" on America; Paul J. Staiti, *Samuel F.B. Morse* (Cambridge University Press, 989), 240–241.

5. Maxine Berg and Pat Hudson, *Slavery, Capitalism and the Industrial Revolution* (Polity Press, 2023), 87–89.

6. Roger Burlingame, *Machines that Built America* (Signet, 1955), 61. Joseph Wickham Roe, *Connecticut Inventors* (Yale University Press, 1934), 3, cited patent statistics to "show that activity of invention has been higher here [Connecticut] than in any other state of the country, and for many years it probably has been higher than in any other similar district in the world."

7. Meyer, *Networked Machinists*, 84–85, 20–21. Re federal armories, see Merges, *American Patent Law*, 53–55. Re US industrial regions, see Neil Smith and Ward Dennis, "The Restructuring of Geographical Scale: Coalescence and Fragmentation of the Northern Core Region," *Economic Geography* 63 (1987): 162–165. Re complex networks, see Peter J. Taylor, *World City Network* (Routledge, 2004), 57–60.

8. Joseph W. Roe, *Connecticut Inventors*, 3, went on to list "climate, land, and social conditions." Roe identification in David A. Hounshell, *From American System to Mass Production 1800–1932* (Johns Hopkins University Press, 1984), 335–336.

9. Chevalier, *Lettres* (1837), 1: xiv–xvi. Re Chevalier's views, see Daniel Hovey Calhoun, *American Civil Engineer* (MIT Press, 1960), 195.

10. Edwin T. Layton, Jr, "European Origins of the America Engineering Style of the Nineteenth Century," in Nathan Reingold and Mark Rothenberg, eds., *Scientific Colonialism* (Smithsonian, 1987),156, 151; Edwin T. Layton Jr., "Millwrights and Engineers, Science, Social Roles, and Evolution of the Turbine in America," in *The Dynamics of Science and Technology: Social Values, Technical Norms and Scientific Criteria in the Development of Knowledge*, ed. Wolfgang Krohn et al. (Riedel, 1978), 78. Hagen attributed the Industrial Revolution to "the British [being] a more innovational people"; Everett E. Hagen, "British Personality and the Industrial Revolution: The Historical Evidence," in *Social Theory and Economic Change*, ed. Tom Burns and S. B. Saul (Tavistock, 1967), 35–66. For a demolition of American stereotyping, see Nathan Reingold, "American Indifference to Basic Research: A Reappraisal," in *Nineteenth-Century Science*, ed. George H. Daniels (Northwestern University, 1972), 39, 45, 46, 49, 50, 59; Irene Cheng, "Structural Racialism in Modern Architectural Theory," in "Irene Cheng et al., *Race and Modern Architecture: A Critical History from the Enlightenment to the Present* (University of Pittsburgh Press, 2020), 142–143. Re ongoing use of "Anglo-Saxon" in France, see Billie Melman, "Claiming the Nation's Past: The Invention of an Anglo-Saxon Tradition," *Journal of Contemporary History* 26 (1991): 575–595; Emile Chabal, "The Rise of the Anglo-Saxon: French Perceptions of the Anglo-American World in the Long Twentieth Century," *French Politics, Culture & Society* 31 (2013): 24–46.

11. Amy Slaton, *Race, Rigor, and Selectivity in U.S. Engineering: The History of an Occupational Color Line* (Harvard University Press, 2010), ix–x; Elizabeth Deans Romariz, "Mobility and Memory: British Architects and Their Tools, ca. 1700–1870," *Material Culture Review* 85 (2017): 35–46.

12. Re genetics, see Noah A. Rosenberg et al., "Interpreting Polygenic Scores, Polygenic Adaptation, and Human Phenotypic Differences," *Evolution, Medicine, and Public Health* 2019 (2018): 26–34. Still, 20 percent of white Americans attribute income inequality to genetics rather than racial exclusion; Ann Morning et al., "Socially Desirable Reporting and the Expression of Biological Concepts of Race," *Dubois Review: Social Science Research on Race* (2019): https://doi.org/10.1017/S1742058X19000195. Re exclusion of women, see Deborah J. Merritt, "Hypatia in the Patent Office: Women Inventors and the Law, 1865–1900," *American Journal of Legal History* 35 (1991): 235–306; Margaret Talbot, "Out of Frame," *New Yorker*, November 4, 2019, 69–73. Re Black inventors, see Brian L. Frye, "Invention of a Slave," *Syracuse Law Review* 68 (2018): 181–229; Kara W. Swanson, "Race and Selective Legal Memory: Reflections on 'Invention of a Slave,'" *Columbia Law Review* 120 (2020): 1077–1118; Lisa D. Cook, "Violence and Economic Activity: Evidence from African American Patents, 1870–1940," *Journal of Economic Growth* 19 (2014): 221–257.

13. Calhoun, *American Civil Engineer*, 177–181, 193–199.

14. *Plan, Durchschnitt und Aufriss der drey merkwerdigsten hoelzernen Bruecken in der Schweiz, nämlich der beruehmten Schaffhauser-Bruecke, ueber den Rhein, jener zu Wettingen, ueber die Limmat* (Mechel, 1803), 14, and A.R. Emy, *Traité de l'art de la charpenterie* (Carilian-Goeury et Anselin, 1837, 1841), 2: 393–394, apparently responding to scientist Christof Jezeler.

15. Roger Hale Newton, *Town and Davis, Architects* (Columbia University Press, 1942), 18–19. Bender, *New York Intellect* (Knopf, 1987), 11–12, 26, 184–191, and throughout.

16. Town also knew painter John Trumbull and Eli Whitney. "Silliman College," Yale, Slavery and Abolition, http://www.yaleslavery.org/WhoYaleHonors/silliman.html. Eric Herschthal, "The Missing Link: Conservative Abolitionists, Slavery, and Yale," *Black Perspectives*, March 31, 2017, https://www.aaihs.org/the-missing-link-conservative-abolitionists-slavery-and-yale. Re Morse's nationalism and anti-Catholicism, see Daniel Walker Howe, *What Hath God Wrought: The Transformation of America, 1815–1848* (Oxford University Press, 2007), 692–693, and Kenneth Silverman, *Lightning Man: Accursed Life of Samuel F. B. Morse* (Knopf, 2002), 132–142. Re Calvinism, see Staiti, *Morse*, 6–8, 223. Re Whitney, see David R. Meyer, *Networked Machinists* (John Hopkins University, 2006), 85–86; Nancy Finley, "North and South: The Legacy of Eli Whitney," Connecticut History, https://connecticuthistory.org/north-and-south-the-legacy-of-eli-whitney. Nathaniel Jocelyn, born in New Haven, knew Town, Whitney, and Morse, was associated with the National Academy of Design, had a New York office during the 1840s, and painted Town's portrait; David B. Dearinger, ed., "Nathaniel Jocelyn," *Paintings and Sculpture in the Collection of the National Academy of Design*, vol. 1: *1826–1925* (Hudson Hills, 2004), 318–319. Re Whitney and Jocelyn, see Chieffo-Reidway, "Nathaniel Jocelyn," 16–17; John Donoghue, "Alexander Jackson Davis, Romantic Architect, 1803–1892" (Ph.D. diss., New York University Press, 1979), 99. Re Morse's view of slavery, see "Morse College," http://www.yaleslavery.org/WhoYaleHonors/morse.html. On Silliman's ties to Whitney and views on slavery, see Eric Herschthal, *The Science of Abolition: How Slaveholders Became the Enemies of Progress* (Yale University Press, 2021), 131–46. Re Trumbull, Jocelyn, et al., see Susan B. Matheson, *Art for Yale: A History of the Yale University Art Gallery* (Yale University Art Gallery, 2001), 11–25. Re Town's social world, see David P. Handlin, "New England Architects in New York, 1820–40," *American Quarterly* 19 (1967): 685–687. Town received in 1825 an honorary master of arts, https://secretary.yale.edu/programs-services/honorary-degrees/since-1702?field_degrees_value=Master+of+Arts&field_year_value=1825&keys=. The *Fayetteville Observer*, October 13, 1825, 3, reported on degree and Town, "formerly of this place and ingenious constructor of Clarendon Bridge." Yale retains no materials on the award; email, Liz Quercia, Assistant Director of Institutional Affairs, Office of Institutional Affairs, Yale University, 1 February 2021. On speculation about Town's early education, see Seymour, "Ithiel Town: Architect," 715. Re individuals cited here, see Antony Dugdale et al., *Yale, Slavery and Abolition* (Amistad Committee, 2001), http://www.yaleslavery.org/YSA.pdf. Town's partner Davis knew Jacob Bigelow of Harvard; Donoghue, "Alexander Jackson Davis," 74–82, and William Dunlap, *History of the Rise and Progress of the Arts of Design in the United States* (Scott, 1834), 2: 410.

17. See chapter 5. "Timothy Dwight College," Yale, Slavery and Abolition, http://www.yaleslavery.org/WhoYaleHonors/dwight1.html; "Naomi, Slave of T. Dwight," Yale, Slavery and Abolition, http://www.yaleslavery.org/WhoYaleHonors/dwight2.html, and "Yale and the South," Yale, Slavery and Abolition, http://www.yaleslavery.org/TownGown/south.html. Nathaniel's father Simeon was known for antislavery views. Town's views on slavery are unknown.

18. Albert H. Marckwardt, "The Chronology and Personnel of the Bread and Cheese Club," *American Literature* 6 (1935): 389–399, and Nelson F. Adkins, "James Fenimore Cooper and the Bread and Cheese Club," *Modern Language Notes* 47 (1932): 71–79. Re Rome and Cooper-Morse

relations in New York and Italy, see Wayne Franklin, "James Fenimore Cooper and American Artists in Europe," in *Transatlantic Romanticism*, ed. Andrew Hemingway and Alan Wallach (University of Massachusetts Press, 2015), 151–155. On Town's social circles, see Newton, *Town and Davis*, 18–19. Re Cole, see Bender, *New York Intellect*, 134–135. Re Morse and Silliman, see Sarah Kate Gillespie, "Morse and 'Mechanical Imitation," in *Samuel F. B. Morse's Gallery of the Louvre and the Art of Invention*, ed. Peter John Brownlee (Terra Foundation for American Art, 2014), 101–102, and Staiti, *Morse*, 6–7. Town's partner Davis also knew Morse and Cole; Amelia Peck, ed., *Alexander Jackson Davis, American Architect 1803–1892* (Rizzoli/Metropolitan Museum of Art, 1992). Re Cooper, see Daniel Walker Howe, *What Hath God Wrought* (Oxford University, 2007), 234. Re Town's connections, see Donoghue, "Alexander Jackson Davis"; re Town's New York library and the cultural elite, see "Architects and Architecture of New York," *Brother Jonathan*, May 20, 1843, 61–62. Re 1820 date of Town's New York office, see Quinn, "Library of Ithiel Town," 22.

19. Moreover, Northeastern individuals and institutions were foisting national character imaginings southward and westward, to the consternation of those with other ideas of "America"; James A. Hijaya, "Why the West Is Lost," *William and Mary Quarterly* 51 (1994): 276–292.

20. Compare the landscape tradition in Jocelyn's *Town's Bridge* (1821), Cole's *River in the Catskills* (1843), and Frederic Church's *New England Scenery* (1851). Miller, *Empire*, 167–168, 190–195, and passim; Thomas Cole, *River in the Catskills*, Museum of Fine Arts Boston, https://collections.mfa.org/objects/33073; Franklin Kelly and Gerald L. Carr, *The Early Landscapes of Frederic Edwin Church, 1845–1854* (Amon Carter Museum, 1987), 64–67.

21. Ellwood C. Parry III, "Thomas Cole's Imagination at Work in The Architect's Dream," *American Art Journal* 12 (1980): 57, 59. Thomas Cole, "Essay on American Scenery," *American Monthly Magazine* n.s. 1 (1836): 1–12 and Wallach, "Thomas Cole and Transatlantic," 210–15. Cole was born in England and trained there as an engraver. Re historic-national issues in his artworks, see Martin Christadler, "Romantic Landscape Painting in America: History as Nature, Nature as History," in *American Icons: Transatlantic Perspectives on Eighteenth- and Nineteenth-Century American Art*, ed. Thomas W. Gaehtgens and Heinz Ickstadt (Getty Center, 1992), 93–99. Also, Tim Barringer et al., *Picturesque and Sublime: Thomas Cole's Trans-Atlantic Inheritance* (Thomas Cole National Historic Site, Yale University, 2018). Rebecca Ayres Schwartz, "Thomas Cole on Architecture: Picturing the Gothic" (PhD diss., University of Delaware, 2016), 26–71; re Cole and Cooper, 41–42; re the dreaming figure, 68–70.

22. Re *Architect's Dream*, see Ellwood C. Parry III, "Thomas Cole's Imagination at Work in The Architect's Dream," *American Art Journal* 12 (Winter 1980): 41–59; William H. Pierson, *American Buildings and Their Architects 2* (Oxford University, 1978), 2:127–129; Randall C. Griffin, "The Untrammeled Vision: Thomas Cole and the Dream of the Artist," *Art Journal* 52 (1993): 66–73; Charles Colbert, "Dreaming Up *The Architect's Dream*," *American Art* 6 (1992): 78–91; Abraham Davidson, *The Eccentrics and Other American Visionary Painters* (Dutton, 1978), 23–24; Julia M. Treuttner, *Aspirations for Excellence: Alexander Jackson Davis and the First Campus Plan for the University of Michigan, 1838* (University of Michigan Press, 2003), 45–46; Barbara Novak, *American Painting of the Nineteenth Century*, 3rd. ed. (Oxford University Press, 2007),

244; Barbara Novak, *Nature and Culture: American Landscape and Painting, 1825–1875*, 3rd ed. (Oxford University Press, 2007), 140–142. Re *Course of Empire*, see Linda S. Ferber, *The Hudson River School: Nature and the American Vision* (New York Historical Society / Skira Rizzoli, 2009), 190–198; Schwartz, "Cole," 56. Also, Alan Wallach, "Thomas Cole's 'River in the Catskills' as Antipastoral," *Art Bulletin* 84 (2002): 340–342; Kenneth W. Maddox, "Thomas Cole and the Railroad: Gentle Maledictions," *Archives of American Art Journal* 30 (1990): 146–154; Elizabeth Kornhauser and Tim Barringer, *Thomas Cole's Journey: Atlantic Crossings* (Metropolitan Museum of Art / Yale University Press, 2018), 221–223.

23. Christine MacLeod, *Heroes of Invention* (Cambridge University, 2007), 13–14; Dell Upton, "Inventing the Metropolis: Civilization and Urbanity in Antebellum New York," in Catherine Hoover Voorsagner and John K. Howat, eds., *Art and the Empire City: New York, 1925–61* (Metropolitan Museum of Art; New Haven: Yale University Press, 2000), 35–36. Re "American" style, see W. Barksdale Maynard, *Architecture in the United States, 1800–1850* (Yale University Press, 2002), 51–71. On fig. 7.3, see Richard Sanders Allen, *Covered Bridges of the Middle West* (Greene, 1970), 7–9 and "Architects and Architecture of New York," *Brother Jonathan*, June 17, 1843, 181.

24. Re moralizing intentions of Town's architecture, see David P. Handlin, "New England Architects in New York, 1820–1840," *American Quarterly* 19 (1967): 681–695. Re Greek Revival, see Maynard, *Architecture in the United States*, 219–279; especially 244–252, on "Americanness"; also chap. 2. Re Greece and slavery, see Winterer, *Culture of Classicism*, 74–76. Re enslaved builders, see Catherine W. Bishir, "Black Builders in Antebellum North Carolina," *North Carolina Historical* Review 61 (1984): 423–461. James A. Glass, "The Architects Town and Davis and the Second Indiana Statehouse," *Indiana Magazine of History* 80 (1984): 336.

25. Mitchell Schwarzer, "The Sources of Architectural Nationalism," in *Nationalism and Architecture*, ed. Raymond Quek et al. (Ashgate, 2012), 19–38.

26. Dan Sousa, "'Northampton Mass' by Enoch Wood & Sons: A Design Source Discovery," *TCC Bulletin* 23/1 (2022): 8–9; Donoghue, "Alexander Jackson Davis," 24–25, 100–101; Jan Cunningham, "Samuel Wadsworth Russell House," US Dept. of the Interior, National Park Service, National Historic Landmark Nomination (February 2000).

27. Residences Town designed, the Bowers House (1827), in Northampton, MA, and the Russell House (built 1828–1830), Middletown, CT, helped establish the Greek Revival. He collaborated with New Haven–based sculptor and inventor Hezekiah Augur on the design of a medal commemorating Thanksgiving and New Haven's Green; Laura R. Macaluso and Laura A. Macaluso, "New Haven's Second Centennial Medal and the Origins of City Identity," *Material Culture* 48 (2016): 1–15.

28. Donoghue, "Alexander Jackson Davis," 113, 234–347, and Dunlap, *History of the Rise*, 2:410. Re Davis' artistic skills, 182; James Gallier, *Autobiography of James Gallier, Architect* (Brière, 1864; reprint, Da Capo, 1973), 18–19. Town (1839), back cover, also announced his intention to publish a book on bridges.

29. Connecticut Office of the State Historian, "July 5: New Haven Builds a New State Capitol Building," Today in Connecticut History, July 5, 2019, https://todayincthistory.com/2019/07

/05/july-5-new-havens-new-state-house/; Derek Strahan, "State House, New Haven, Connecticut," Lost New England, July 14, 2019, https://lostnewengland.com/2019/07/state-house-new-haven-connecticut/. The upper exterior walls of this building (and perhaps the columns) were built of brick and covered with stucco painted white to resemble the marble on the lower portion of the walls; J. S. Buckingham, *The Eastern and Western States of America* (Fisher, [1842]), 1:387. Also, Office of the House Clerk, Connecticut General Assembly, "The State Houses," https://www.cga.ct.gov/hco/his-places.asp. For an account of the demolition, see "Gone . . . But Not Forgotten," *Nineteenth Century* (Autumn 1976): 4–5.

30. For the "French engineering tradition" and "continental tradition," see Edwin T. Layton, Jr., "Millwrights," 63, 77. Eric Hobsbawm, "Introduction: Inventing Traditions," in *The Invention of Tradition*, ed. Eric Hobsbawm and Terence Ranger (Cambridge University Press, 1983), 1–14.

31. Fred Kniffen, "American Covered Bridge," *Geographic Review* (1951), 119.

32. Joseph D. Conwill, "Builders and Practices," in Christianson and Marston, *Covered Bridges* (HAER, 2015), 161, 163.

33. "Striking differences in behavior seem completely out of proportion to the small divergences in the situation of these three closely similar regions"; Julius Rubin, "Canal or Railroad: Imitation and Innovation in the Response to the Erie Canal in Philadelphia, Baltimore, and Boston," *Transactions of American Philosophical Society* n.s., 51/7 (1961), 9. Francis T. Evans, "Roads, Railways, and Canals: Technical Choices in 19th-Century Britain," *Technology and Culture* 22 (1981), 1–34; Edmund N. Todd, "A Tale of Three Cities: Electrification and the Structure of Choice in the Ruhr, 1886–1900," *Social Studies of Science* 17 (1987), 387–412; James C. Burke, "North Carolina's First Railroads: A Study in Historical Geography" (Ph.D. diss., University of North Carolina, 2008).

34. Thomas C. Cochran, "Cotton Textiles and Industrialization," in Randolph Shipley Klein, ed., *Science and Society in Early America* (American Philosophical Society, 1986), 268–269, and Thomas Cochran, *Frontiers of Change* (Oxford University Press, 1981), 14–15. See also Sarah Kate Gillespie, "Morse and 'Mechanical Imitation,'" in *Samuel F.B. Morse's Gallery of the Louvre and the Art of Invention*, ed. Peter John Brownlee (Terra Foundation for American Art, 2014), 108n8, and AnnaLee Saxenian, *Regional Advantage: Culture and Competition in Silicon Valley and Route 128* (Harvard University Press, 1994). Meyer, *Networked Machinists*, 1–2. Re constraints, see Timothy J. Sturgeon, "How Silicon Valley Came to Be," in *Understanding Silicon Valley: The Anatomy of an Entrepreneurial Region*, ed. Martin Keeney (Stanford University Press, 2000), 16–17. Also "Silicon Valley Habitat," in *The Silicon Valley Edge: A Habitat for Innovation and Entrepreneurship*, ed. Chong-Moon Lee et al. (Stanford University Press, 2000), 2–15.

35. J. Parker Snow, chief engineer of Boston and Maine, noted that the wooden lattice bridge was "frequently referred to as peculiarly unscientific and wasteful of timber. It is, however, the best of the purely wooden bridges, and its present survival here and its economy over all other types disproves its wastefulness." Snow, "Wooden Bridge Construction," 36; "It is the only kind that can be built in competition with iron at present prices" (39). Of Boston and Maine's 1,561 bridges 1,085 were composed of wood and bolts in 1895 (31). See also, Richard Sanders

Allen, *Covered Bridges of the Northeast*, 99, and Fletcher and Snow, "History," 333. Re Snow's high-rise work, see J. Parker Snow in Henry H. Quimby, "Wind Bracing in High Buildings," *ASCE Transactions* 27 (1892): 228–229. The railroad's former names were Contoocook Valley and Concord & Claremont; from 1887, Boston and Maine. Contoocook Railroad Museum, "World's Oldest Surviving Railroad Bridge," https://www.contoocookdepot.org/bridge.html.

36. Thomas Rodd, "Memoir of Felician Slataper," *ASCE Transactions* 58 (1907): 540–541; "Bridge over the Alleghany," *Journal of the Franklin Institute* 3rd ser., 55 (1868): 30–34.

37. Board of Railroad Commissioners of the State of New York, *Strains on Railroad Bridges of the State*, ed. Charles F. Stowell (Weed, Parsons, 1891), passim, contains calculations and drawings for every span of every railroad bridge in New York; "lattice bridge" in this publication includes structures in which diagonals intersect only once; I counted as lattice bridges only with two or more intersections. My figures are approximate; note that the number of spans and length per span varies per bridge. See also George E. Gray, "Notes on Early Practice in Bridge Building," *ASCE Transactions* 37 (1897): 1–16.

38. Boller, cited in "Historical Sketch of the Development of American Bridge Specifications," *Bulletin, American Railway Engineering Association* 25 (October 1923): 20–22, reprinted from *AREA Proceedings* 6 (1905). According to Boller, "schools" of "accidental" formation. During an 1892 discussion of bridge design, an engineer asserted, "With few exceptions, designers accustomed to one form of bridge cannot properly design the other"; A. J. Swift, in J. A. L. Waddell, "Some Disputed Points in Railway Bridge Designing," *ASCE Transactions* 26, no. 1 (1892): 222–223. George H. Thomson, in Gray, "Notes," 5, says, "[Howard] Carroll designed in riveted lattice."

39. J. G. Van Zandt, "Outline History of Railway Bridge Building in the U.S.," *Railway Engineering and Maintenance of Way* 8 (1912): 453–454. Re Pennsylvania Railroad bridge design in the 1860s and 1870s, see Dario A. Gasparini, "Joseph M. Wilson, Henry Petit and the Iron Truss Bridges of the Pennsylvania Railroad," in Mascarenhas-Mateus and Pires, *History of Construction Cultures: Proceedings of 7th International Congress on Construction History*, ed. Joao Mascarenhas-Mateus and Ana Paula Pires, eds., (CRC, 2021), 379–386.

40. George E. Gray in J. E. Greiner, "American Railroad Viaduct—Its Origin and Evolution," *ASCE Transactions* 24 (1891): 269; S. Whipple, "Extracts from a Paper Read by S. Whipple, Civil Engineer, Member of the Society, On Truss Bridge Building," *ASCE Transactions* 1 (1872): 240–241.

41. Squire Whipple, *A Work on Bridge Building: Consisting of Two Essays, the One Elementary and General, the Other Giving Original Plans and Practical Details for Iron and Wooden Bridges* (Curtiss, 1847), iii; Whipple in Greiner, "American Railroad Viaduct," 363–364. It's unlikely Whipple's work remained within the United States; his publication was publicized through *American Railroad Journal* articles by Whipple and his friend, engineer E. F. Johnson. See chapter 6; Whipple devoted two pages (119–120) of his book to the Dee Bridge collapse in England earlier in the year, indicating access to transatlantic networks.

42. Census according to Wisely, *American Civil Engineer 1852–1974* (ASCE, 1974), 5–7, who notes that forty people tried to form a professional organization in Baltimore in 1839. Calhoun

notes a total of 997 West Point graduates working in the United States in 1850 as civil engineers; 634 were in the army; Calhoun, *American Civil Engineer*, 206–208. There were about 300 engineers in New York, with one third members of "New York Institute of Civil Engineers," *Scientific American* 5 (February 1850), 157.

43. William J. Gilmore, *Reading Becomes a Necessity of Life* (University of Tennessee Press, 1989), 22–27; according to Gilmore, literacy conditions in the Northeast were "an utter anomaly" (27).

44. Christine Lehmann and Bertram Maurer, *Karl Culmann und die graphische Statik* (Ernst & Sohn, 2006), 83, note that Whipple's publication was not found in Culmann's library. See note 41 above.

45. "Memoir of George Edward Gray," *ASCE Transactions* 76 (1913): 2; "New York Central Railroad," https://en.wikipedia.org/wiki/New_York_Central_Railroad.

46. Smaller bridges were built previously; some included cast-iron, which NYCRR did not allow. See Gray, "Notes," 2; re Rider Bridge, see chapter 4.

47. Gray in Greiner, "American Railroad Viaduct," 369. The discussion in his explicitly "American" (357) article mentions the Dirschau, Nogat, and Cologne bridges. C. Shaler Smith, *Comparative Analysis of the Fink, Murphy, Bollman & Triangular Trusses* (Woods, 1865; reprint, 1866); addenda (1865) and 51 (1866).

48. Gray in Greiner, "American Railroad Viaduct," 369–371; Gray, "Notes," 16.

49. "Historical Sketch," 20; Boller in Greiner, "American Railroad Viaduct," 364; Gray in Greiner, "American Railroad Viaduct," 369–370; "Historical Sketch of the Development of American Bridge Specifications," in Report of Committee on No. VII—On Wooden Bridges and Trestles, in *Proceedings of the Sixth Annual Convention of American Railway Engineering and Maintenance of Way Association* 6 (1905): 200. Thompson claimed that Gray was not directly involved in bridge design; George Thomson, letter to Theodore Cooper, January 10, 1889, Cooper Bridge Plan Collection, 1887–1889: series I, box 1 (Correspondence), Division of Rare and Manuscript Collections, Cornell University Library, https://hdl.handle.net/1813/56233. Thomson reported eight other bridges in a list he sent Cooper; Thompson, letter to Cooper, January 7, 1889, Cooper Bridge Plan Collection, 1887–1889: series I, box 1 (Correspondence), Division of Rare and Manuscript Collections, Cornell University Library, https://hdl.handle.net/1813/56230. The Mohawk Bridge dates are from George H. Thomson, "Riveted Lattice for Railroad Bridges of Maximum Span: A Plea for a Return to Rational Design," *Engineering News* 59, January 23, 1908, 91.

50. Gray, "Notes," 16; Gray was correcting Cooper's claim in "American Railway Bridge," 16. Engineers A. S. C. Würtele, S. J. Fields, and John F. Alden subsequently learned from Hilton; George H. Thomson in Gray, "Notes," 5–6. Thomson had incorrectly told Cooper that Hilton was Carroll's pupil. George Thomson, letter to Theodore Cooper, January 10, 1889, Cooper Bridge Plan Collection, 1887–1889: series I, box 1 (Correspondence), Division of Rare and Manuscript Collections, Cornell University Library, https://hdl.handle.net/1813/56233. See also David Guise, "Lattice Configurations," *Structure* (August 2011): 22–26; David Guise, "American Lattice-Truss Bridge and the Hilton Truss," part 2, *Structure* (October 2011): 30–32. Re relationships among the engineers, see Report of Committee No. XV—On Iron and Steel

Structures, "Historical Sketch of the Development of American Bridge Specifications," in *Proceedings of Sixth Annual Convention of American Railway Engineering and Maintenance of Way Association* 6 (1905): 199–217.

51. J. E. Greiner, "What is the Life of an Iron Railroad Bridge?" *ASCE Transactions* 34 (1895): 335; T. C. Clarke, 309; Walter Katté, 318–319; George L. Vose, *Manual for Railroad Engineer and Engineering Students* (Lee and Shepard, 1874), 253–262, pl. 22; Louis-Antoine Comolli, *Les ponts de l'Amérique du Nord: Étude, calcul, description de ces ponts* (Ambrose Lefèvre, 1879), 2:48–57, pl. 25. Also, "Table of Bridges in New York State" (undated), Cooper Bridge Plan Collection, 1887–1889: series I, box 1 (Correspondence), Division of Rare and Manuscript Collections, Cornell University Library, https://hdl.handle.net/1813/56234, and "Double Lattice Wrought Iron Bridge at Springfield, MA," *Scientific American* 32 (1875): 21. Mead & Hunt, Engineers, *Contextual Study of New York State's Pre-1961 Bridges* (New York State Department of Transportation, 1999), 93.

52. No single source indicates the number or location of riveted lattice bridges; on debates around riveted versus pin-connected bridges, see Dario Gasparini and David Simmons, "American Truss Bridge Connections in the 19th Century II: 1850–1900," *Journal of Performance of Constructed Facilities* 11 (1997): 138–139. See also Théophile Seyrig, "Different Modes of Erecting Iron Bridges," *Institution of Civil Engineers: Minutes of Proceedings* 63 (1881): 160, 199, and [Ernest Pontzen], *Verrugas Viaduct* (Railroad Gazette, 1876), including observations by W. W. Evans and Charles Bender. The characterization of pinned bridges as American appeared widely in the engineering literature of the period. "The favorite iron bridge in Europe, even at the present day, is the lattice truss"; W. Howard White, "European Railways—As They Appear to an American Engineer," *ASCE Transactions* 3 (1874): 63.

53. Theodore Cooper, "American Railroad Bridges," *ASCE Transactions* 21 (1889): 35–38; Paul Sibly, "The Prediction of Structural Failures" (PhD diss., University of London, 1977), 108–112; William Addis, *Structural Engineering* (Ellis Howard, 1990), 70–71, 162–163.

54. Charles Stowell in Cooper, "American Railway Bridges," 586. Gasparini and Simmons, "American Truss Bridge Connections in the 19th Century II," 130–140.

55. Bridge types included Linville, Post, and Schneider; "Historical Sketch," 20; Cooper, "American Railroad Bridges," 20.

56. E. B. Cushing in Gray, "Notes," 15, and Charles Stowell in Cooper, "American Railway Bridges," 586–587.

57. Thomson, "Riveted Lattice," 91. From the 1850s to the 1870s, prominent engineers recommended pinned connections over indeterminate, prestressed, adjustable forms like the Howe; on changes in design and analysis, see Gasparini and Simmons, "American Truss Bridge Connections I: 1829–1850," *Journal of Performance of Constructed Facilities* 11 (1997): 127–128. Gasparini and Simmons, "American Truss Bridge Connections in the 19th Century II," 132, 136, refer to examples in Gregory K. Dreicer, "The Long Span. Intercultural Exchange in Building Technology: Development and Industrialization of the Framed Beam in Western Europe and the United States, 1820–1870" (Ph.D. diss., Cornell University, 1993), 425–428. Bridge types are discussed in Thomas Curtis Clarke, "Design Generally of Iron Bridges of Very Large Span for Railway Traffic," *Institution of Civil Engineers: Minutes of Proceedings* 54 (1878): 179–247.

58. Édouard Lavoinne and Ernest Pontzen, *Les chemins de fer en Amérique* (Dunod, 1880), 1:177–178, 234–245.

59. Gasparini and Simmons, "American Truss Bridge Connections in the 19th Century II," 132–136; Sibly, "Prediction of Structural Failures," 112; re riveting, see Richard Byrom, "William Fairbairn—Experimental Engineer and Mill-Builder" (Ph.D. diss., University of Huddersfield, 2015).

60. D.A. Gasparini and Caterina Provost, "Early Nineteenth Century Developments in Truss Design in Britain, France and the United States," *Construction History* 5 (1989): 22–33; Stefan M. Holzer, "Polonceau Roof and Its Analysis," *International Journal for the History of Engineering & Technology* 80 (2013): 22–54. Pontifex, "European v. American Bridges," *Engineering* 51 (1891): 495; Kirti Gandhi, "Verrugas Viaduct and its Reconstruction, Peru, South America," in *Safety and Reliability of Bridge Structures*, ed. Khaled M. Mahmoud (CRC, 2009), 461–479.

61. Gasparini and Simmons, "American Truss Bridge Connections in the 19th Century II," 134–139. Fink and Bollman bridges belong to "a very cheap and scientific class of bridges," per "Historical Sketch of the Development of American Bridge Specifications," in Report of Committee on No. VII—On Wooden Bridges and Trestles, in *Proceedings of Sixth Annual Convention of American Railway Engineering and Maintenance of Way Association* 6 (1905): 201.

62. H. Blakeley in Gray, "Notes on Early Practice," 9. British engineer Ewing Matheson asserted that "he did not think any Government Inspector in England, Germany, or France would allow such a bridge to be constructed," i.e., a bridge without a deck that would stop a train from plunging through it; Thomas Curtis Clarke, "Design Generally of Iron Bridges of Very Large Span for Railway Traffic," *Institution of Civil Engineers: Minutes of Proceedings* 54 (1878): 229.

63. P. Alex Peterson, "Chaudière Railway Bridge," *The Gazette* (Montreal), August 7, 1880, 2.

64. Thomson started at New York Central Railroad around 1872. A.W. Carpenter, "Memoir of George Huntington Thomson," *ASCE Transactions* 71 (1911): 438–441.

65. George H. Thomson, "American Bridge Failures," *Engineering* 46 (1888): 273–274, and Thomson, "Riveted Lattice," 91; read in front of British Association for Advancement of Science; Carpenter, "Memoir of George Huntington Thomson," 440, apparently incorrectly reported that Thomson had delivered the talk.

66. "Collapsed Bridge in Williamsville," University of Vermont, Landscape Change Program, https://landscape.uvm.edu/landscape_new/search/details.php?ls=07716&sequence=000&set_seq=7&imageSet=1617384557-6067546d41472&AddRel=0.

67. "The highway bridge on the left is a wooden lattice, with four wooden pins in each intersection . . . like the first one built in America by Mr. Towne, about 1816–1820." Thomson, "American Bridge Failures," 253. See also Thomson, "Riveted Lattice," 93.

68. Thomson may have worked with Hilton for more than five years. Thomson in Gray, "Notes," 7; Thomson, "American Bridge Failures," 253; Thomson, "Riveted Lattice," 91. Substantial changes in pin-connected bridge design likely accounts for the positive evaluation in Lavoinne and Pontzen, *Chemins de fer en Amérique*, 248–259. Re profusion of riveted bridges, lattice or pin-connected, Thomson noted in 1897, "The ideas of engineers change.

Twenty-three years ago but one shop in the United States built lattice bridges"; Thomson in Gray, "Notes," 7. For an extended discussion of historical change in and diverse perspectives on bridge design, see J.A.L. Waddell, "Some Disputed Points in Railway Bridge Designing," 77–282; Waddell notes at the start an ongoing convergence of design standards "based on the principle of survival of the fittest."

69. Thomson, "American Bridge Failures," 294.

70. Thomson, "Riveted Lattice," 91–93.

71. [Editorial], *Engineering News* 59 (1908): 101–102. Earlier, an engineer called Thomson's advocacy of lattice rather than pin-connected bridges "retrograde" and listed the lattice's deficiencies, many relating to the "ambiguity" of the stresses (indeterminacy); J. A. L. Waddell, in Cooper, "American Railroad Bridges," 598–599.

72. C. A. P. Turner, "Riveted Lattice or Pin-Connection for Railroad Bridges: An Answer to the Lattice Argument," *Engineering News* 59 (1908): 229–230. In 1898, engineer Edmund T. D. Myers, who had begun his career in 1848, mentioned "the condition that we poor fellows were in then, when we knew nothing in science better than a Towne's [*sic*] lattice bridge"; *Proceedings of the Eighth Annual Convention of the Association of Railway Superintendents of Bridges and Buildings* (Rumford, 1898): 31.

73. Thomson, "Riveted Lattice," 93.

74. Thomson, "Riveted Lattice," 92–93. Two years later, in 1910, Thomson died from an injury sustained in an 1892 railway accident; Carpenter, "Memoir of George Huntington Thomson," 438–439.

75. Theodore Cooper, "American Railroad Bridges," *ASCE Transactions* 21 (1889): 1–58, 566–607; reprinted without the discussion as a book, Theodore Cooper, *American Railroad Bridges* (Engineering News, [1889]). See also Mark Aldrich, *Death Rode the Rails* (John Hopkins University Press, 2006), 148–150.

76. Gustave Eiffel, "Eiffel Tower" (from *New Review*), in *Annual Report of Board of Regents of the Smithsonian Institution . . . to July, 1889* (GPO, 1890), 730, cited in Jill Jonnes, *Eiffel's Tower* (Viking, 2009), 51.

77. Cooper, "American Railroad Bridges," 1, 18–19, 22. "By a process of elimination based upon the principle of the survival of the fittest, certain types of bridges for the different governing conditions have almost become standard"; Waddell, "Some Disputed Points," 77–78.

78. Cooper, "American Railroad Bridges," 41. The first engineer who contributed to the discussion noted that Cooper had vastly understated the significance of riveted lattice bridges, especially in comparison to pin-connected bridges; Harry D. Bush in "American Railroad Bridges," 568, 570.

79. Stowell in Cooper, "American Railroad Bridges," 586; Thomson, "American Bridge Failures," 253, Thomson, "Riveted Lattice," 91. Thomson's concern about the combining of riveted lattice and pin-connected design seems to be supported by Eda Kranakis, "Fixing the Blame: Organizational Culture and the Quebec Bridge Collapse," *Technology & Culture* 45 (2004): 500 and 510n63 (also, Sibly, "Prediction of Structural Failures," 106). In 1897, Thomson noted

that the pins-versus-rivets debate was over and that pin-connected-bridge designers had incorporated lattice bridge features; by 1905, the president of the ASCE observed that "no vital difference exists between American and European bridges of moderate spans"; Thomson in Gray, "Notes," 7–8; Charles C. Schneider, "Evolution of the Practice of American Bridge Building," *ASCE Transactions* 54 (1905): 223.

80. Addis, *Structural Engineering*, 162. Re methods of analysis applied to the lattice bridge, see Lola Bennett and Dorottya Makay, "Contoocook Railroad Bridge," 13–44, and Dorottya Makay, "Addendum to Contoocook Railroad Bridge," Written Historical and Descriptive Data, HAER, National Park Service, US Department of Interior, 2003, from Prints and Photographs Division, Library of Congress (HAER NH-38), 6–35.

81. Cooper doesn't name Thomson, but there's no ambiguity regarding his target; Cooper, "American Railway Bridges," 47–52. It is astonishing to discover that Thomson was likely present; the clue is an excerpt of his reply, in "American Society of Civil Engineers—Convention of 1889," *Railroad Gazette*, June 28, 1889, 424, which appeared the week after Cooper presented his paper. Thomson provided numbers demonstrating that there were many more long lattice bridges in the United States than Cooper indicated. Thomson discussed the lattice in 1897 in Gray, "Notes," 5–8. He had provided information to Cooper as Cooper was preparing his paper! Cooper Bridge Plan Collection, 1887–1889: series I, box 1 (Correspondence), Division of Rare and Manuscript Collections, Cornell University Library, https://rmc.library.cornell.edu/EAD/htmldocs/RMM04709.html#info_for_users.

82. Sibly believes that "tactless" and sensational presentations of safety issues made engineers less open to hearing them; Sibly, "Prediction of Structural Failures," 109. An investigation into the accuracy of the statistics is beyond the scope of Aldrich, *Death Rode*; awaiting a definitive, balanced inquiry.

83. Cooper, "American Railroad Bridges," 48.

84. Aldrich, *Death Rode*, 133–134, 145–146; Stowell showed it again in 1886 (148–150). Aldrich (152) believes that metal bridges were safer than wooden bridges, a conclusion requiring substantially more evidence and analysis; "rust" appears only once in Aldrich within a quote on bridge failure (141), though a reference to the effect of corrosion is significant in engineering literature; e.g., Theodore Cooper in Henry D. Blunden, "On the Care and Maintenance of Iron Bridges," *ASCE Transactions* 11 (1882): 424–428, and T. Kennard Thompson in Gray, "Notes," 12.

85. It was pointed out by a couple of engineers, including Stowell, in Cooper, "American Railroad Bridges," 585: on Hilton's specs, see "Historical Sketch," 30 and J. N. Clary, "History of Early Bridge Specifications," in Wallace T. McKeel, Jr. and Ann B. Miller, *History of Early Bridge Specifications: A Reprint of a Paper by J. N. Clary,* Final Report VTRC 07-R10 (Virginia Transportation Research Council, 2006). Apparently, Thomson was not present for this discussion; there is no written reply.

86. Cooper, "American Railroad Bridges," 50.

87. Thanks to Amy Slaton for pointing out this distinction.

88. Cooper, "American Railroad Bridges," 49–50. Cooper bragged that his specs constituted "the first paper on bridge construction in which the relic of ignorance, 'the factor of safety,' was entirely omitted"; Cooper, "American Railroad Bridges," 22; Addis, *Structural Engineering,* 163.

89. Gasparini, "Joseph M. Wilson, Henry Petit and the Iron Truss Bridges of the Pennsylvania Railroad," 382–385.

90. Cooper, "American Railroad Bridges," 50; William Cawthorne Unwin, "Illustrations of the Use of Theory in the Work of the Engineer" [address to the Junior Engineering Society], *The Engineer* 66 (1888): 419; Henry Petroski, *Engineers of Dreams* (Vintage, 1996), 98.

91. Unwin, "Illustrations," 429.

92. Re a contemporary comparison, see Petroski, *Engineers of Dreams*, 92. "Le Grand Pont de Forth en Écosse," *La Nature* 16, February 25, 1888, 197, points out that civil engineers internationally were watching these structures.

93. Theodore Cooper, "Bridge Design," *Proceedings of the Engineers' Society of Western Pennsylvania* 7 (1891): 102–103. His words, slightly altered, appeared in *Engineering News* 58 (1907): 391, after the collapse of the Quebec Bridge; cited in Petroski, *Engineers of Dreams*, 91–92.

94. Carpenter, "Memoir of George Huntington Thomson, *ASCE Transactions* 71 (1911): 440.

95. "William Cawthorne Unwin 1838–1933," *Obituary Notices of Fellows of the Royal Society* 1 (December 1934): 169–171. The notorious collapse of the Tay Bridge in 1879, the year following its completion, was widely reported; Kranakis, "Fixing the Blame," 503. Re Cooper's and others on the Forth, see Björn Åkesson, *Understanding Bridge Collapses* (Taylor & Francis, 2008), 55, 59–60; Petroski, *Engineers of Dreams*, 56. Re Bouch, see William Weaver Tomlinson, *The North Eastern Railway: Its Rise and Development* (Andrew Reid, 1915), 617, 734–735.

96. Petroski, *Engineers of Dreams*, 102; Cynthia Pearson and Norbert Delatte, "Collapse of the Quebec Bridge, 1907," *Journal of Performance of Constructed Facilities* 20 (2006): 89–90. See the devastating assessment in Sibly, "Prediction of Structural Failures," 95.

97. Petroski, *Engineers of Dreams*, 102; Kranakis, "Fixing the Blame," 495. US engineers were aware that the longest bridge in the world would not be American; e.g., Waddell, *System of Iron Railroad Bridges*, 2. In a twenty-first-century nationalist twist, Petroski appears to portray the Quebec Bridge as a Canadian rather than US story; Henry Petroski, *To Forgive Design: Understanding Failure* (Belknap, 2012), 175–167, 199. Although viewed as American, the Canadian government intended it to be a Canadian symbol; Sibly, "Prediction of Structural Failures," 91.

98. Thomson, "Riveted Lattice," 93.

99. Re craniology, see Penn Museum, "A History of Craniology in Race Science and Physical Anthropology," https://www.penn.museum/sites/morton/craniology.php#_ftnref28. The related science of phrenology was also widely known. On how theorists and evolutionists regarded anatomy and brain size, see John P. Jackson Jr., and Nadine M. Weidman, *Race, Racism, and Science* (ABC/CLIO, 2004), 39–52; 80–84. See also Michael Adas, *Machines as the Measure of Men* (Cornell University Press, 1989), 292–297.

100. Åkesson, *Understanding Bridge Collapses*, 55, 59–60; Petroski, *Engineers of Dreams*, 109–10.

101. See Introduction. Debunking of failure in Addis, *Structural Engineering*, 180–189.

102. Cooper, "American Railroad Bridges," 49.

103. Petroski, *Engineers of Dreams*, 116–117.

104. Cooper, "American Railroad Bridges," 22–24.

105. Sibly, "Prediction of Structural Failures," 110, 125–26; Addis, *Structural Engineering,* 12.

106. Kranakis, "Fixing the Blame," 508–509, 511; analysis of collapse, Sibly, "Prediction of Structural Failures," e.g. 99–103.

107. Charles Ingels pointed this out in 1944; Wilfred G. Lockett, "Lesson of the Quebec Bridge," *Scientia Canadensis* 11 (1987): 87; Kranakis, "Fixing the Blame," 489, 496, 498–499.

108. Sibly, "Prediction of Structural Failures," 111–115.

109. "Le Grand Pont de Forth en Écosse," *La Nature* 16, February 25, 1888, 198. On the Eiffel systematization of construction, see Peters, *Building the Nineteenth Century*, 264–66.

110. Re organizational culture, see Kranakis, "Fixing the Blame," 487–492. Kranakis believes Cooper could not have altered workflow and emphasizes organizational structures (505–08); targeting Cooper took the blame off the "engineering profession" (517–518).

111. William D. Middleton, *The Bridge at Québec* (Indiana University Press, 2001), 78–79; "Why the Quebec Bridge Failed," *Scientific American* 97 (October 12, 1907): 257.

112. "Construction is sublime male poetry," she continued; Camille Paglia, *Sexual Personae: Art and Decadence from Nefertiti to Emily Dickinson* (Vintage, 1991), 37–38.

113. Regine Bendl et al., "Diversity Management Discourse Meets Queer Theory," *Gender in Management* 23 (2008): 382–394. Rather than the result of fixed internal nature, the expression of national identity *is* the identity (385–386).

114. Ruth Oldenziel, *Making Technology Masculine: Men, Women and Modern Machines in America, 1870–1945* (Amsterdam University Press, 1999). Michele Speitz, "Infrastructural Sublime and Imperial Landscape Aesthetics: Robert Southey, Poet Laureate, and Thomas Telford, Father of Civil Engineering," *European Romantic Review* 32 (2021): 44–46.

115. Andreas Chatzidakis et al., *The Care Manifesto: The Politics of Interdependence* (Verso, 2020), 9–10, ascribes "collateral damage" to neoliberalism.

116. Sibly, "Prediction of Structural Failures," 76. Safety information is shared online via, e.g., Collaborative Reporting for Safer Structures US (CROSS-US), https://www.cross-safety.org/us, with UK and Australian sites.

117. The paragraph continues, "The tension members consisting of eye-bars and the compression members of rectangular latticed sections, built up of plates and angle bars, the whole bridge being pin-connected. The skeleton design shows also the distinctive American features of wide panels, great depth of truss, and a resulting apparent lightness of the individual members. Furthermore, the consulting engineer of the bridge is perhaps the most distinguished bridge

engineer in this country, his 'Specifications for Railroad Bridges' having been for many years the standard authority on that subject"; "A Portentous Bridge Disaster," *Scientific American* 97 (1907): 182.

118. Sibly, "Prediction of Structural Failures," 95, 98–100; "Rectangular Lattice Compression Members in Bridges," *Scientific American* 98 (1908): 406; Henry S. Prichard, "Latticed Compression Members in Bridges," *Scientific American* 99 (1908): 43.

119. "Kahnawake Mohawks Mark 1907 Quebec Bridge Disaster," CBC News, August 30, 2006, https://www.cbc.ca/news/canada/kahnawake-mohawks-mark-1907-quebec-bridge-disaster-1.623180. Labor included US, French-Canadian, and Irish workers; Cheryl Cornacchia, "When Quebec Bridge Came Crashing Down," *Montreal Gazette*, August 30, 2006, https://www.pressreader.com/canada/montreal-gazette/20060830/281522221564792. During the 1916 construction of a replacement bridge, the center span fell while being hoisted, killing an additional thirteen workers.

120. "The Briggers: The Story of the Men Who Built the Forth Bridge," http://www.briggers.com/; "Forth Bridge," https://en.wikipedia.org/wiki/Forth_Bridge#cite_note-b86-70; and Elspeth Wills, *The Briggers: The Story of the Men Who Built the Forth Bridge* (Casemate, 2009; 2021), which I was unable to access. "One log book of accidents and sickness had 26,000 entries"; National Museums Scotland, "Forth Bridge Paint Mixer," https://www.nms.ac.uk/explore-our-collections/stories/science-and-technology/forth-bridge-paint-mixer/.

121. David Coates, *Capitalism: The Basics* (Routledge, 2016), 53–80.

122. Re the repair person's perspective see, for example, Susan Strasser, *Waste and Want: A Social History of Trash* (Holt, 1999), 10–11.

123. Petroski, *Engineers of Dreams*, 98; the paragraph continues with the equivocation, "The line between too little and too much safety is not always very clear and distinct, however, and that is what makes the best engineering also the most difficult. It is also what can lead to the worst engineering."

124. Concern was evident in the 1850s, but meaningful change took two decades; Aldrich, *Death Rode*, 141, 306. According to Sibly, "Prediction of Structural Failures," it is significantly less costly to systematically collect and analyze data that may prevent accidents than not do so. Re complexities of assessing, see George L. Vose, "Safety in Railway Travel," *North American Review* 135 (1882): 374–377.

125. Elizabeth V. Spelman, *Repair: The Impulse to Restore in a Fragile World* (Beacon, 2002).

126. "Bridges," in *2021 Report Card for America's Infrastructure* (ASCE), 18–25.

127. Thomson, "American Bridge Failures," 294.

128. Thomson, "American Bridge Failures," 294.

129. Institution of Civil Engineers, *Royal Charter and By-laws* (*Institution of Civil Engineers*, 2020), 5; "Institution of Civil Engineers," https://en.wikipedia.org/wiki/Institution_of_Civil_Engineers.

CHAPTER 8

1. On the early high-rise, see Gerald R. Larson, "The Elevator, the Iron Skeleton Frame, and the Early Skyscrapers: Part 1," *International Journal of High-Rise Buildings* 9 (2020): 1–15; Condit, "Two Centuries of Technical Evolution," 11–24.

2. Joanna Merwood-Salisbury, "First Chicago School and the Ideology of the Skyscraper," in Peggy Deamer, ed., *Architecture and Capitalism* (Routledge, 2014), 25–47; re "foreigners," 38–40.

3. Merwood-Salisbury, "First Chicago School," 42.

4. Merwood-Salisbury, "First Chicago School," 39–42; Joanna Merwood-Salisbury, *Chicago 1890* (University of Chicago Press, 2009), 25–26. "Gothic" could denote an architectural style or a construction philosophy.

5. Re development of bridge piers leading to the Eiffel Tower, see Roberto Serrano-López et al., "La adaptación de la forma a las cargas en el arte structural: Desde las pilas y el arco, a la torre / Adaptation of the shape to loads on structural art: from piers and arch, to the tower," *Informes de la Construcción* 66 (2014): 1–9. Also, Bertrand Lemoine, "L'Entreprise Eiffel," *Histoire, Économie et Société: Entreprises et Entrepreneurs du Batiment et des Travaux Publics (XVIIIe–XXe Siècles)*, ed. Dominique Barjot, 14 (1995): 281. Re skyscraper walls, see Gregory K. Dreicer, "High-Rise Building Façade, 1880–1930: Construction, Repair, Protective Legislation" (master's thesis, Columbia University, 1985), 131–147. Henry H. Quimby, "Wind Bracing in High Buildings," *ASCE Transactions* 27 (1892): 222, noted that "a writer on the subject recently called [exterior walls] 'mere curtains to shield the exterior'" (225). Donald Friedman, "Hidden Intricacies: The Development of Modern Building Skeletons," *APT Bulletin: Journal of Preservation Technology* 43:4 (2012): 13–21. Thomas Leslie, "Built Like Bridges: Iron, Steel and Rivets in the Nineteenth-century Skyscraper," *Journal of Society of Architectural Historians* 69 (2010): 234–261; the title suggests a discussion of the relationship between bridge and skyscraper but the article references only the Tay Bridge collapse, 1879.

6. Corydon T. Purdy, "Use of Steel in Large Buildings," *Journal of Association of Engineering Societies* 14 (1895): 183–184; of the building frame, he writes, "In any truss of this kind, if we may call it by that name, the columns become the chords, one being in compression and the other in tension" (195).

7. Conrad Matschoss, *Männer der Technik* (1925; reprint, VDI Verlag, 1985), 191; Victor Röll, ed., *Encyklopädie des gesamten Eisenbahnwesens*, 2nd ed. (Vienna, 1915), 7:364–365; *Österreichisches Biographisches Lexikon*, 143; *Österreichische Eisenbhanzeitung* 31 (1908): 255. In December 1858, Nordling legally changed his name; *Bulletin des lois de la République francaise* (Imprimerie Impériale, 1859), 87.

8. Nordling, *Documents*, 10, 59; see full citation below, note 10. Re rail lines, see "Civil Engineering in the Paris Exhibition—Iron Viaducts of La Cere, Busseau d'Ahun and du Midi," *Engineer* 23 (1867): 386.

9. The viaduct at Rouzat, between the stations of Saint-Bonnet-de-Rocehfort and Gannat; over Bellon, near the stations of Louroux-de-Bouble and Echassières; over Bouble, between the stations of Lapeyrouse et Louroux-de-Bouble, near Echassières. Eiffel contributed to enhanced

fabrication in the foundry of wrought- to cast-iron connections for the tower piers his firm built; Bertrand Lemoine, *Gustave Eiffel* (Hazan, 1984), 32, 34; Nordling, *Mémoire*, 15 (see below, note 10), and Nordling, *Documents*, 65; Devance, *Gustave Eiffel*, 126. He also modified the design of rollers used in launching; Henri Loyrette, *Gustave Eiffel* (Rizzoli, 1985), 54–57. Re construction time, see Ach. Ferary, "Viaduc de la Cère au Ribérès (Cantal): Travaux en cours d'exécution," *L'Illustration*, August 19, 1865, 121.

10. Quote on p. 1, Wilhelm Nordling, "Mémoire sur les piles en charpente métallique des grands viaducs," *Annales des ponts et chaussées* 4/2 (1864): 1–126. Note: references in this chapter refer to the offprint, as Nordling, *Mémoire*). The second paper is Wilhelm Nordling, "Documents relatifs aux viaducs métalliques de la ligne de Commentry à Gannat," *Annales des ponts et chaussées* 4/1 (1870): 125–187 (similarly, published separately: *Documents relatifs aux viaducs métalliques de la ligne de Commentry à Gannat* [Dunod, 1870]. A series, begining with some of Nordling's "Mémoire sur les piles" of 1864 without attribution, presents an extensive discussion of Nordling's articles and viaducts, including wind-bracing design: "Civil Engineering in the Paris Exhibition—Iron Viaducts of La Cere, Busseau d'Ahun and du Midi," [and various titles], *Engineer* 23 (1867): 384–387, 398, 430–431, 462–463, 488–489, 512–513; "Great Iron Viaducts of the Commentry and Gannat Railway (France) II," *The Engineer* 26 (1868): 409–410; re Nordling's prose; his writing "proved its author's mastery of science as well as of practice, and his sound sense that wastes not a line upon needless scientific display."

11. Nordling, *Mémoire*, 9.

12. X-bracing and lattice were equally valid options, per Nordling, *Documents*, 12. Eiffel's choice likely reflected his experience at bridge over Garonne in Bordeaux; Lemoine, *Gustave Eiffel*, 23–26.

13. Nordling, *Mémoire*, e.g., 7–8.

14. R. Serrano-López et al., "La adaptación," 8. Previously, piers were strongly anchored into foundations to prevent overturning; adding cables was considered but deemed too challenging; Nordling, *Documents*, 7–10. Carl W. Condit, "Wind Bracing of Buildings," *Scientific American* 230 (February 1974): 96. Engineers would cease to use cast iron in bridge structures because, despite advantages in compression (appropriate for columns), it was weak in tension. It was also challenging to make effective, rigid connections among cast-iron column elements and between cast-iron columns and iron and steel members. Cast iron is brittle; it can fail without warning under high loads or as the result of invisible manufacturing flaws. Catastrophic accidents encouraged engineers to limit the use of cast iron. Re Nordling's choice of metal, see Nordling, *Mémoire*, 119–120. By around 1880, cast iron was no longer used in the United States (or France) in pier construction; engineers made designs "such as to eliminate tensile strains in the pier columns at the specified wind pressure," according to C. Shaler Smith, "Wind Pressure upon Bridges," *ASCE Transactions* 10 (1881): 145, 157. See Corydon T. Purdy, "Use of Steel in Large Buildings,"191–201 and Robins Fleming, *Six Monographs on Wind Stresses* (Engineering News, 1915), 27–41, 43–52.

15. Re pyramid and stability, see Nordling, *Mémoire*, 15; re use of metal, Nordling, *Mémoire*, 119–20 and Nordling, *Documents*, 65. Apparently Nordling, perhaps with Moreau, designed the Bellon Viaduct's curving pier. This preceded two bridges built by Eiffel's firm; Louis Devance,

Gustave Eiffel, 126. Portage trestles were described as "massive pyramids of timber," in "Great Railroad Bridge," *American Railroad Journal* 25 (1852), 486.

16. Nordling, *Mémoire*, 116, 121. The wind-driven catastrophe that Nordling feared happened (Tay Bridge, 1879), but metal piers were established by then.

17. Nordling, *Document*, 8, 23. Piers of the latter four viaducts were, per Nordling, simpler, more stable, and cheaper.

18. Nordling, *Mémoire* (1864), 2–6. "Civil Engineering in the Paris Exhibition—Iron Viaducts of La Cere, Busseau d'Ahun and du Midi," *Engineer* 23 (1867): 386, further discusses precedents. During the following decades, there was no systematic study of structures and wind, though the challenge was recognized as significant; Sibly, "Prediction of Structural Failures," 71–73.

19. This sample list does not include all journals in which the news appeared, sometimes duplicated, sometimes commented upon. "Railroad Bridge at Portageville," *Wyoming County Mirror*, September 16, 1851; thanks to Daniel Fink for typescript, from Madden Collection, Canal Society of New York State. "A Stately Bridge," *Scientific American* 7 (1852): 385 (from *Wyoming County Mirror*); "Great Railroad Bridge," *American Railroad Journal* 25 (1852): 486–467 (from *N. Y. Famer and Mechanic)*; "'High Bridge,' Portage, New York," *Canadian Journal*, 68–69 and plate; "'High Bridge,' Portage, New York," *Civil Engineer and Architects Journal*, 65–66, pl. 8; "Hohe Brücke bei Portage," *Allgemeine Bauzeitung* 18 (1853): 338; 341; "High Bridge, Portage, N.Y.," *Appleton's Mechanics' Magazine* 3 (1853): 98–99, pl. 1; "'High Bridge' at Portage, New York" *Mechanics' Magazine* 58 (London, 1853): 169–170 (from *Canadian Journal*); "'Hohe Brücke' in Portage bei Neuyork," *Polytechnisches Centralblatt* NF7 (1853): 1083 (from *Canadian Journal* via *Civil Engineer and Architects Journal*); "Hohe Brücke bei Portage, New York," *Zeitschrift des Architekten- und Ingenieur-Vereins für das Königreich Hannover* 1 (1855): 419–420; "Haut Pont de Partage [*sic*] dans l'État de New-York," *Nouvelles Annales de la Construction* 1 (1855): 3, pl. 24; "Hohe Brücke bei Portage, New-York," *Dingler's Polytechnisches Journal* 139 (1856): 76–77 (from *Canadian Journal*); "Hohe Brücke bei Portage, New York," *Zeitschrift des Oesterreichischen Ingenieur-Vereines* 8 (1856): 186–87 (from *Canadian Journal*). Re bridge opening, see *New-York Daily Tribune*, August 16, 1852, 8. Nordling, *Mémoire*, 2, cited the regularly updated book by Auguste Perdonnet et Camille Polonceau, *Nouveau portefeuille de l'ingénieur des chemins de fer* (Lacroix-Comon, 1857), 37–38, pl. M.

20. Edward Harold Mott, *Between the Ocean and the Lakes: The Story of Erie* (Collins, 1899), 362.

21. "A Stately Bridge," *Scientific American* 7 (1852): 385.

22. Wisely, *American Civil Engineer*, 5.

23. Wisely, *American Civil Engineer*, 13–16.

24. "Beiträge zu den Holzkonstrukzionen," *Allgemeine Bauzeitung* 10 (1845): 11, pl. DCXLVI646; also in Duggan, *Specimens of the Stone, Iron and Wood Bridges, Viaducts, Tunnels, Culverts, &c. of the United States Rail Roads* (Appleton, 1850).

25. *First Annual Report of S. W. Roberts, Chief Engineer of the Little Schuylkill and Susquehanna Rail Road Company* (Clark, 1839), 5. Re measurements, see *Second Annual Report of S. W. Roberts, Chief Engineer of the Little Schuylkill and Susquehanna Rail Road Company* (Clark, 1840), 6. Thanks

to Julie Herrada, Special Collections Research Center, University of Michigan Library. Moncure Robinson, original engineer on the line, was a lattice bridge builder (see chapter 3) and was involved in 1839 when Roberts and Miller with a group attempted to establish an engineering society; "Address of Committee Appointed at Convention of Civil Engineers, Baltimore, Md.," *American Railroad Journal* 8 (1839): 25–27. Also "Proceedings of Institution of Civil Engineers, for the Formation of a Society of Civil Engineers in the United States," *American Railroad Journal* 8 (1839): 153–57; Wisely, *American Civil Engineer*, 6–13; Daniel Hovey Calhoun, *American Civil Engineer* (MIT Press, 1960), 182–185. Wendler, *Friedrich List as a Railway Pioneer in the USA: The Little Schuylkill Railroad in Pennsylvania* (Springer, 2020), 47–55.

26. *First Annual Report of S. W. Roberts*, 5.

27. Conversation with Grazia Cione, June 28, 2024.

28. "Beiträge zu den Holzkonstrukzionen," pl. 646, and Duggan, *Specimens*.

29. *Second Annual Report of S. W. Roberts*, 5–6; *First Annual Report of S. W. Roberts*, 4.

30. Stone was "undoubtedly preferable"; Stephen H. Long, *Description of Col. Long's Bridges, Together with a Series of Directions to Bridge Builders* (Brown, 1836), 29–30.

31. The claim that the pier was a precedent for the Portage appeared many years later: Edwin F. Smith, "Some Early Engineering Works in Pennsylvania," *Proceedings of the Engineers' Club of Philadelphia* 21 (1904): 76–77, and "Bibliography," in Report of Committee on No. VII—On Wooden Bridges and Trestles, in *Proceedings of the Sixth Annual Convention of American Railway Engineering and Maintenance of Way Association* 6 (1905): 39.

32. "Civil Engineering in the Paris Exhibition," 386; see note 10 above for additional page numbers. *France, Universal Exhibition at Philadelphia in 1876: Notes on Models, Charts and Drawings Relating to Works of the Ponts et Chaussées and the Mines* (Government Printing Office, 1876), 22–31; drawings of Bouble Viaduct were also displayed, 45–48. Re image count (apparently not including technical journals) and how it shaped what non-visitors saw, see Thomas Smits, *European Illustrated Press and Emergence of a Transnational Visual Culture of the News, 1842–1870* (Routledge, 2020), 14, 171–217.

33. Nordling, *Documents*, 5, 63. Uncounted were Thomas Bouch's viaducts in England: Tees Valley (1860, demolished 1971) and Gaunless (1863, demolished 1964), both on stone piers; Belah (1860, demolished 1963) and Deepdale (1861, demolished 1963) viaducts. William Weaver Tomlinson, *North Eastern Railway: Its Rise and Development* (Newcastle-upon-Tyne: Andrew Reid, 1915), 597–598, 604. Bouch's work became suspect when his Tay Bridge failed in 1879 (617, 734–735).

34. Shaler Smith, "Wind Pressure upon Bridges," *ASCE Transactions* 10 (1881): 152, 156; 145–146 for a discussion of design and materials of US, "French, Swiss, and Austrian" viaduct piers. Generalizations about "European" bridges were exploded by Gustave Bouscaren, who noted that the bridges discussed appeared during an early phase of viaduct design (156–57); also in C. Shaler Smith et al., "Wind Pressure upon Bridges," *Engineering News* 8 (1881): 395–400.

35. Jules Gaudard, "Resistance of Viaducts to Sudden Gusts of Wind," *Institution of Civil Engineers: Minutes of Proceedings* 69 (1882): 130–133, 216.

36. Alfred Picard, *Exposition Universelle Internationale de 1889 à Paris: Rapport Général 2: Travaux de l'Exposition Universelle de 1889* (Imprimerie Nationale, 1891), 273–279. "Le Fer," in J.-K. Huysmans, *Certains* (Tresse & Stock, 1889), 178. Gustave Eiffel, *Travaux scientifiques exécutés à la Tour de trois cents mètres de 1889 à 1900* (Maretheux, 1900), 4; quotation, 229–230. Eiffel, "Eiffel Tower," (from *New Review*), in *Annual Report of the Board of Regents of Smithsonian Institution . . .* to July, 1889 (GPO, 1890), 729–730; Eiffel may have made it appear that the outward-curving pier design was his idea; evidence does not indicate this is accurate; see notes above for his contribution.

37. Eda Kranakis, "Social Determinants of Engineering Practice: A Comparative View of France and America in the Nineteenth Century," *Social Studies of History* 19 (1989): 16. David P. Billington, *The Tower and the Bridge* (Basic, 1983), 93. Billington incorrectly states (66–67) that Eiffel built the four Commentry-Gannat line viaducts; he attributes Nordling's tower designs to Eiffel; he believes "that a work of structural art comes from the mind of a single artist" (93). Antoine Picon, *L'invention de l'ingénieur modern* (Presses de l'école des ponts et chaussées, 1992), 465. Lavoinne and Pontzen, *Les Chemins de Fer en Amérique* 1 (Dunod, 1880), 217. Carl Condit recognized Nordling's team's work on wind resistance, but attributed it to Eiffel; Carl W. Condit, "Wind Bracing of Buildings," *Scientific American* 230 (February 1974): 96–98.

38. W. de Nordling, "Nécrologie: Notice sur la Vie et sur les Travaux de M. Thirion," *Revue générale des chemins de fer* 10 (1897): 94. Perhaps Nordling, like Michel Chevalier, was tainted by association with Napoleon III; Drolet, "Industry, Class and Society," 1237–1238. Attitudes in France about Germany (emulation rather than revenge) may explain why Nordling was able to move back to France; Rachel Chrastil, "Who Lost the Franco-Prussian War? Blame, Politics, and Citizenship in the 1870s," *Journal of Western Society for French History* 32 (2004): http://hdl.handle.net/2027/spo.0642292.0032.016; Norimitsu Onishsi, "France Seeks Lessons in Book About Fall of the Nation in 1940," *New York Times*, June 10, 2020, A9; Karine Varley, *Under the Shadow of Defeat: The War of 1870–71 in French Memory* (Palgrave Macmillan, 2008).

39. Lemoine, *Gustave Eiffel*, 43–44. Re opening of the Orléans line, see "Un Exploit Technique et Humain," Les Amis des Viaducs (2019), https://amisdesviaducs.wordpress.com.

40. Re Nordling, see Lemoine, *Gustave Eiffel*, 26, 32, 34–35, 42; Bertrand Lemoine, *L'architecture du fer, France: XIXe siècle* (Champ Vallon, 1986), 102–103, 107–108; Bernard Marrey, *Les ponts modernes: 18e-19e siècles* (Picard, 1990), 210–212, 310.

41. Cédric Lamarsaude et al., "Diagnosis and Restoration of Four Historical Eiffel-Type Rail Viaducts," *Tomorrow's Megastructures*, 40th IABSE Symposium, Nantes, Frances, September 19–21, 2018, S21-1-S21–8 and *Structural Engineering International* 28 (2018): 150–156.

42. "Why Philosophers Care about Dehumanization" (Maria Kronfeldner and Alexander Reutlinger)," September 30, 2021, Socially Engaged Philosophy, YouTube, 1 hr, 0 min., 41 sec., https://www.youtube.com/watch?v=UPVjeUnYxAw.

43. Durbach, Etzel (see chapter 5), Jacqmin and Nordling, "Reconstruction du viaduct de Grandfey," *Bulletin Technique de la Suisse Romande* 51 (1925): 301.

44. Emile Deschanel, "Inauguration de chemin de fer de Fribourg à Lausanne," *Journal des Débats*, September 13, 1862, 2–3. Also [Viaduc de Grandfey], *La Lorgnette* (Bordeaux), July 20, 1862, 3. Freiburg postcards featured cathedral and bridges. "Les artistes contre la tour Eiffel," *Le Temps*, April 14, 1887, 2–3. Also, David E. Nye, *American Technological Sublime* (MIT Press, 1994), 28–29. On industrial fears, see Hollis Clayson, "The Ornamented Eiffel Tower: Awareness and Denial," *nonsite.org* 27 (2019), https://nonsite.org/the-ornamented-eiffel-tower/#foot_56-11530.

45. "Rekonstruktion des Viadukts von Grandfey," *Berner Woche in Wort und Bild* 17 (1927): 175. Consulting engineer: Robert Maillart. "Umbau des Grandfey-Viadukts der Schweizerischen Bundesbahnen: Entwurf und Bauleitung," *Schweizerische Bauzeitung* 87/88 (1926): 217–221, 232–37 and "Reconstruction du viaduct de Grandfey," *Bulletin Technique de la Suisse Romande* 51 (1925): 301–307.

46. Re "thicket" (*fourré*), see Nordling, *Mémoire*, 13. Re "symmetry," see Nordling, *Documents*, 64.

47. Sanghun Lee, "Technology and Form: Iron Construction and Transformation of Architectural Ideals in Nineteenth Century France, 1830–1889" (PhD diss., MIT, 1996) throughout; re buildings identified here, 216–221.

48. Nordling, *Mémoire*, 13. Nordling noted that this experience was possible with temporary scaffoldings of wood, which were left uncovered. He believed that the latticework on the Grandfey Viaduct piers did little structurally and was visually disturbing; Nordling, *Mémoire*, 5, 13–15.

49. B. Dupuis, "Viaduc de la Sarine à Fribourg (Suisse)," *L'Illustration*, June 21, 1862, 392, 394; Ferary, "Viaduc de la Cère," 121–122.

50. Letter, November 23, 1869, cited in Bermond, *Eiffel*, 154. He used a Sioule pier photograph when promoting his company; https://gallica.bnf.fr/ark:/12148/btv1b104825035?rk=21459;2#.

51. Re the tower's public and professional acceptance, see Lemoine, *Architecture du fer*, 282, 288. Re Koechlin and Nouguier, see Lemoine, "L'Entreprise Eiffel," 281, and "Dipl. Ing. Maurice Koechlin und der Eiffelturm," Schriftenreihe der ETH-Bibliothek 27 (ETH-Bibliothek,1990), https://www.research-collection.ethz.ch/handle/20.500.11850/20447. Re tower's lattice system, see Roderic Lakes, "Materials with Structural Hierarchy," *Nature* 361 (1993): 511–515, and M. Meenakshi Sundaram and G. K. Ananthasuresh, "Gustave Eiffel and his Optimal Structures," *Resonance* (2009): 849–865. Eiffel quote from Charles Van Büren, *Der 300-Meter-Turm* (1988), cited in Pfammatter, *Making of the Modern Architectand Engineer* (Birkhäuser, 2000), 164–166. Pfammater notes the tower incorporated approaches of the École Centrale and Karlsruhe and Zurich engineering schools. On perceptions of and changes to Eiffel Tower, see Hollis Clayson, "Ornamented Eiffel Tower."

52. Erie published its first tourist guide in 1854. Tom Breslin and Tom Cook, "Portage Bridge," *Exploring Letchworth Park History*, http://www.letchworthparkhistory.com/bridge.html and

www.letchworthparkhistory.com/Photo_Index.html; "Erie Railway Bridge at Portage," *Scientific American* 17 (September 7, 1867), 148; Herbert Gottfried, *"Erie Railway Tourist," 1854–1886* (Lehigh University Press, 2018), 5. Re bridge experience, see "Great Portage Railroad Bridge," *Kenosha Telegraph* (Wisconsin), August 5, 1853, 1. Irene Cheng, "Structural Racialism in Modern Architectural Theory," in *Irene Cheng et al., Race and Modern Architecture: A Critical History from the Enlightenment to the Present* (University of Pittsburgh Press, 2020), 147–48; Shelly Hales, "The History of Human Habitation," in Katharine T. von Stackelberg and Elizabeth Macaulay-Lewis, eds., *Housing the New Romans: Architectural Reception and Classical Style in the Modern World* (Oxford University Press), 92–118.

53. Eiffel was unable to pass the entrance exams to the École Polytechnique. His family name change was legalized in 1879. Lemoine, *Gustave Eiffel*, 14–15.

54. Théophile Seyrig, for example, designed several of the firm's most prominent bridges. After his conflicted departure, he could be found sharing knowledge in London: Seyrig, "Different Modes of Erecting Iron Bridges," *Institution of Civil Engineers: Minutes of Proceedings* 63 (1881): 151–216; summary in "Erection of Iron Bridges," *Engineering News* 8 (1881): 17–18.

55. Frederick Brown, "Eiffel's Tower," *New England Review* 29 (2008): 19–21.

56. There is no body of evidence showing that technologists based in France designed differently than innovators in other lands. Peter Lundgreen, "Engineering Education in Europe and U.S.A., 1750–1930," *Annals of Science* 47 (1989), 39–41. Capsule biographies of engineers and companies, Lemoine, *Architecture du fer*, 298–305; Marrey, *Ponts modernes*, 304–312. Dominique Barjot, "Innovation et travaux publics," 403–414. French engineers could leave government service, work in private industry, then return (*pantouflage*); Alan Mitchell, *The Great Train Race: Railways and the Franco-German Rivalry, 1815–1914* (Berghahn, 2000), 91. Re engineering hierarchy in France, see Picon, *Invention*, 139–145, 457–464. Re technical schools' "fundamental characteristics," see Peter Lundgreen, "Organization of Science and Technology in France: A German Perspective," in *Organization of Science and Technology in France 1808–1914*, ed. Robert Fox and George Weisz (Cambridge University Press, 1980), 323. Also, Kranakis, *Constructing a Bridge*, 261–262; Terry Shinn, "From 'Corps' to 'Profession': the Emergence and Definition of Industrial Engineering in Modern France," in Fox and Weisz, *Organization*, 185, 207.

57. Harold Callender, "Gandhi Dissects the Ford Idea," *New York Times Magazine*, November 8, 1931, 2; David A. Hounshell, *From American System to Mass Production 1800–1932* (Johns Hopkins University Press, 1984), 322.

58. A. Buchanan, in Seyrig, "Different Modes," Discussion, 192; also, Iron Bridges," from *London Quarterly Review* (1858), reprinted in *American Railroad Journal* 31 (1858): 586. Lemoine, *Gustave Eiffel*, 37; J. G. James, "The Origins and Worldwide Spread of Warren-Truss Bridges in the Mid-Nineteenth Century, Part 1: Origins and Early Examples in the UK," *History of Technology* 11 (1986): 109; James, "Overseas Railways." Theodore Cooper noted that bridge-building companies were associated with specific styles, locations, or railway companies; Dario Gasparini and David Simmons, "American Truss Bridge Connections in the 19th Century II: 1850–1900," *Journal of Performance of Constructed Facilities* 11 (1997): 130–131. Also, Kranakis,

"Fixing the Blame," 487–518. Kennard exported structures fabricated in its factory adjacent to the Crumlin Viaduct; Henry N. Maynard, *Handbook to the Crumlin Viaduct* (Wilson, 1862); 20–21, 102–106.

59. See chapter 3.

60. Michael Kammen, *Spheres of Liberty: Changing Perceptions of Liberty in American Culture* (University Press of Mississippi, 1986), and Henry Steele Commager, "The Idea Behind the Symbol: Changing Meaning of Liberty in American History," in *Liberty: The French-American Statue in Art and History* (Perennial Library / Harper and Row, 1986), 12–19; "Abolition," *Statue of Liberty*, National Park Service, https://www.nps.gov/stli/learn/historyculture/abolition.htm.

61. Lemoine, *Gustave Eiffel*, 74–76.

62. Francesca Lidia Viano, *Sentinel: The Unlikely Origins of the Statue of Liberty* (Harvard University Press, 2018), notes the role of Saint-Simonism, e.g., 62–65.

AFTERWORD

1. Benedict Anderson, *Imagined Communities* (Verso, 1983; rev. ed. 2006); Colin Woodard, *American Nations: A History of the Eleven Rival Regional Cultures of North America* (Viking, 2011).

2. Based on experience with public and experts in my professional role, Chicago Architecture Foundation, 2007–2013.

3. Gerald R. Larson and Roula Mouroudellis Geraniotis, "Toward a Better Understanding of the Evolution of the Iron Skeleton Frame in Chicago," *Journal of Society of Architectural Historians* 46 (1987): 39–48; Condit, "Two Centuries of Technical Evolution Underlying the Skyscraper," in *Second Century of the Skyscraper*, ed. Lynn S. Beedle, ed., Council on Tall Buildings and Urban Habitat, *Second Century of the Skyscraper*, (Van Nostrand Reinhold, 1988), 11–24. Early skyscraper designer LeRoy S. Buffington designed columns "braced diagonally, after the manner of lattice bridge girders"; "Twenty-Eight Story Building," *Chicago Tribune*, September 2, 1888, 27; Gerald R. Larson, "The Tragic History of Buffington's Patent," *Chicago School of Architecture*, April 5, 2021, https://thearchitectureprofessor.com/2021/04/05/3-17-the-tragic-history-of-buffingtons-patent/.

4. E.g., "Rhine Bridge," https://en.wikipedia.org/wiki/Rhine_Bridge. Some bridges (e.g., lattice) were harder to destroy; Georges Ribeill, "Vie et mort des ouvrages d'art: L'exemple des ponts des chemins de fer," *Culture Technique* 26 (1992): 101–112.

INDEX

Page numbers followed by *f* denote a figure.
Index does not refer to notes.

Publisher contact:
The MIT Press
Massachusetts Institute of Technology
77 Massachusetts Avenue, Cambridge, MA 02139
mitpress.mit.edu

EU Authorised Representative:
Easy Access System Europe, Mustamäe tee 50,
10621 Tallinn, Estonia
gpsr.requests@easproject.com

Printed by Integrated Books International,
United States of America